Nordic Climate Histories

Nordic Climate Histories

Impacts, Pathways, Narratives

edited by

Dominik Collet, Ingar Mørkestøl Gundersen,
Heli Huhtamaa, Fredrik Charpentier Ljungqvist,
Astrid E.J. Ogilvie and Sam White

Copyright © 2025
The White Horse Press, The Old Vicarage, Main Street, Winwick, Cambridgeshire PE28 5PN, UK

Set in 10.5 point Adobe Caslon Pro

OPEN ACCESS CC BY-NC 4.0
Reuse of illustrations may not be permitted.

doi: 10.63308/63881023874820.book

British Library Cataloguing in Publication Data
A catalogue record for this book is available from the British Library

ISBN 978-1-912186-98-3 (PB); 978-1-912186-99-0 (Open Access ebook)

A Note on the Cover Illustration

The cover of the book shows an engraving from *Vies des savants illustres* by Louis Figuier depicting 'Maupertuis Measuring Meridian Arc in the Arctic Circle'. Pierre-Louis Moreau de Maupertuis (1698–1759) was a French polymath who led an expedition to northern Finland in order to determine the shape of the earth. As noted by Lundstad et al. in this volume, the originator of the Celsius temperature scale, Anders Celsius (1701–44), also took part in this 'Lapland Expedition', as it became known.

Contents

List of Figures and Maps vii

Introduction. Integrating, Connecting and Narrating Nordic Climate Histories
Dominik Collet, Ingar Mørkestøl Gundersen, Heli Huhtamaa, Fredrik Charpentier Ljungqvist, A. E.J. Ogilvie and Sam White 1

Chapter 1. The Development of Meteorological Institutions and Early Instrumental Climate Data in the Nordic Countries
Elin Lundstad, Stefan Norrgård and A. E.J. Ogilvie 29

Ancient and Medieval Climate

Chapter 2. Cold or Culture? Effects of Mid-Holocene Temperatures on Forager and Early Farmer Demographics in Southern Norway
Svein Vatsvåg Nielsen 59

Chapter 3. A Series of Unfortunate Events: Two Central Norwegian Settlements Facing the Climatic Downturn after AD 536–540
Ingrid Ystgaard and Raymond Sauvage 77

Chapter 4. Volcanic Vulnerability in Medieval Iceland
Carina Damm 103

Chapter 5. The Moving Manors and Adaptation in Sixteenth-Century Denmark
Sarah Kerr 123

Chapter 6. Architectural Climate Change Adaptations in Little Ice Age Norway c. 1300–1550
Kristian Reinfjord 147

Little Ice Age Climate

Chapter 7. The Impact of Wildfire and Climate on the Resilience and Vulnerability of Peasant Communities in Seventeenth-Century Finland
Jakob Starlander 167

Chapter 8. Northern Iceland Temperature Variations and Sea-Ice Incidence c. AD 1600–1850
A. E.J. Ogilvie and M.W. Miles 186

Chapter 9. Integrating Agricultural Vulnerability and Climate Extremes. Eighteenth-Century Norway through the Works of Jacob Nicolaj Wilse (1735–1801)
Ingar Mørkestøl Gundersen 209

Chapter 10. 'An Ice Breakup as in the Good Old Days'. Ice Jams in the Aura River, Turku, Southwest Finland, 1739–2024
Stefan Norrgård 237

Narrating Climate Histories

Chapter 11. Climate Narratives in Norwegian Public Histories
Eivind Heldaas Seland 261

Chapter 12. Glacier Poetry in Norwegian Literary Historiography
Kristine Kleveland 275

Chapter 13. Through a Mirror, Darkly: Bringing Deep Environmental History into the Museum
Felix Riede 294

Chapter 14. Back to the Future: Weaving Climate History into Nordic National Museum Narratives
Natália Melo, Bergsveinn Þórsson, Felix Riede and Stefan Norrgård 317

Index. 337

List of Figures and Maps

Introduction

Figure 1. Temperatures past and present.. 5

Figure 2. A simplified model of climate–society interactions. 13

Chapter 1

Map 1. Overview of early meteorological stations in the
    Nordic countries. 31

Figure 1. The earliest and longest annual temperature trends
    from four Nordic locations. 41

Figure 2. Historical annual temperature trends for Ilulissat
    and Longyearbyen. 42

Figure 3. Pairwise scatter plots of annual temperature time
    records for five Nordic capitals. 44

Figure 4. The temperature time series for the five capitals of
    the Nordic region... 45

Figure 5. Annual mean temperature trends from 1829 to
    2024 for five Nordic capitals.. 48

Chapter 2

Figure 1. Comparison of pre-Bronze Age cold LIA-like events............ 64

Map 1. Spatial visualisation of test results from the simulation analysis.65

Map 2. Spatial distribution of Early and Middle Neolithic
    artefacts and site types. 69

Figure 2. Early Neolithic flint axes found at Krågevoll in
    Klepp, southwestern Norway. 70

Chapter 3

Map 1. The location of Vik and Vinjeøra, central Norway.................... 78

List of Figures and Maps

Figure 1. Fragments of a glass beaker, manufactured within the Western Roman Empire. 82

Figure 2. Insular bronze fitting, from the ninth century A.D. 82

Map 2. The Ørland peninsula showing the excavation site at Vik. 83

Map 3. The shoreline at Ørland with the excavated areas at Vik marked. 84

Figure 3. Summed probability distribution (SPD) and Kernel density estimate (KDE) of radiocarbon dates from Vik archaeological excavations. 86

Figure 4: Last dates from Vik, areas A and C. 87

Map 4. The Vinjefjord area showing sites mentioned in the text. 89

Figure 5. Summed probability distribution (SPD) plots and Kernel density estimates (KDE) of radiocarbon dates from Vinjefjord archaeological excavations. 90

Figure 6. Settlement feature dates from Skeiet 2. 91

Chapter 5

Map 1. Map of Jutland with location of the current Nørre Vosborg manor house. 128

Map 2. Simplified drawing showing the position of the four (1–4) Vosborg manor houses. 129

Figures 1 and 2. Reconstruction drawings of Vosborgs 1 and 2 from the publication of the excavations. 131

Figure 3. Reconstruction drawing showing Vosborg 1 destroyed after the 1593 storm surge. 133

Figure 4. Aerial image showing the remains of Vosborg 2 in cropmarks. 135

Figure 5. Photograph of the extant Vosborg 3. 138

List of Figures and Maps

Chapter 6.

Figure 1. Climate as an actant in a network of architectural technology.. 149

Figure 2. Inserting drains down into cobblestone floors to secure a drier courtyard, an architectural adaption of the sixteenth century.. 153

Figure 3. Secondary built closed drains could have been a good way to achieve a drier indoor climate............................... 155

Figure 4. In eastern Norway, brick use in rebuilding contributed to warmer dwellings. 157

Figure 5. High-class home-owners installed tile ovens in their dwellings. 159

Figure 6. Building underground rooms enabled frost-free storage........ 161

Chapter 7

Figure 1. Number of reported forest fires in North Ostrobothnia and Lower Satakunta during the seventeenth century... 171

Figure 2. Mean temperature during JJA (June, July, August) of 1670–1677 and 1678... 172

Figure 3. Mean temperature during JJA (June, July, August) of 1690–1694 and 1695–1699. 173

Figure 4. Number of fire support applications in North Ostrobothnia and Lower Satakunta during the seventeenth century... 179

Chapter 8

Map 1. Map of Iceland's location with surrounding North Atlantic Ocean currents and sea-ice limits. 188

Map 2. Map of Iceland and its 22 traditional counties (*sýsla/* plural *sýslur*)... 191

Figure 1. Temperature and sea ice indices for Iceland........................... 195

List of Figures and Maps

Chapter 9

Map 1. Map by Jacob Nicolaj Wilse, 1768, covering Spydeberg rectory and associated fields. 211

Map 2. Location of Wilse's SMP weather stations in Østfold county. 214

Figure 1. Yield ratio at Spydeberg rectory for main crops 1769–1779.. 215

Map 3. GDD-model for Indre Østfold based on 1783 weather data.. 222

Map 4. GDD-model for Indre Østfold based on 1784 weather data.. 223

Figure 2. Population growth rates for Akershus County, Norway (excluding Akershus), and Øvre Borgesyssel deanery... 225

Figure 3. Number of births and deaths in Spydeberg (1696–1779) and Eidsberg (1765–1794). 226

Chapter 10

Map 1. City map of Turku. 240

Figure 1a. The mid-winter ice jam on 27 February 1887, taken from the Tuomiokirkkosilta Bridge. 242

Figure 1b. A 2025 photo taken from Tuomiokirkkosilta Bridge, showing the height of the stone quays from the water level between the bridges. 243

Figure 2. The Aura River ice jam index, 1739–2025, in Turku. 245

Figure 3. The ice-off dates in the Aura River, 1749–2025 252

Chapter 11.

Figure 1. Use of the terms 'Little Ice Age' and 'Medieval Warm Period' in English language books. 269

Chapter 13.

Figure 1. Visitor numbers for museums and cultural and natural history sites in Denmark for 2016 to 2023. 298

Figure 2. A climate historical timeline of the waning ice age. 300

List of Figures and Maps

Figure 3. The Laacher See today. 301

Figure 4. 'After the Apocalypse' in the making. 303

Figure 5A. Poster content, as positioned in the exhibition: a climate historical timeline of the closing millennia of the Pleistocene and its transition into the Holocene. 304

Figure 5B. Poster content, as positioned in the exhibition: a climate historical timeline for the transition from the Holocene to the Anthropocene. 305

Figure 6. A collage of images from the cabinets in the 'After the Apocalypse' exhibition 307

Chapter 14

Figure 1. National museums can take a practical and incremental approach to integrating climate narratives into permanent exhibitions. 329

Figure 2. National museums can foster critical reflection, community engagement and informed action for climate justice. 330

Introduction

INTEGRATING, CONNECTING AND NARRATING NORDIC CLIMATE HISTORIES

Dominik Collet, Ingar Mørkestøl Gundersen, Heli Huhtamaa, Fredrik Charpentier Ljungqvist, A.E.J. Ogilvie and Sam White

Rediscovering Climate History

The rapidly accelerating global warming of our time has reawakened interest in past periods of climatic change. While the scale and speed of twenty-first-century anthropogenic warming are unprecedented, historical experiences hold important information to situate the present. The past does not just provide the indispensable baselines for forecasting the future; it holds the available inventory of human responses to 'socialise climate'. Crucially, it also helps us connect the abstract forecasts of global computer models to 'lived' regional experiences.

The Nordic countries, specifically, have long been imagined in relation to their climatic setting. From the sixteenth-century descriptions of Olaus Magnus of the icy North and the idealisations of nineteenth-century romantic nationalism, their cold, harsh, snowy climate has been employed to set the Nordic countries apart (Sörlin 2024). Yet, just as elsewhere, interest in climate history has waxed and waned over time in accordance with perceived risks. The traditional entanglement of climate and society that characterised earlier natural histories fragmented during the nineteenth century with the proliferation of specialised academic disciplines. Attempts in the early twentieth century to enlist climate for the legitimation of colonialism and racism discredited the field for decades. Only during the 1970s did an integrated study of 'historical climatology' start to reemerge, building on the availability of new datasets. However, it was not until the current climate change became pressingly obvious, that the dispersed research communities of geographers, historians, archaeologists and other social and natural scientists once again galvanised into a coherent field.

Today, climate history is characterised by its ambitious 'big interdisciplinarity'. This connects the natural sciences and the humanities. Research in this

doi: 10.63308/63881023874820.intro

area explores both 'archives of nature' (tree-rings, ice cores, etc.) and 'archives of societies' (historical records, early observations, material culture). The collaboration among natural scientists, historians and climate communicators remains both challenging and constitutive. These communities bring their own methods, vocabulary and work routines (Haldon et al. 2018). As a result, climate history is characterised by plurality, heterogeneity and occasionally eclecticism. Quantitative and qualitative approaches coexist alongside local, regional and global perspectives. Different areas of research emphasise different 'proxies' – indirect sources of information – ranging from tree-ring data and sediment cores to agricultural records, poetry and songs (Pfister et al. 2018). Some climate histories explore single climatic periods or anomalies in detail, while others follow regional trajectories across centuries or even millennia. What unites this diverse field is the ambition to answer similar questions: What can the reconstruction of climatic trends and shocks add to our understanding of the past? What can these earlier climate–society interactions tell us about future scenarios, both biophysical and societal? What lessons can be learned from earlier failures and successes in climate change adaptation?

Nordic environments offer exceptional opportunities for climate histories. They are both particularly sensitive to changes in climate and richly documented. Nordic countries offer a consummate range of climate proxies, some unique to the region. Trees in high latitudes provide exceptionally accurate reconstructions of summer temperatures. Other proxies, such as sea- or lake-ice break-up dates, are much richer across Fennoscandia than elsewhere. In addition, the natural archives of Nordic countries, such as glaciers and freezing lakes and rivers, are located close to human populations, reducing the need to interpolate societal impacts over large distances. Societies in many parts of the Nordic countries also persisted in rather marginal ecological environments, where small changes in climate could have disproportionate impacts. They cultivated a broad repertoire of responses to climatic stress, including specific foodways, housing and heating practices as well as cultural and communal adaptations. As a result, climatic changes are often visible in the history of the North before they are seen to have affected warmer regions.

Paradoxically, this abundance of information has reduced the need for interdisciplinary collaboration. Scholars in Nordic countries have felt less pressure to reach across academic and national borders to supplement sparse data. Research into past climate has often been compartmentalised, with a lack of integration across natural and societal archives. Attempts to (re-)connect these data through a combined socio-natural approach remain rare. Compared to Central Europe and East Asia, climate history in the Nordic countries commands little institutional structure and support. Integrative research is often pursued by individual researchers or temporary research projects. As a result, the visibility of these approaches and their impact on public debate have been limited.

This collection aims to present climate histories from across the Nordic countries in a single volume. It was conceived during an international conference at the University of Oslo in May 2024. The meeting connected leading and emerging scholars in the field with practitioners in climate communication and governance. The papers in this volume have been selected to represent the whole scope of recent developments in the vibrant field of climate history. The collection is deliberately interdisciplinary in format, connecting climatologists, geographers, archaeologists, historians and museologists. It presents historiographical and methodological research, as well as histories from every Nordic country, from prehistory to the present. It includes in-depth case studies as well as reflexive meta-histories of longer pathways. These cover the period from the mid-Holocene to the twentieth century. The papers also draw on the whole range of available records from glaciology to material culture and poetry. Crucially, the collection also includes several texts that debate how these climate histories can be communicated. These investigate how museums and literature can bring these histories into conversation with a current audience looking for lived experiences of climate adaptation. Inevitably, some gaps remain. Denmark is covered only peripherally, and research with a strong (palaeo)climatological focus has not been included. However, the selected texts make extensive use of the available palaeoclimatic data and results.

The resulting volume is intended to provide an overview of the rich and rapidly growing scholarship in Nordic climate history and its connection to public debate in the Nordic countries and beyond. Its essays speak to various academic communities (climatology, archaeology, history, literature) and stakeholders (museum, climate communication and advocacy practitioners) as well as a wider public interested in the vibrant debate on climate adaptation and experience. Taken together, they explore timely questions: How did Nordic societies cope with climate variability in the past? What responses did the Medieval Climate Anomaly or the Little Ice Age trigger in the different societies they affected? Can these past experiences help to identify current shortcomings and reductions?

Variations in Climate in the Nordic Countries

The Nordic countries share a range of climatic characteristics. However, they also exhibit substantial internal variation in climatic conditions, not just over time, but also geographically. Norway is mountainous as opposed to Denmark and most of Sweden and Finland. Coastal Norway is also exceptionally mild for its latitude in winter – and wet year-round – as a consequence of prevailing westerly winds and the relatively warm North Atlantic currents. Southwestern Sweden and Denmark also have mild winters, but with far less precipitation than in Norway, and much warmer growing seasons. Conversely, interior northern Sweden and Finland have the coldest and longest winters. The Lake Mälaren region of east-central Sweden (around Stockholm), southeastern Sweden, and southeastern Finland enjoy the warmest – and also driest – summers, shorter in length but with temperatures temperatures resembling continental climate. Summers along the Norwegian west coast are cool and precipitation-rich due to a strong oceanic influence. The annual mean precipitation reaches 2,500–3,000 millimetres in parts of coastal Norway, with rainfall averages across the Nordic agricultural areas decreasing strongly from west to east. In the drier regions, mainly located in southeastern Sweden, late spring and early summer drought is a major challenge. Here spring precipitation is particularly low and irregular (Wastenson et al. 1995; Tveit et al. 2001).

For centuries, these climatological and ecological frameworks have influenced the livelihoods that supported European pre-industrial societies. During the 1961–1990 reference period, the average length of the growing season (daily mean temperature exceeding 5 degrees Celsius) exceeded 200 days in almost all of Denmark and in southernmost Sweden and surpassed 175 days in the major agricultural regions of Sweden and Norway. In almost all of Finland, the growing season was shorter. At the northern edge of agriculture in the Nordic countries, this is reduced to only 120 days and even today a warm summer is necessary for good harvests. This crucial variable not only shows large inter-annual and inter-decadal fluctuation – it has also varied over time with climatic change. During the climax of the Little Ice Age in the late sixteenth and early seventeenth centuries (Wanner et al. 2022) it was approximately six weeks shorter due to late and cold springs. In addition, in inland regions – due to a large day-to-night temperature range – the nominal growing season is often considerably shorter because of the occurrence of frost (Tveit et al. 2001). In these environments climatic shifts could have disproportionate agricultural effects.

Climate histories have identified some major trends in the past climate of the Nordic regions (Figure 1). Tree-ring based summer (growing season) temperature reconstructions reveal sharp and very strong volcanically-induced cooling, following the large AD 536 and AD 540 eruptions, lasting from 536 to at least c. 545, commonly called the Late Antique Little Ice Age (LALIA) (Büntgen et al. 2016). Summers were mainly warm, slightly exceeding mean 1961–1990 conditions, between c. 850 and 1050 and again, but less so, between c. 1150 and 1250. These warm periods are commonly collectively named the Medieval Warm Period (MWP) or Medieval Climate Anomaly (MCA) (Ljungqvist et al. 2016; Wang et al. 2023). Summers became generally colder, and possibly more variable,

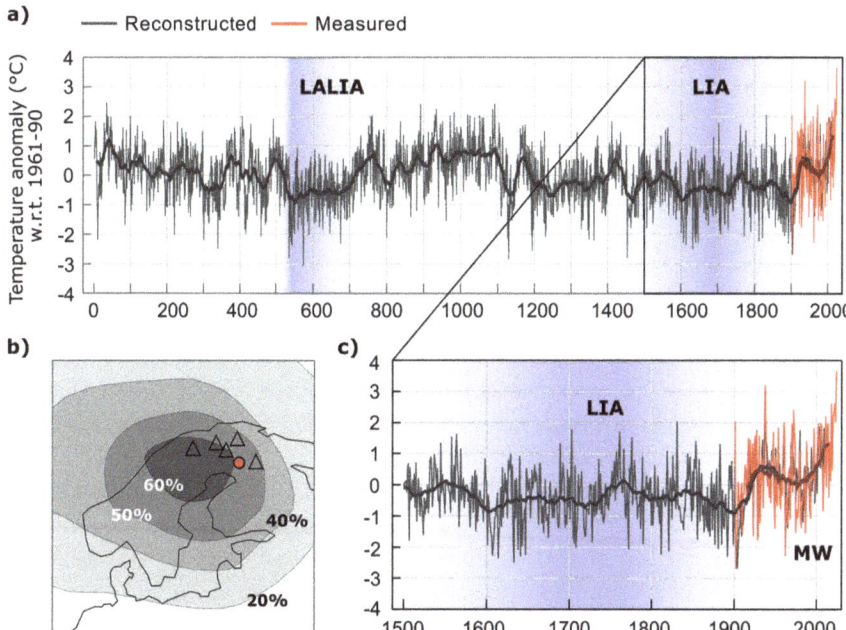

Figure 1. Temperatures past and present. (a) Tree-ring based reconstruction of summer (June, July, August) temperature anomalies (with respect to 1961–90 mean) over northern Fennoscandia in grey, and Sodankylä weather station recordings over the same months in red. (b) Locations of the tree-ring sites (triangles) used for the reconstruction and Sodankylä weather station (red dot). The percentages indicates how well the reconstruction explains the observed temperature variability over the Nordic regions. (c) Close up over the periods of the Little Ice Age (LIA) and Modern Warming (MW). Data sources: Matskovsky and Helama 2014; Morice et al. 2021; Finnish Meteorological Institute (Accessed 11 Nov. 2024 via https://www.ilmatieteenlaitos.fi/havaintojen-lataus).

after the mid-thirteenth century, marking an early onset of the Little Ice Age (LIA) (Wanner et al. 2022). The longest period of very cold summers occurred in the decades around 1600 and during parts of the nineteenth century. Little is known about the temperature conditions in the Nordic countries for other seasons prior to the sixteenth century. The longest period of very cold and long winters occurred in the second half of the sixteenth century and first half of the seventeenth century. In the mid-eighteenth century, winter temperatures, on the other hand, were as mild as in the late twentieth century (Leijonhufvud et al. 2010). Post-2000 winters and springs have most likely been the warmest for at least one, possibly over two, millennia in the Nordic countries. It is less certain whether recent summer warming has also been unprecedented. Importantly, the amplitude of temperature variability, across all timescales, appears to have been about three times larger in winter and early spring than during summer and autumn. Regarding hydroclimate, very dry summers seem to have prevailed during much of the twelfth century, the second half of the fifteenth century and early nineteenth century (Seftigen et al. 2017).

Because of its island location in the North Atlantic, the climate of Iceland is very different to that of the other Nordic countries. A key feature is its variability, due to its proximity to both warm and cold ocean and air currents. In the past, Iceland's climate was much affected by the sea ice that drifted southward to its shores on the East Greenland Current. There is considerable evidence that the climate was relatively mild around the time of the Viking Age settlement at the end of the ninth century, but the picture is complex. Historical records of climate change appear in the late twelfth century. Iceland is known for its excellent historical records of climate information. However, although they first appear in the late twelfth century, they do not become prolific until the beginning of the seventeenth century. From that time onwards the variability of the climate is evident. Yet frequent sea-ice incidence added to the severity of the climate. A warming trend began in the early twentieth century that has for the most part continued to the present day (Ogilvie and Miles, this volume).

The Historiography of Climate History in the Nordic Region

Writings on the impacts of climate change and variability in the Nordic countries may be found as early as the mid-sixteenth century. In his great opus of 1555, *Historia de Gentibus Septentrionalibus* (Description of the Northern Peoples) the Swedish writer and cartographer Olaus Magnus (1490–1557) refers to topics that include climatology, meteorology and hydrology. Icelanders were particularly precocious in their interest in the topic. In the late-sixteenth century, the scholars Arngrímur Jónsson and Oddur Einarsson wrote treatises on sea ice and climate that were followed in the seventeenth century onwards by similar accounts (Ogilvie 2005; 2022). One of the earliest works anywhere on what would now be termed climate-impact studies must be that by the Icelandic Bishop, Hannes Finnsson (1739–1796) *Mannfækkun af Hallærum* (Loss of Life as a Result of Dearth Years) first published in 1796 (and in Danish translation, 1831; Ogilvie 2005).

As regards early instrumental observations, scientists from mainland Scandinavia were among the pioneers in this field. A notable example is Anders Celsius (1701–1744) an astronomer, physicist and mathematician from Uppsala in Sweden. The temperature scale he invented is the principal one in use in the world today (Lundstad et al., this volume). The Enlightenment movement and the attendant 'Age of Reason' of the seventeenth and eighteenth centuries brought an increase in knowledge in all spheres and provided a basis for an increasing interest in climate and its effects (Gundersen, this volume). To some extent, the study of climate was the province primarily of practitioners of medicine who were interested in the influence of climate on diseases and epidemics. Related to this, and in parallel with the development of early meteorological observations, came an interest in gathering historical documentary information on climate. Such information was assembled in compilations which later on were much used by the pioneers of historical climatology. Some fifteen such compilations were published in different countries, with three of them relating to Nordic regions.[1] Generally, as no attempt was made at historical source analysis, the compilations contain a mixture of reliable and unreliable information (Bell and Ogilvie 1978).

Discussions on historical climate change in the North Atlantic and Nordic regions had begun by the early nineteenth century. An example is the account entitled *Om Climaternas rörlighet* ('On Climatic Variations') published in 1824 by the Swedish scholar and diplomat, F.V. von Ehrenheim (discussed in Ogilvie and Jónsson 2021). In the early years of the twentieth century, a discussion among academics in Iceland, Norway and Sweden focused on the climate of Scandinavia in medieval times. In 1913, the Swedish oceanographer Otto Pettersson published a paper entitled 'Climatic Variations in Historic and Prehistoric time' (Pettersson 1913, 1914). He asserted not only that climates vary but also that climate was an important causal factor in Scandinavia's economic and demographic decline in late medieval times. In spite of his pioneering work on the past climate of Iceland, Thorvaldur Thoroddsen did not agree (Thoroddsen 1914; Ogilvie and Miles this volume). The Norwegian historian, Edvard Bull, then entered the debate, calling for a more comprehensive analysis. Bull showed himself to be ahead of his time by making a plea for accurate and reliable sources in historical climate analyses (Bull 1915). The debate continued with, among others, Speerschneider, in his compilation on ice conditions, and Arnold Norlind, who in 1915 published *Till frågan om det historiske klimatet*,

---

1. Amongst these are *Einige Bemerkungen über das Klima der historischen Zeit nebst einem Verzeichnis mittelatlerlicher Witterungserscheinungen* ('Some Remarks on the Climate of Historical Times, together with a List of Medieval Weather Phenomena') by the Swedish historical geographer Arnold Norlind (1783–1929) published in 1914 and *Om Isforholdende i Danske Farvande i aeldre og nyere tid, aarene 690–1860* ('On Ice Conditions in Danish Waters in Early and Recent times, the Years 690–1860') by the Danish meteorologist C.I.H. Speerschneider (1915). For Iceland, the compilation *Árferði á Íslandi í púsund ár* ('The Seasons in Iceland in One Thousand Years') was published in 1916–17 by the geologist and geographer Thorvaldur Thoroddsen (1855–1921) (Ogilvie and Miles, this volume).

särskilt i Nord- och Mellaneuropa ('With Regard to the Question on Historical Climate, especially in northern and central Europe'). In the early 1920s, the Norwegian humanitarian, scientist and explorer Fridtjof Nansen (1861–1930) took up the debate. Nansen believed that sea ice had been extensive in medieval times and remained sceptical of warm conditions in the centuries around AD 1000. What ended this discussion, however, was the growing realisation that the climate was actually changing. This point was brought home by another Scandinavian scientist, the Danish geologist, Lauge Koch, in his major work on Arctic sea ice, *The East Greenland Ice* (1945).[2]

In spite of this early pioneering scholarship on historical climates in the North Atlantic, the possible role of the effects of climate variability fell out of favour in mainstream Scandinavian discourse from around the 1930s. For example, the Norwegian agrarian historian Sigvald Hasund suggested that the Black Death pandemic alone was responsible for the late medieval crisis (Hasund 1934). This became the prevailing view for decades.

A new era may be said to have begun in the early 1970s with a renewed international interest in what was then becoming known as 'historical climatology' (see, e.g., Ingram et al. 1978). A pioneering contribution to this field was made by the British climatologist Hubert Lamb (1913–1997), the first Director and founder of the Climatic Research Unit (CRU) at the University of East Anglia in the UK. Lamb coined the term the 'Medieval Warm Period' and began to use the 'Little Ice Age' to refer to a period from around AD 1400 to 1800. Lamb spoke Norwegian and had a high regard for his Scandinavian colleagues, whom he encouraged and collaborated with. These included Erik Wishman from Norway who, together with colleagues at the Stavanger Museum of Archaeology, established a 'national historical-climatological database' supported by the Norwegian Research Council.[3] Knud Frydendahl of the Danish Meteorological Institute established a project to search for documentary climate information and collaborated closely with Lamb, including on a compendium of historical North Sea storms (Lamb and Frydendahl 1991). A student of Lamb's, Astrid Ogilvie, decided instead to focus on Iceland because of the wealth of historical information on climate and climate impacts (Ogilvie 1982 et seq., Ogilvie and Miles, this volume). She continued the collaboration established by Lamb with Scandinavian colleagues such as with the meteorologist Øyvind Nordli (see, e.g., Nordli et al.) but in particular with Icelanders, including the meteorologist Trausti Jónsson who has undertaken pioneering work in early instrumental data (Jónsson and Garðarson 2001).

---

2. An expanded discussion on these early debates may be found in Ogilvie and Jónsson 2001; Huhtamaa and Ljungqvist 2021.
3. https://uis.brage.unit.no/uis-xmlui/handle/11250/216733?locale-attribute=en

A key development within the Nordic region during the 1970s included the Scandinavian Research Project on Deserted Farms and Villages (*Det nordiske ødegårdsprosjekt*). As early as 1951 the archaeologist Axel Steenberg had suggested that climate could have played a part in the abandonment of farms in Denmark during the fourteenth century. The 1970s project on Deserted Farms and Villages, however, represented a larger collaborative project among Nordic historians to systematically map the desertion and abandonment of farms (*gårdsbruk*) in the Nordic countries during the late Middle Ages. Although the possibility of effects from climate variability was a significant part of the project, no firm conclusions were made as to the possible role of climate for the late medieval crises in any Nordic country (Holmsen 1978; Sandnes 1978; Sandnes and Salvesen 1978). Other scholars later revisited the issue. For example, the Norwegian historian Audun Dybdahl wrote on climate, dearth years and crises in Norway during the past 1,000 years (Dybdahl 2016) and suggested that historians since Hasund's days had probably underestimated the role of climatic cooling in Norway's late medieval crises. The Danish historian Nils Hybel also implicated climate variability in Denmark's famines during the medieval period (Hybel 1997, 2002).

An important theme in the study of climate impacts in Nordic history studies has been the effects of climate on agriculture. In the 1950s, Swedish economic historian Gustaf Utterström undertook pioneering studies focused on the early modern period. Utterström (1954, 1955) tentatively linked changing climatic conditions to harvest variations and demographic fluctuations, emphasising that the coldest years and decades generally coincided with large-scale harvest failures in early modern Sweden. As in Sweden, early studies on climate history in Finland focused primarily on the early modern period. In 1975, Finnish historian Eino Jutikkala undertook one of the first studies of the adverse effects of colder climatic periods for Finnish agriculture and society. His study highlighted the use of data from the natural sciences to understand past climate, as well as the impact of hunger-related infectious diseases on mortality following harvest shortfalls (Jutikkala 1975; see also Jutikkala 1994, 2003). In the 1980s and 1990s, quantitative examination was established as one of the main approaches within climate history research, as well as a closer following of the state-of-the-art research in palaeoclimatology. Subsequently, concepts such as the Little Ice Age were introduced in historical research (Seland, this volume; Tornberg 1989, 1992).

More recently, historian Sven Lilja reassessed links between early modern Swedish climate, harvests and demography (Lilja 2006, 2012), but his conclusions were tentative due to a lack of quantitative approaches and interdisciplinary collaboration. A study by Edvinsson et al. (2009) presented quantitative estimates of climate–harvest relationships in 1724–1955, concluding that summer temperatures were of major importance only in the sparsely populated northern two-thirds of Sweden. In the main population areas of Sweden, drought would have been the major hazard for agriculture instead. This was confirmed by Skoglund (2022, 2023, 2024), although Ljungqvist et al. (2023) showed that cold conditions, in addition to drought, had an adverse effect on grain yields in most major agricultural areas in Sweden.

The past decade has brought a growing number of interdisciplinary investigations of climate history, with emphasis on identifying societal vulnerabilities, resilience and adaptation strategies to climate. Heli Huhtamaa has studied the effects of climatic variability on crop yields in pre-modern Finland, from medieval times to the nineteenth century, as well as socio-economic and demographic consequences of harvest failures (Huhtamaa et al. 2015, Huhtamaa 2018). Her research has uncovered important differences in resilience to harvest failures among different social groups in Little Ice Age Finland (Huhtamaa and Helama 2017; Huhtamaa et al. 2022). Furthermore, Skoglund showed that eighteenth- and nineteenth-century agriculture in southern-most Sweden could be notably resilient to climate shocks, as even severe drought years had relatively modest impacts on grain agriculture in this fertile area (Skoglund et al. 2022, 2024). However, for environmentally marginal locations, such as Iceland, it is easily demonstrable that variations in climate had a substantial impact on vegetation growth, famines and related issues; and considerable research has been undertaken for this location (see, e.g., Bergthórsson 1985; Ogilvie 2001, 2020; Júlíusson 2021; Jónsson, G. 2023; Ogilvie and Sigurðardóttir, forthcoming).

More scholars have come to recognise the potential benefits of interdisciplinary research into the human dimensions of past climate change, and the value of local Indigenous knowledge has become more apparent. Although beyond the scope of this introduction, there are now numerous examples of how the use of palaeoclimatic data from the natural sciences can be used together with historical documentary data and archaeological data in order to assess climate impacts in the past. A current venture in the discipline of historical climatology and related studies is the NORLIA project.[4] Furthermore, recent advances in the disciplines of both history, archaeology and palaeoclimatic studies offer new opportunities to bring such investigations to fruition. This volume is a contribution to current debates on climate changes and adaptation through a connected, socio-natural, approach.

---

4. Center for Advanced Study Oslo, The Nordic Little Ice Age (1300–1900): Lessons from Past Climate Change: https://cas-nor.no/intranet/project/nordic-little-ice-age-1300-1900-norlia.

Methods and Approaches

Beyond well-established short-run impacts of climate and weather on agriculture and prices lie difficult questions about the long-term influence of climatic variability and change in Nordic history more generally. How should climate figure (if it figures at all) in the stories of economic and demographic developments, military victories and defeats, political crises and revolutions, or cultural and religious movements that fill the pages of conventional national history books (see Seland, this volume)? This question raises challenging issues that frame the contributions of this volume.

Scholars often broach the question of the long-run influences of climate in the context of powerful global climate events, usually in the wake of large volcanic eruptions, which historians have associated with episodes of crisis and transformation in world history. These include the 'Late Antique Little Ice Age' (LALIA) and crisis of the sixth century (Büntgen et al. 2016); famine, plague, and the 'Great Transformation' of the fourteenth century (Campbell 2016); Europe's crisis of the 1590s (Clark 1985) and the following 'General Crisis' of the seventeenth century (Parker 2013); as well as social and cultural turmoil, such as witchcraft trials and famine catastrophes during the Little Ice Age (Ljungqvist et al. 2021, 2024).

These world events may contextualise historical developments in Nordic countries and provide a framework for investigating possible long-term roles of climate variability and change in the region. At the same time, Nordic history may well present distinct features that test and refine theories and narratives developed in other parts of the world. For example, 'maximalist' interpretations of catastrophic impacts during the LALIA and the crisis of the sixth century have become hotly contested among scholars of the eastern Mediterranean, where much of the original research has focused. These debates may caution Nordic scholars against *a priori* assumptions of climate-driven catastrophe, but they hardly invalidate evidence of settlement abandonment, migration and social reorganisation in Nordic regions (Anagnostou et al. 2024). Moreover, these debates highlight the value of further local and comparative studies that help explain why some societies survived the sixth century intact and others faced crises (such as Ystgaard and Sauvage, this volume). A number of recent studies on the sixth century in Scandinavia clearly demonstrate how climate change, although a global phenomenon, can have very different local consequences, depending on both environmental and social factors, such as topography and food strategies (see Gundersen and van Dijk, forthcoming).

This raises the question of why some communities, even within the same climate zone, were more susceptible to climate shocks than others. It also contests the idea of past climate crises as something uniform and highlights the need to integrate both environmental and social data in analyses when trying to understand the wider societal ramifications of climate change in the past. Abandonment of Nordic farms, both before and after the Black Death, has been an important piece of evidence for the role of climate change in the Late Medieval Crisis in Europe, centred around the fourteenth century. Yet local tree-ring evidence indicates that northeastern Europe did not experience the same cold, wet summers that characterised the 'Great Famine' in Western Europe during the 1310s (Huhtamaa 2020). Utterström's classic (1955) study on climatic cooling and harvest failures in early modern Nordic countries helped draw attention to the 'European crisis of the 1590s'. Yet recent studies have emphasised that the crisis during this period – in the sense of a turning point in political and economic fortunes – focused on the Spanish and Ottoman Empires rather than parts of northern or western Europe usually considered more vulnerable to climatic cooling (Parker 2018; White 2017).

Within a longer perspective, the consequences of climatic impacts also varied among and within the Nordic countries. Health and prosperity in all Northern Europe depended to some degree on the seasons well into the industrial era; nevertheless, the continuing vulnerability of Iceland and Finland to weather-related famines as late as the nineteenth century constitutes a significant element in those countries' economic and demographic development compared to Sweden and Denmark (Huhtamaa 2018; Ponzi 1995; Ogilvie and Miles, this volume). Within these divergent national climate histories are also diverse local stories: sites of higher or lower mortality from hunger or disease as they related to diverse systems of taxation and landholding or different strategies of land use and migration (e.g., Huhtamaa et al. 2022; Starlander, this volume).

These problems of scale highlight the need for clear causal analysis when assessing the roles of climate in Nordic history. Climate alone did not 'make history' in Nordic countries or anywhere else. Yet some *distinctions* about past climate and weather made significant *differences* in Nordic history. For example, climate alone does not explain why there were famines and witch hunts in the early modern Nordic countries. However, the timing and severity of the worst cold spells during the Little Ice Age can help explain the timing and severity of the worst famines and some Nordic witch hunts, as well as why those witch trials focused on accusations of weather magic. Rather than offering arguments that climate did (or did not) determine Nordic history, studies in this volume relate features of variability and change in past climate to differences in demographic, economic or cultural developments across diverse scales, from the local to the regional and from years to centuries.

As new studies emphasise, causation in climate history also calls for consideration of historical context, structures and agency. Climate and weather

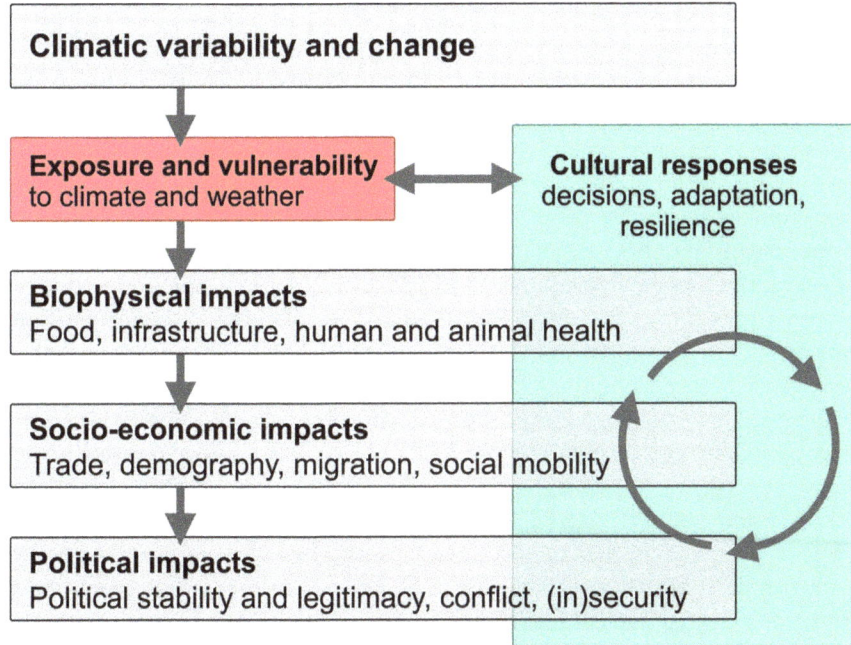

Figure 2. A simplified model of climate-society interactions often used in climate history. It conceptualises climate impacts cascading through the various societal levels mediated through pre-existing vulnerabilities, mitigated by cultural arrangements on all levels.

could be unpredictable yet consequential contingencies at critical moments in history: for example, Denmark's stunning military defeat and political transformation following the invasion of Swedish soldiers into Zeeland across the frozen North Sea straits in the winter of 1657–58. Nevertheless, the frequency and severity of most climate and weather impacts – historically, as in the present day – were a function of exposure and vulnerability to weather and to meteorological hazards such as frosts and floods. In short, they were not just about the weather itself. Cultural responses, including individual and political decisions and the adaptability and resilience of institutions, also mediated the scope and consequences of climate- and weather-related disasters and the risks of cascading effects such as migration or conflict (Figure 2). Studies in this volume highlight context and cultural responses associated with the influence of climate in Nordic history – as well as possibilities and challenges in using historical examples to explain current issues of climate, causation and agency to a wider public (Melo et al. and Riede, this volume).

Climatic variability and change affect societies mainly through their exposure and vulnerability to extreme weather and meteorological hazards, such as droughts, floods, frosts and heatwaves. Experience of extremes and hazards also generates cultural responses: individual and collective decisions and adaptation measures, including improvements to infrastructure and public or private disaster relief. These cultural responses, in turn, influence future exposure and vulnerabilities. The most direct impacts of extreme weather and hazards were usually biophysical, affecting food production, physical infrastructure, and health (Ljungqvist et al. 2021). Biophysical impacts could, in turn, affect economies (e.g., prices and employment) and populations (e.g., mortality and migration); and socioeconomic disruption could, in its turn, produce political effects, including disruptions to political stability and legitimacy or crime and violent conflict. The worst crises of the Little Ice Age (e.g., White 2011) involved feedback loops among biophysical, socioeconomic, and political impacts, which brought recurring disasters, population loss and displacement, and long-term conflict and political instability. However, these 'cascading effects' from one type of impact to another are mediated through cultural responses. Timely and effective decision-making, as well as resilient political, economic and cultural institutions could prevent disasters from spiralling into crises.

Ways of Knowing Nordic Climate

The contents of this book can be read in the chronological order they are presented, an arrangement chosen for the sake of convenience. Such a deep dive reveals long pathways and the continuity of exposure that characterises the Nordic world. Yet the collected papers also disclose an array of common themes. They contain the persistent dynamic of hazards and adaptation in response to ice, cold, floods, fires and storms. Instead of fixed responses, the studies unveil plural and creative choices that might not usually be associated with pre-industrial societies. They also highlight how these entanglements over time condensed into 'ways' (Richter 2020) of knowing climate. The papers illustrate how these experiences were expressed through building techniques, trading networks and settlement patterns, but also through lifestyles, poetry, and stories.

Opening the volume, Lundstad et al. provide the first major overview of early Nordic meteorology and discuss how prevailing theories on climate and weather influenced how weather data were observed and analysed. In the second section, they introduce the data available and analyse how it can be used to reconstruct the temperature increase in the Nordic capitals up to this day. Similarly, Ogilvie and Miles compile evidence for sea-ice occurrence off the coast of Iceland in the early modern period by using contemporary historical records.

They stress that Iceland is characterised by distinct climate conditions separate from those in continental Europe – a key finding of Lundstad et al. as well.

Two papers focus on architectural adaptations in response to Little Ice Age climate conditions – a topic that is receiving renewed attention in the light of current challenges. Both focus on late medieval high-status residences. Kerr draws our attention to the severe storm surges that ravaged coastal Denmark during a key period of the Little Ice Age and the evidence for estate relocation. According to Kerr, the moving manors demonstrate resilience among the ruling elite to climate extremes. Reinfjord focuses on the buildings themselves and explores how they were modified to accommodate colder and wetter weather. He focuses on prominent alterations in late medieval buildings in Norway, such as drains, tile stoves, stone cellars and building materials, and argues that these improvements were entangled with climatic stress.

The dynamic 'socialisation' of climate hazards is explored in three more papers. Norrgård investigates historical ice jams in the Aura River, Turku, Finland. He compiles an ice-jam index from historical records and newspapers. This tool helps him to analyse not just the severity of the incidents. He can also show how the anthropogenic built environment contributed to their development. Starlander explores the impact of wildfires and climate on early modern Finnish peasant communities. He highlights that building resilience among the Finnish rural population against these hazards depended on dynamic cooperation, between the peasants, communities and authorities. Damm traces similar interactions in response to volcanism in medieval Iceland. Using 'vulnerability' as an analytical lens, she situates her study at the intersection of external shocks and communal responses. Her research unveils the careful balance between practical measures and religious interpretations. All three studies illustrate how societal vulnerability to hazards depends not just on environmental stress, such as the climate, but also on human practices and responses.

Foodways and agricultural responses permeate many papers. The contribution by Nielsen is a strong reminder that the impact of Little Ice Age type events depended on the coexistence of risk drivers across the natural and human domains. His reflexive paper hypothesises that a mid-Holocene cold event should have had a negative impact on human demography in southern Norway. However, the reconstructed population trajectory shows little influence, prompting him to explore alternative pathways and explanations. Ystgaard and Sauvage use 'vulnerability' and 'resilience' to unlock these dynamics in two Iron Age farming settlements in mid-Norway. Both settlements seem to have experienced a profound crisis during the mid-sixth century LALIA climate anomaly. Yet, their analysis also explores the long-term impacts beyond the immediate disaster. Their research stresses plurality, revealing how different social and environmental conditions resulted in one settlement becoming permanently abandoned, while the other seems to have been able to reorganise and reoccupy farming lands after the climate shock. While recognising the sixth century as a time of crisis, their interpretations support the recent challenge to a maximalist approach by highlighting diversity in disaster impact and human responses (see Anagnostou et al. 2024; Gundersen and van Dijk, forthcoming). A similar approach can be found in Gundersen's socio-natural study of harvest failures in eighteenth-century Norway. He deploys a growing degree days (GDD) model to critically investigate the entanglements of meteorological and agricultural data. His study flags that local farmers were locked into a narrow farming strategy that made them highly vulnerable to climatic shocks. In such a built ecology, the development of a crisis depended not just on local climatic but also on regional societal factors. Conversely, due to this linkage, climatic stress could (and did) result in socio-political reform and rupture in late-eighteenth-century Norway.

Two more papers investigate how these persistent shocks and their influence on Nordic lifeworlds have often been marginalised. Kleveland revisits a long-neglected part of Norwegian literature, glacier poetry. As a motif, a symbol and even an agent, glaciers have a long tradition in Norwegian poetry that has not been previously acknowledged. This is reminiscent of how the climate was discussed in public histories as late as the twentieth century. Seland's historiographic examination of multi-volume popular histories of Norway shows that climate was recognised early on by archaeologists and medieval historians, but often as a stage for human action rather than an active element in human and environmental history. Modern-era historians devoted even less attention to climate. Many treated it as a static backdrop. The growing awareness of past climate change among scientists, historians, and archaeologists alike, failed to surface even in the latest editions. Only in the twenty-first century did climate find a reluctant place (again) in historical narratives.

The complexities of narrating climate are unfolded at the culmination of this volume. The contribution by Melo, Þórsson, Riede and Norrgård studies how climate history is framed and presented in Nordic national museums. Their permanent exhibitions demonstrate the willingness to incorporate climate. Yet its impact is rarely integrated into their storytelling, often appearing as isolated fragments disconnected to overall narratives and storylines. In response, the authors propose six key strategies to guide and inspire critical engagements and collective action. They suggest tools to help museums incorporate climate narratives, including the use of the past to situate future-oriented perspectives. Riede's paper departs in a similar direction. He reflects on his exhibition covering the consequential Lacher See eruption (~13,000 years BP). The museum intervention used the disruption of Paleolithic populations to envision critical vulnerabilities of modern societies experiencing anthropogenic global warming. The exhibition, named 'After the Apocalypse', sought to highlight the potential to learn from past existential risk. Both these papers explore main themes of the volume – the potential of past events to situate, contextualise, challenge and to enrich our debates on human–climate interactions in the present.

The Future of Past Climate

Three critical points lend relevance to climate histories today. First, they help improve and ground our *understanding of past societies*. Studying past communities, particularly pre-industrial ones, without their climatic context ignores a crucial variable shaping their choices and pathways. Marginalising past climatic environments also reinforces the modern fallacy of neglecting ecological context when imagining societal change. Reconnecting climatic and societal development can help to make sense of the timing and regional coherence of historical changes.

Second, climate histories advance the development of *scenarios for the future*. Highly resolved climate histories provide baselines for predictions and planning for both biophysical and societal trends. The combination of natural and societal archives produces more precise climate and weather reconstructions and models. Their integration also allows higher accuracy in calculating societal impacts and the range of human responses (Adamson et al. 2018; Haldon et al. 2024).

Third, they provide a way to *orient ourselves* in regard to current climate change. Technical scenarios expressing climatic change in global degree changes or CO_2 gigatons, can be challenging to process for non-scientists. Climate histories ground these scenarios in human experiences. They connect forecasts with actual events and situate current challenges within collective memories and heritage. Climate histories provide narratives that make abstract models actionable, answering questions such as: What might a 1.5 degrees Celsius change entail? Which type of responses were effective and which failed? Crucially, regional climate histories, such as the ones presented here, reduce planetary abstractions and bring climatic challenges closer to home.

In this way, climate histories have the potential to challenge the persistent gap between what we know and what we do. However, to succeed in these tasks, research into the field of climate history must overcome substantial challenges. We identify six areas where improvement is necessary and desirable:

- **Achieving true interdisciplinarity.** The impacts of climate change cross boundaries and reach every aspect of our lives, while our research infrastructures remain fragmented geographically and institutionally. Collaborating across the entrenched divisions of current universities and research centres requires more than merely reading each other's papers. Understanding the methods, biases, and gaps of other fields demands sustained and intense forms of co-working and co-creation (Haldon et al. 2018; van Bavel et al. 2019; Degroot et al. 2022; Ogilvie et al. 2024). Currently, the Nordic countries offer limited support for 'big interdisciplinarity' across the sciences and the humanities. Few climate-focused centres host such research.[5] The design of Nordic research calls and expected career tracks often remain within disciplinary limits. At the same time, a range of contemporary collaborative projects demonstrates the potential of tightly integrated climate histories.[6] Their experience suggests that successful interdisciplinarity in this field requires the close personal and spatial integration of research and work best when embedded in the research design from the start.

- **Pathways rather than determinisms.** Earlier climate histories often posited direct relationships of cause-and-effect between climatic impacts and societal developments. Current research instead highlights the *interaction* of environment and society. The papers in this volume illustrate how effects of climatic shifts can cascade through the various levels of society from agriculture to the economy, then on to politics and culture, often with decreasing intensity (Figure 2). At the same time, societal constellations and traditions moderate how and if these effects gain influence. In close-up studies, climatic effects often serve as a catalyst rather than a cause. They seem to accelerate change that is already occurring or escalate tensions already in place. Climate histories have developed specific interaction models to safeguard against 'determinism' and 'reductionism'

---

5. Cf. the Oslo Center for Environmental Humanities (OCEH), the Greenhouse Centre for the Environmental Humanities at the University of Stavanger, the Bolin Centre for Climate Research or the KTH Environmental Humanities Laboratory in Stockholm, the Centre for Environmental Humanities at Aarhus University or The Stefansson Arctic Institute, Akureyri, Iceland.

6. Cf. the ClimateCultures and VIKINGS projects at the University of Oslo (https://www.hf.uio.no/iakh/english/research/projects/climatecultures/; https://www.mn.uio.no/geo/english/research/projects/vikings/), the VICES (Volcanic Impacts on Climate, Environment and Society) team at Bern (https://www.hist.unibe.ch/forschung/forschungsprojekte/forschung_wirtschafts_sozial_und_umwelt_geschichte/vices/index_ger.html), the Nordic ARCPATH (Arctic Climate Predictions: Pathways to Resilient, Sustainable Societies: https://arcpath.nersc.no/) or the Icelandic MYSEAC (The Mývatn District of Iceland: Sustainability, Environment and Change ca. AD 1700 to 1950) projects and the Adaptations to Climate Change in the Northern Baltic Region ca 1500–1900 study at Stockholm University (https://www.su.se/english/research/research-projects/adaptations-to-climate-change-in-the-northern-baltic-region-ca-1500-1900).

(Hulme 2011). This includes the reflexive use of terminology such as 'collapse', 'resilience' or 'sustainability' borrowed and translated from ecological sciences to social settings often without critical evaluation of their conceptual baggage. Future research should be careful to heed these qualifications despite the current demand for single-factor solutionism. Highlighting the ecological guardrails and historical pathways in which climate effects take indirect or delayed action can help to advance current debates.

- **Regional dynamics to qualify global trends.** Climate science has often focused on global perspectives and planetary averages. While these abstractions helped to identify the large-scale trends of anthropogenic warming and facilitated international political action, they also severed climatic shifts from personal experience and regional heritage (Adamson and Rapson 2024). Climate histories have in part mirrored this trend, with earlier research taking a larger view, often productively challenging the national foci of historical research. However, the proliferation of more highly resolved data has now made it possible to go beyond planetary abstractions and identify regional trends. One example is the intense debate around the Little Ice Age and its uneven regional duration, amplitude, seasonality and societal effects (Pfister and Wanner 2021; Wanner et al. 2022). The focus on more regionally resolved archives such as tree-rings has helped to highlight spatial variations. This trend to regional scopes has not only helped qualify current models, both past and present. It also creates the opportunity to challenge the myth that consequences are limited to distant areas and to reconnect climate histories with local traditions, collective memory and cultural heritage instead. This link has been identified as crucial in translating planetary science into local action (Adamson et al. 2018). We suggest that regional climate histories have an important role to play in establishing these relationships.

- **Identifying risk drivers.** Climate variability can have very different consequences both within and between societies. Local climates, topography, food strategies, networks, foreign relations, socio-economic structures, past experiences, decision-making processes and the complexity of the climate system itself are a few examples of key variables that contribute to different types of impact and human responses. Understanding these variables is particularly important for understanding the spatiotemporal variability present in the archaeological and historical records, but also for our ability to draw lessons from the past to inform future solutions. It is crucial for future research in climate history to address why some situations developed into crisis, and others did not, and identify the various risk drivers that significantly influenced the course of events.

- **Broadening the repertoire for action.** Current debates on how to cope with a changing climate often stay close to established precedents. In the midst of turbulence, it can be difficult to envision transformation outside familiar trajectories. Climate histories can challenge the technology-centred, top-down debates that dominate our present. Past societies developed a range of responses that could not rely on international accords or techno-fixes (Haldon et al. 2018). The papers in this volume highlight a broad mix of mitigating reactions that range from new foodways, improved legal codes, advanced risk-sharing arrangements, broadened political participation to new forms of cultural memory and communication of hazards. From a modern perspective, these climate histories can highlight that the current focus on technologies might be too restrictive and that adaptations need to be flanked by social transformation to be effective.

- **Narrating climate.** The climate sciences often stop short of providing a narrative for their results. This can be the consequence of disciplinary traditions of evidencing, an abstract systems-oriented perspective or a lack of information-sharing when designing future scenarios (Daniels et al. 2009). Similarly, earlier historical work on past climate often relied on laden tropes and emplotments to provide meaning. Imaginaries of collapse, decline and disaster often served as the silent referents of many older works (Carey 2012). The studies in this volume illustrate that a broader, reflexive approach to narration increases both relevance and impact. This can include creative entanglements of abstract data and individual historical experience or storyline approaches that visualise research in the form of exhibitions or public interventions. They challenge the reductionism of declensionist tropes and include creative, place-based ways of 'socialising climate'. The narration of complex cascades of interaction between climatic stress and societal responses can now draw on a range of recent storytelling and interaction designs (Ljungqvist et al. 2021; Melo 2023; White et al. 2023). Including them has the potential to overcome the fatalism and passivity that the scale and abstraction of planetary models often instill in audiences.

Climate histories that creatively embrace interdisciplinarity, that explore dynamic pathways rather than rigid causalities, that operate on an accessible regional scale, that challenge the limitations of current mitigation discourses, and that reflexively employ narrative strategies can fill important gaps. These include the trajectories of *past* societies, the pathways for the *future* and the communication of climate challenges in the *present*. In this sense, the future of past climate can and should be an integral supplement to our presentist discourses.

Acknowledgements

The editors would like to thank *UiO: Energy and Environment* for funding the initial meeting in Oslo; and the *Norwegian Research Council* (project nr. 315441) and the *Centre for Advanced Study (CAS) at the Norwegian Academy of Science and Letters* (NORLIA) for supporting its publication. Our heartfelt thanks go to Daniel Rogoża-Žuklys for his help in finalising the manuscripts.

Bibliography

Adamson, G. and J. Rapson. 2024. 'Weather, heritage, and memory'. *WIREs Climate Change* **15** (6): e913. https://doi.org/10.1002/wcc.913.

Adamson, G., M. Hannaford and E. Rohland. 2018. 'Re-thinking the present: The role of a historical focus in climate change adaptation research'. *Global Environmental Change* 48: 195–205. https://doi.org/10.1016/j.gloenvcha.2017.12.003.

Anagnostou, E., J. Linderholm and K. Lidén. 2024. 'The AD 536/540 climate event in Sweden – a review'. *Boreas* **54** (1): 1–13. https://doi.org/10.1111/bor.12672.

Bell, W.T. and A.E.J. Ogilvie. 1978. 'Weather compilations as a source of data for the reconstruction of European climate during the medieval period'. *Climatic Change* 1: 331–348. https://doi.org/10.1007/BF00135154.

Bergthórsson, P. 1985. 'Sensitivity of Icelandic agriculture to climatic variations'. *Climatic Change* 7: 111–127. https://doi.org/10.1007/BF00139444.

Bull, E. 1915. 'Islands klima i Oldtiden'. *Geografisk Tidsskrift* 23: 1–5. https://tidsskrift.dk/geografisktidsskrift/article/view/46893.

Büntgen, U., V. Myglan, F.C. Ljungqvist et al. 2016. 'Cooling and societal change during the Late Antique Little Ice Age from 536 to around 660 AD'. *Nature Geoscience* 9: 231–236. https://doi.org/10.1038/ngeo265.

Carey, M. 2012. 'Climate and history: A critical review of historical climatology and climate change historiography'. *WIREs Climate Change* 3: 233–49. https://doi.org/10.1002/wcc.171.

Daniels, S. and G.H. Endfield. 2009. 'Narratives of climate change: introduction'. *Journal of Historical Geography* **35** (2): 215–222. https://doi.org/10.1016/j.jhg.2008.09.005.

Degroot, D. et al. 2022. 'The History of climate and society: A review of the influence of climate change on the human past'. *Environmental Research Letters* **17** (10): 103001. https://doi.org/10.1088/1748-9326/ac8faa.

Dybdahl, A. 2012. 'Climate and demographic crises in Norway in medieval and early modern times'. *The Holocene* **22** (10): 1159–1167. https://doi.org/10.1177/0959683612441843.

Edvinsson, R., L. Leijonhufvud and J. Söderberg. 2009. 'Väder, skördar och priser i Sverige'. In B. Liljewall, I.A. Flygare, U. Lange, L. Ljunggren and J. Söderberg (eds), *Agrarhistoria på många sätt: 28 studier om människan och jorden. Festskrift till Janken Myrdal på hans 60-årsdag*. Stockholm: Kungl. Skogs- och lantbruksakademien. pp. 115–136.

von Ehrenheim, F.V. 1824. *Om Climaternes rörlighet*. Stockholm: Kongl. Vetenskaps Akademien.

Eyþórsson, J. 1926. 'Um loftlagsbreytingar á Íslandi og Grænlandi siðan á landnámsöld' ['On climatic changes in Iceland and Greenland since settlement times']. *Skírnir* 100: 113–133.

Finnsson, H. 1831. *Om Folkemængdens Formindskelse ved Uaar i Island*. Trans. by Haldor Einarsen. Copenhagen: I Commission i den Schubotheske Boghandling.

Gundersen, I.M. and E. van Dijk. Forthcoming. 'Interlocking climate and society: Climate and harvest failure in sixth-century Scandinavia'. In A. Franklin-Lyons and T. Newfield (eds), *Contextualizing Medieval Food Shortages: Causes, Definitions, and Historiography*. Amsterdam: Amsterdam University Press.

Haldon, J., L. Mordechai, T.P. Newfield et al. 2018. 'History meets palaeoscience: Consilience and collaboration in studying past societal responses to environmental change'. *Proceedings of the National Academy of Sciences* 115: 3210–3218. https://doi.org/10.1073/pnas.1716912115.

Haldon, J., L. Mordechai, A. Dugmore et al. 2025. 'Past answers to present concerns. The relevance of the premodern past for 21st century policy planners: Comments on the state of the field'. *WIREs Climate Change* 16: e923. https://doi.org/10.1002/wcc.923.

Hasund, S. 1934. *Det norske folks liv og historie gjennem tidene: Tidsrummet 1280 til omkring 1500*. Oslo: Aschehoug.

Huhtamaa, H. 2018. '"Kewät kolkko, talwi tuima" – Ilmasto, sää ja sadot nälkävuosien taustalla'. In T. Jussila and L. Rantanen (eds), *Nälkävuodet 1867–1868*. Helsinki: Suomalaisen Kirjallisuuden Seura. pp. 33–65.

Huhtamaa, H. and S. Helama. 2017. 'Distant impact: Tropical volcanic eruptions and climate-driven agricultural crises in seventeenth-century Ostrobothnia, Finland'. *Journal of Historical Geography* 57: 40–51. https://doi.org/10.1016/j.jhg.2017.05.011.

Huhtamaa, H. and F.C. Ljungqvist. 2021. 'Climate in Nordic historical research – a research review and future perspectives'. *Scandinavian Journal of History* 46 (5): 665–695. https://doi.org/10.1080/03468755.2021.1929455.

Huhtamaa, H. 2020. 'Climate and the crises of the early fourteenth century in northeastern Europe'. In G.J. Schenk and M. Bauch (eds), *The Crisis of the 14th Century: Teleconnections Between Environmental and Societal Change*. Berlin: De Gruyter. pp. 80–99.

Huhtamaa, H., S. Helama, J. Holopainen, C. Rethorn and C. Rohr. 2015. 'Crop yield responses to temperature fluctuations in 19th century Finland: Provincial variation in relation to climate and tree rings'. *Boreal Environment Research* **20** (6): 707–723.

Huhtamaa, H., M. Stoffel and C. Corona. 2022. 'Recession or resilience? Long-range socioeconomic consequences of the 17th century volcanic eruptions in northern Fennoscandia'. *Climate of the Past* **18** (9): 2077–2092. https://doi.org/10.5194/cp-18-2077-2022.

Hulme, M. 2011. 'Reducing the future to climate: A story of climate determinism and reductionism'. *Osiris* **26**: 245–66. https://doi.org/10.1086/661274.

Hybel, N. 1997. 'Klima, misvækst og hungersnød i Danmark 1311–1319'. *Historisk Tidsskrift* **97** (1): 29–40. https://tidsskrift.dk/historisktidsskrift/article/view/53825.

Hybel, N. 2002. 'Klima og hungersnød i middelalderen'. *Historisk Tidsskrift* **102** (2): 265–281. https://tidsskrift.dk/historisktidsskrift/article/view/56005.

Ingram, M.J., D.J. Underhill and T.M.L. Wigley. 1978. 'Historical climatology'. *Nature* **276**: 329–334. https://doi.org/10.1038/276329a0.

Jónsson, Guðmundur. 2023. 'Hve Margir Dóu? Hungursneyð Átjánda Aldar og Nítjándu Aldar og Mannfall í Þeim' ['How many died? Famines in the eighteenth and nineteenth centuries and loss of life in them']. *Saga Tímarit Sögufélags* **61** (2): 53–88.

Jónsson, T. and H. Garðarsson. 2001. 'Early instrumental meteorological observations in Iceland'. In A.E.J. Ogilvie and T. Jónsson (eds), *The Iceberg in the Mist: Northern Research in Pursuit of a 'Little Ice Age'*. Dordrecht: Kluwer Academic Publishers. pp. 169–187.

Júlíusson, Á.D. 2021. 'Agricultural growth in a cold climate: The case of Iceland in 1800–1850'. *Scandinavian Economic History Review* **69**: 217–232. https://doi.org/10.1080/03585522.2020.1788985.

Jutikkala, E. 1994. 'Ilmaston muutokset ja historia'. In P. Karonen (ed.) *'Pane leipään puolet petäjäistä' Nälkä ja pulavuodet Suomen historiassa*. Jyväskylä: University of Jyväskylä. pp. 9–24.

Jutikkala, E. 1975. *Ilmaston muutosten vaikutuksia pohjoismaiden väestö- ja asutushistoriaan*. Helsinki: Societas Scientiarum Fennica.

Jutikkala, E. 1989. *Kuolemalla on aina syynsä: maailman väestöhistorian ääriviivoja*. Helsinki: WSOY.

King, L.A. and A.E.J. Ogilvie. 2021. 'The challenge of synthesis: Lessons from arctic climate predictions: Pathways to resilient, sustainable societies (ARCPATH)'. In D.C. Nord (ed.) *Nordic Perspectives on the Responsible Development of the Arctic: Pathways to Action*. Cham: Springer Nature Switzerland. pp. 393–412. https://doi.org/10.1007/978-3-030-52324-4.

Koch, L. 1945. *The East Greenland Ice. Meddelelser om Grønland 130: 3*. Copenhagen: Kommissionen for Videnskabelige Undersøgelser i Grønland.

Lamb, H.H. and K. Frydendahl. 1991. *Historic Storms of the North Sea, British Isles, and Northwest Europe*. Cambridge: Cambridge University Press.

Leijonhufvud, L., R. Wilson, A. Moberg, J. Söderberg, D. Retsö and U. Söderlind. 2010. 'Five centuries of Stockholm winter/spring temperatures reconstructed from documentary evidence and instrumental observations'. *Climatic Change* 101: 109–141. https://doi.org/10.1007/s10584-009-9650-y.

Lilja, S. 2006. 'Lokala klimatkriser och kronans intressen – en fallstudie av Södertörns kustsocknar ca 1570–1620'. In S. Lilja (ed.) *Människan anpassaren – människan överskridaren: Natur, bebyggelse och resursutnyttjande från sen järnålder till 1700-tal med särskild hänsyn till östra Mellansverige och Södermanlands kust*. Huddinge: Södertörns högskola. pp. 95–144.

Lilja, S. 2012. 'Klimat och skördar ca 1530–1820'. In S. Lilja (ed.) *Fiske, jordbruk och klimat i Östersjöregionen under förmodern tid*. Huddinge: Södertörns högskola. pp. 59–119.

Ljungqvist, F.C., A. Seim and H. Huhtamaa. 2021. 'Climate and society in European history'. *WIREs Climate Change* 12: e691. https://doi.org/10.1002/wcc.691.

Ljungqvist, F.C., B. Christiansen, J. Esper et al. 2023. 'Climatic signatures in early modern European grain harvest yields'. *Climate of the Past* 19: 2463–2491. https://doi.org/10.5194/cp-19-2463-2023.

Ljungqvist, F.C., P.J. Krusic, H.S. Sundqvist, E. Zorita, G. Brattström and D. Frank. 2016. 'Northern Hemisphere hydroclimatic variability over the past twelve centuries'. *Nature* 532: 94–98. https://doi.org/10.1038/nature17418.

Ljungqvist F.C., A. Seim and D. Collet. 2024. 'Famines in medieval and early modern Europe – Connecting climate and society'. *Wiley Interdisciplinary Reviews: Climate Change* 15 (1): e859924. https://doi.org/10.1002/wcc.859.

Magnus, O. 1555. *Historia de Gentibus Septentrionalibus*. Farnham: Ashgate Publishing.

Matskovsky, V.V. and S. Helama. 2014. 'Testing long-term summer temperature reconstruction based on maximum density chronologies obtained by reanalysis of tree-ring data sets from northernmost Sweden and Finland'. *Climate of the Past* 10 (4): 1473–87.

Melo, N. 2023. *Museums Exhibitions: Showcasing Climate in the 21st Century*. Évora: University of Évora.

Morice, C.P., J.J. Kennedy, N.A. Rayner et al. 2021. 'An updated assessment of near-surface temperature change from 1850: The HadCRUT5 data set'. *Journal of Geophysical Research: Atmospheres* 126 (3): e2019JD032361.

Nansen, F. 1925. 'Klimat-vekslinger i Nordens Historie'. *Avhandlinger utgitt av Det norske videnskabs-akademi, 1. Matematisk-naturvidenskapelig klasse* 3: 1–63.

Nordli, Ø., R. Przybylak, A.E.J. Ogilvie and K. Isaksen. 2014. 'Long-term temperature trends and variability on Spitsbergen: The extended Svalbard Airport temperature series, 1898–2012'. *Polar Research* 33: 21349. https://doi.org/10.3402/polar.v33.21349.

Norlind, G.A. 1914. 'Einige Bemerkungen über das Klima der historischen Zeit nebst einem Verzeichnis mittelatlerlicher Witterungserscheinungen'. *Lunds Universitets Arsskrift, N. F. AFD. 1* **10** (1).

Norlind, G.A. 1915. 'Till frågan om det historiske klimatet, särskilt i Nord- och Mellaneuropa'. *Ymer* 35: 83–97.

Ogilvie, A.E.J. 1982. *Climate and Society in Iceland From the Medieval Period to the Late Eighteenth Century.* Unpublished Ph.D. thesis. Norwich: School of Environmental Sciences, University of East Anglia.

Ogilvie, A.E.J. 2001. 'Climate and farming in northern Iceland, ca. 1700-1850'. In I. Sigurðsson and J. Skaptason (eds), *Aspects of Arctic and Sub-Arctic History.* Reykjavík: University of Iceland Press. pp. 289–299.

Ogilvie, A.E.J. 2005. 'Local knowledge and travellers' tales: A selection of climatic observations in Iceland'. In C. Caseldine, A. Russell, J. Harðardóttir and O. Knudsen. *Iceland – Modern Processes and Past Environments, Developments in Quaternary Science 5.* Amsterdam: Elsevier. pp. 257–287.

Ogilvie, A.E.J. 2010. 'Historical climatology, *Climatic Change*, and implications for climate science in the 21st century'. *Climatic Change* 100: 33-47. https://doi.org/10.1007/s10584-010-9854-1.

Ogilvie, A.E.J. 2020. 'Famines, mortality, livestock deaths and scholarship: Environmental stress in Iceland c. 1500–1700.' In A. Kiss and K. Prybil (eds), *The Dance of Death. Environmental Stress, Mortality and Social Response in Late Medieval and Renaissance Europe.* London: Routledge. pp. 9–24. https://doi.org/10.4324/9780429491085.

Ogilvie, A.E.J. 2022. 'Writing on sea ice: Early modern Icelandic scholars'. In K. Dodds and S. Sörlin (eds), *Ice Humanities: Living, Working, and Thinking in a Melting World.* Manchester: Manchester University Press. pp. 37–56. https://doi.org/10.7765/9781526157782.

Ogilvie, A.E.J. et al. 2024. 'Recent ventures in interdisciplinary Arctic research: The ARCPATH project'. *Advances in Atmospheric Sciences* 41: 1559–1568. https://doi.org/10.1007/s00376-023-3333-x.

Ogilvie, A.E.J. and T. Jónsson. 2001. '"Little Ice Age" research: A perspective from Iceland'. *Climatic Change* 48: 9–52. https://doi.org/10.1023/A:1005625729889.

Ogilvie, A.E.J. and R. Sigurðardóttir. Forthcoming. 'Living with ice and fire: Responses to natural hazards in early modern Iceland'. In D. Degroot, J.R. McNeill and A. Hessl (eds), *The Oxford Handbook of Climate Resilience.* Oxford: Oxford University Press.

Ponzi, F. 1995. *Ísland Fyrir Aldamót – Iceland – The Dire Years.* Mosfellsbær: Brennholt.

Pettersson, O. 1913. 'Klimatförandringar i historisk och forhistorisk tid'. *Kungl. Svenska Vetenskapsakademiens Handlingar* **51** (2).

Pettersson, O. 1914. 'Climatic variations in historic and prehistoric time'. *Svenska Hydrografisk-Biologiska Kommissionens Skrifter* 5: 1–26.

Pfister, C., S. White and F. Mauelshagen. 2018. 'General introduction: Weather, climate, and human history'. In S. White, C. Pfister and F. Mauelshagen (eds), *The Palgrave Handbook of Climate History*. London: Palgrave Macmillan. pp. 1–18. https://doi.org/10.1057/978-1-137-43020-5_1.

Pfister, C. and H. Wanner. 2021. *Climate and Society in Europe: The Last Thousand Years*. Bern: Haupt Verlag.

Richter, L. 2020. 'Forms of meteorological knowledge 1750–1850 in German countries and beyond'. *Wiley Interdisciplinary Reviews. Climate Change* 11 (4): e651. https://doi.org/10.1002/wcc.651.

Sandnes, J. 1978. *Ødegårdstid i Norge. Det nordiske ødegårdsprosjekts norske undersøkelser*. Oslo: Universitetsforlaget.

Sandnes, J. and H. Salvesen. 1978. *Ødegårdstid i Norge: det nordiske ødegårdsprosjekts norske undersøkelser*. Oslo: Universitetsforlaget.

Seftigen, K., H. Goosse, F. Klein and D. Chen. 2017. 'Hydroclimate variability in Scandinavia over the last millennium – insights from a climate model–proxy data comparison'. *Climate of the Past* 13: 1831–1850. https://doi.org/10.5194/cp-13-1831-2017.

Skoglund, M.K. 2022. 'Climate variability and grain production in Scania, 1702–1911'. *Climate of the Past* 18: 405–433. https://doi.org/10.5194/cp-18-405-2022.

Skoglund, M.K. 2024. 'The impact of drought on northern European pre-industrial agriculture'. *The Holocene* 34: 120–135. https://doi.org/10.1177/09596836231200431.

Skoglund, M.K., 2023. 'Farming at the margin: Climatic impacts on harvest yields and agricultural practices in Central Scandinavia, c. 1560–1920'. *Agricultural History Review* 71: 203–233.

Sörlin, S. 2024. *Snö. En historia*. Stockholm: Volante.

Speerschneider, C.I.H. 1915. 'Om Isforholdende i Danske Farvande i aeldre og nyere tid, aarene 690-1860'. *Publikationer fra Det Danske Meteorologiske Institut, Meddelelser 2*.

Steensberg, A. 1951. 'Archæological dating of the climatic change about A.D. 1300'. *Nature* 168: 672–674. https://doi.org/10.1038/168672a0.

Thoroddsen, Þ. 1914. 'Islands klima i oldtiden'. *Geografisk Tidskrift* 22: 204–216.

Thoroddsen, Þorvaldur. 1916–17. *Árferði á Íslandi í púsund ár*. [The Seasons in Iceland in 1000 Years]. Copenhagen: Hið Íslenska Fræðafélag.

Tornberg, M. 1992. 'Ilmastohistoria – nyt'. In E. Kuparinen (ed.) *Muuttuva maailmamme: ympäristöongelmia eilen ja tänään*. Turku: Univeristy of Turku. pp. 47–82.

Tornberg, M. 1989. 'Ilmaston- ja sadonvaihtelut Lounais-Suomessa 1550-luvulta 1860-luvulle'. *Turun Historiallinen Arkisto* 44: 58–87.

Tveit, O.E., E.J. Førland, H. Alexandersson et al. 2001. *Nordic Climate Maps*. Blindern: The Norwegian Meteorological Institute.

Utterström, G. 1954. 'Some population problems in pre-industrial Sweden'. *Scandinavian Economic History Review* 2: 103–165.

Utterström, G., 1955. 'Climatic fluctuations and population problems in early modern history'. *Scandinavian Economic History Review* 3: 3–47.

van Bavel, B., D.R. Curtis, M.J. Hannaford, M. Moatsos, J. Roosen and T. Soens. 2019. 'Climate and society in long-term perspective: Opportunities and pitfalls in the use of historical datasets'. *WIREs Climate Change* 10: e611. https://doi.org/10.1002/wcc.611.

Wang, J., B. Yang, M. Fang, Z. Wang, J. Liu and S. Kang. 2023. 'Synchronization of summer peak temperatures in the Medieval Climate Anomaly and Little Ice Age across the Northern Hemisphere varies with space and time scales'. *Climate Dynamics* 60: 3455–3470. https://doi.org/10.1007/s00382-022-06524-6.

White, S., Q. Pei, K. Kleemann, L. Dolák, H. Huhtamaa and C. Camenisch. 2023. 'New perspectives on historical climatology'. *WIREs Climate Change* 14: e808. https://doi.org/10.1002/wcc.808.

Wastenson, L., B. Raab and H. Vedin. 1995. *Sveriges nationalatlas: Klimat, sjöar och vattendrag*, Stockholm: Sveriges nationalatlas.

The Authors

Dominik Collet is Professor of Climate and Environmental History at the University of Oslo, Norway. He is the PI of *ClimateCultures – Socionatural entanglement in Little Ice Age Norway (1500–1800)* as well as the thematic research group *Nordic Climate History*. He also leads the project *The Nordic Little Ice Age (1300–1900) Lessons from Past Climate Change (NORLIA)* at the Centre for Advanced Study at the Norwegian Academy of Science and Letters. His research focuses on the historical entanglements of climate and culture both in their material and mental configurations.

Ingar Mørkestøl Gundersen is a Postdoctoral Research Fellow in the Department of Archaeology, Conservation, and History at the University of Oslo. He has published on a range of topics on climate history and the Scandinavian Iron Age. Gundersen received his Ph.D. in 2022 with the thesis 'Iron Age Vulnerability', which investigated the archaeological evidence for a sixth-century climate crisis in eastern Norway. His doctoral research was part of the VIKINGS project (Volcanic Eruptions and their Impacts on Climate, Environment, and Viking Society in 500–1250 CE). Together with Dr Manon Bajard, he received the Inter Circle U. prize 2022 for outstanding examples of cross-disciplinary research. He is currently part of two research projects on the Nordic Little Ice Age (ClimateCultures, University of Oslo and *The Nordic Little Ice Age (1300–1900) Lessons from Past Climate Change (NORLIA)* at the Centre for Advanced Study at the Norwegian Academy of Science and Letters.

Heli Huhtamaa is a climate and environmental historian. Her research interests include human consequences of the Little Ice Age and pre-industrial Nordic history. She focuses on interdisciplinary approaches concerning both historical and climate sciences. She is an Assistant Professor at the University of Bern, Switzerland, where she leads a research project on volcanic impacts on climate, environment and society.

Fredrik Charpentier Ljungqvist is Professor of History, in particular Historical Geography, at Stockholm University, Sweden. He also holds the title of Associate Professor of Physical Geography at the same university. Ljungqvist was in 2022 by the Royal Swedish Academy of Letters, History and Antiquities awarded the Rettig Prize for "interdisciplinary works concerning climate and diseases in a long-term perspective". He was a Pro Futura Scientia Fellow at the Swedish Collegium for Advanced Study from 2019 to 2024 and has been a visiting researcher at the University of Cambridge, Lanzhou University, University of Bern and the Freiburg Institute for Advanced Studies (FRIAS).

Astrid E.J. Ogilvie is a Research Professor at the Institute of Arctic and Alpine Research at the University of Colorado and a Senior Associate Scientist at the Stefansson Arctic Institute in Akureyri, Iceland. Her research focuses on the broader issues of climatic change and contemporary Arctic issues, as well as the environmental humanities. Her interdisciplinary, international projects have included leadership of the NordForsk Nordic Centre of Excellence project: *Arctic Climate Predictions: Pathways to Resilient, Sustainable Societies (ARCPATH)*; and *The Natural World in Literary and Historical Sources from Iceland ca. AD 800 to 1800 (ICECHANGE)*. She is the author of some 100 scientific papers and has three edited books to her credit.

Sam White is Professor of Political History at the University of Helsinki, author of *A Cold Welcome* (Harvard University Press, 2017) and editor of the *Palgrave Handbook of Climate History* (Palgrave, 2018).

All authors are currently Fellows of the project *The Nordic Little Ice Age (1300–1900) Lessons from Past Climate Change (NORLIA)* at the Centre for Advanced Study at the Norwegian Academy of Science and Letters.

Chapter 1

THE DEVELOPMENT OF METEOROLOGICAL INSTITUTIONS AND EARLY INSTRUMENTAL CLIMATE DATA IN THE NORDIC COUNTRIES

Elin Lundstad, Stefan Norrgård and A.E.J. Ogilvie

Abstract

The Nordic countries share much, in terms of political and economic histories, and geographically all countries border or lie within the Arctic Circle. This study focuses on their shared history with regard to meteorological observations and provides insights into this development from the 1700s until the establishment of national meteorological institutes in the latter half of the 1800s. An overview of the founding of these institutes is included. To our knowledge, this is the first study to discuss and present the history of meteorological observations across all Nordic countries together. Beginning in the 1700s, the study explores how prevailing theories on climate and weather influenced the recording and analysis of meteorological observations. Temperature records for each country are presented using a novel approach to illustrate temperature increases up to the present day. A cross-correlation analysis of temperature data indicates a strong correlation between all Nordic capitals, except Reykjavik, highlighting Iceland's distinct climatic conditions even within the Nordic context. Finally, using the Mann-Kendall trend analysis, we found that Copenhagen exhibits the highest temperature trend among the Nordic capitals.

Introduction

The aim of this study is to provide an overview of the development of meteorological observations and institutions in the Nordic countries from the early eighteenth century to the establishment of national meteorological institutes in the latter half of the 1800s. The primary focus is on observations of air temperature which started in the early 1700s and the development of the Hauksbee (1708), Fahrenheit (1724) and Celsius (1740) thermometers. This study comple-

doi: 10.63308/63881023874820.ch01

ments and supplements earlier related studies (e.g., Kington 1972; Lamb 1977; Moberg and Bergström 1997; Jónsson and Garðarsson 2001; Ogilvie and Jónsson 2001ab; Pfister et al. 2018; Huhtamaaa and Ljungqvist 2021; Norrgård 2024). Considering the scope and aim of this article, descriptive weather journals are not discussed. A reason for considering the Nordic countries as a group, apart from geographical proximity, excluding Iceland and Greenland, and their shared political and economic histories, is that all countries border or lie within the Arctic Circle. The Nordic countries are often referred to as one region; however, they span a vast geographical area with very different landscapes and topographies. Stretching from Karelia in eastern Finland, to Svalbard in the northeast, to Iceland and Greenland in the west and Denmark in the south, their climatic regimes vary greatly. Moreover, it is easy to forget that Finland, Norway and Sweden are three of the eight largest countries in Europe and Greenland is the world's largest island.

Climatic regimes vary significantly across the region due to differences in latitude, proximity to the ocean and topography. They are generally characterised by long, cold winters and short, mild to warm summers, but local variations create distinct climatic zones. Norway, Denmark and Iceland have a temperate maritime climate, especially along the coast. The North Atlantic Current brings milder winters and relatively cool summers. Western Norway and Iceland experience high precipitation due to moist Atlantic air. Sweden, Finland and parts of Norway have a more continental climate, with colder winters and warmer summers. Inland areas experience greater seasonal temperature variations, with harsh winters and warm summers. Northern parts of Finland, Norway, Sweden and much of Iceland fall under the subarctic and Arctic climate zones. The Nordic climate is diverse, shaped by oceanic influences, latitude and elevation. While traditionally known for cold and snowy winters, climate change is altering long-term patterns, making winters milder and increasing extreme weather events across the region.

The main motivation for this study is that it would appear that, apart from the valuable and seminal work of Kington (1972), there has been no specific study of the history of instrumental meteorological observations for the Nordic countries as a whole. While the focus is on key developments in each country, a caveat is that it has not been possible to discuss every single available meteorological record. Emphasis has been placed on the length of the observations and their possible value for climate reconstruction purposes. Map 1 shows the Nordic countries and highlights the meteorological stations begun before 1900, the end of the early instrumental period, and is based on data derived from the global historical climate database HCLIM, which comprises 12,452 records from 118 countries (Lundstad et al. 2022, 2023; Valler et al. 2024).

The Development of Meteorological Institutions

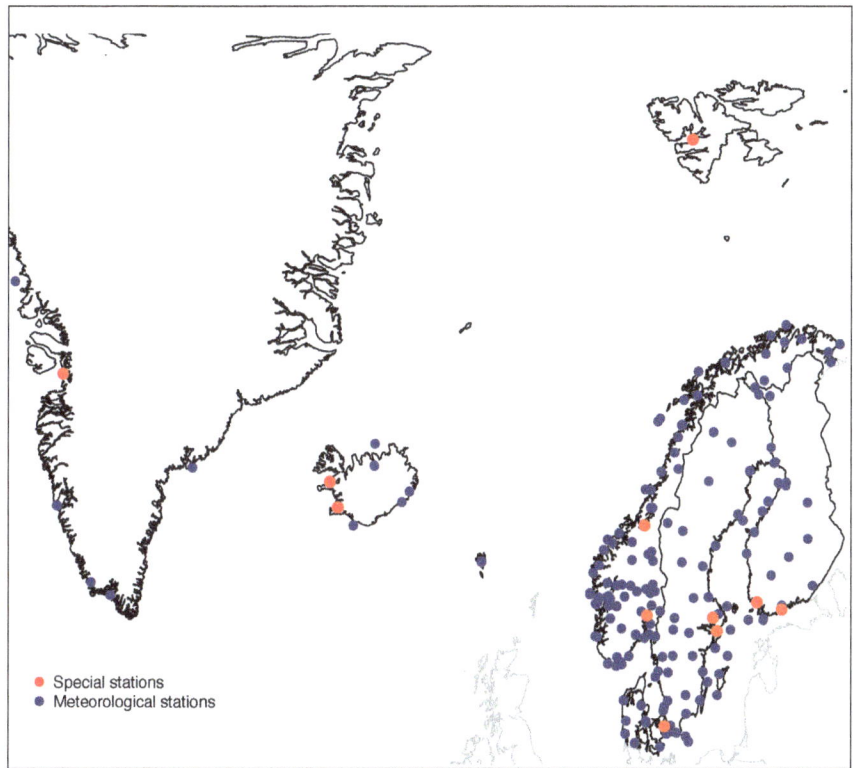

Map 1. Overview of early meteorological stations established before 1900 in the Nordic countries (dark blue), including Norway, Sweden, Finland, Denmark, Iceland, Svalbard and Greenland. Special stations are highlighted in red, e.g. Ilulissat, Reykjavik, Stykkishólmur, Oslo, Trondheim, Longyearbyen, Copenhagen, Uppsala, Stockholm and Helsinki. Data source: HCLIM database (Lundstad et al. 2022).

The Origin of Instrumental Meteorological Observations: 1700–1740

The Enlightenment gave birth to empirical science in Europe. Concerning meteorological observations, the invention of the barometer by Evangelista Torricelli (1608–1647) in 1643 was followed by the development of several of the current thermometers and scales during the first half of the 1700s. Although this marked a shift from qualitative descriptions of weather and the beginning of instrumental climate records (Hiebl 2006) these continued to be produced, even up to recent times. The earliest instrumental observations from the Nordic countries are those made by the Danish astronomer Ole Rømer (1644–1710). Rømer's observations cover the first ten years of the 1700s and include the extremely cold winter 1708/09. It is of interest that he also hosted Daniel Gabriel Fahrenheit (1686–1736), who visited Rømer in 1708, after which he developed the Fahrenheit thermometer and scale, which saw daylight in 1724. There are several records and meteorological journals that have not been analysed, and there may be instrumental records that have not been uncovered. Nevertheless, the records following Rømer's depended much on the development of the thermometer and whether it was possible to acquire such an instrument.

In the 1720s, a milestone in the development of meteorological observations was James Jurin's (1684–1750) invitation in 1723 to submit such observations to the Royal Society in London. Although James Jurin's invitation has often been labelled as one of the first networks, it can hardly be regarded as such. Jurin was interested in understanding wind and weather, as he believed that weather determined people's health. His purpose, therefore, was presumably not so much to create a network as to try to solve a medical query. However, Jurin's invitation was important in that he provided an example of how to keep meteorological journals, divided into rows and columns, and thus influenced the structure of such journals in the following decades. His invitation went to many observers and most meteorological journals kept during the greater part of the 1700s followed his suggested structure. The invitation inspired several observers in the Nordic countries with a majority of these coming from Sweden. The measurements of these early observers were presented in the *Philosophical Transactions* in a series of articles published in the 1730s and 1740s. The first records to be submitted from the Nordic countries were observations made by Conrad Quensel (1676–1732) in Lund for the year 1724. These were followed by observations made in Bettna, Bygdeå, Lund, Piteå and Uppsala covering 1726. However, only the temperature observations from Uppsala, made with the Hauksbee thermometer, and Lund, were published. The others were disregarded as the scale and thermometers used were unknown. This remained a problem with the later observations and was a general dilemma related to several observations submitted to London. In addition to the Swedish observations, one year of observations was also sent from Christiania (Oslo), in Norway, covering 1727. In this case, the observations were submitted by a Pehr Kink (Derham 1734a, 1734b) and the observer also remains unknown. For 1728, the list of Swedish observations continued to expand with records from Svenåker, Risinge, Hudiksvall and Härnösand. From Finland, only one series was sent to London, which was that made by Hermann Spöring (1701–1747) in Turku in 1731.

Except for the Uppsala observations, none of the above-mentioned series survived the passage of time. Initially, the Uppsala series was kept by the astronomer Anders Celsius (1701–1744) who invented the Celsius thermometer and scale in 1740. Celsius is also noteworthy for his participation in a scientific

The Development of Meteorological Institutions

expedition in 1736/37 organised by the French Academy of Sciences to Torneå in northern Finland. This expedition, known as 'The Lapland Expedition', led to one of the first and longest records from one of the northernmost cities in the world, Torneå, with a set of observations that covers the 1737–1749 period (Holopainen and Vesajoki, 2001). Celsius also shared Jurin's instructions with others, such as Johan Leche (1704–1764), professor of medicine, who made meteorological observations in Lund between 1742 and 1744 (Norrgård 2024). The connection resulted in another long-term record as Leche made observations in Turku in Finland from 1748 to the 1820s. This was the first long-term record from Finland made with the Celsius thermometer.

Meteorological Observations and Scientific Societies: 1740–1800

The Celsius scale was born at a critical stage during the Enlightenment and, as the structure of the meteorological journals became to some extent formalised, and thermometers became increasingly common, measurements were made across multiple sites. However, it was not the only thermometer and scale in use, and especially not in the other Nordic countries. The fact that observers such as Johan Daniel Berlin (1711–1781) used Reaumur's thermometer and the zoologist and minister Hans Strøm (1726–1797) used a Fahrenheit thermometer shows that the instruments used varied. For example, in Oulu (Uleåborg), the pharmacists Jacob Karlberg (1730–1782) and Johan Julin (1752–1820) made observations between 1776 and 1803 using at first the Florentine and then the Celsius thermometer (Vesajoki and Holopainen 2000).

An important part of the development of observations was the establishment of scientific societies. The Royal Society of Sciences in Uppsala (*Kungliga Vetenskaps-Societeten i Uppsala*) was established in 1710. However, the Royal Swedish Society of Sciences (*Kungliga Vetenskapsakademien*), established in 1739, became more influential, not least because it published research in Swedish rather than Latin. The Royal Danish Academy of Sciences, also known as the Danish Scientific Society (*Det Kongelige Danske Videnskabernes Selskab*), was founded in 1742 and the Royal Norwegian Society of Science and Letters (*Det Kongelige Norske Videnskabers Selskab*) in 1767, although it was first established under a different name in Trondheim in 1760. In Iceland, the Icelandic Society of Letters, also translated as the Icelandic Literary Society (*Hið Íslenzka Bókmenntafélag*), was founded somewhat later in 1816, with a focus not just on literature but on general scholarship. The purpose of these societies was to promote science and to benefit the development of their respective countries. Meteorological observations became an important part of this process. These societies all still exist today. During the period under discussion, Norway, Iceland and Greenland were under the control of the Danish Crown and the Danish Scientific Society played a significant part, in particular in the establishment of observations in Iceland, as well in Denmark. By the late eighteenth century, the interest in climatic mapping led to the efforts of the *Societas Meteorologica Palatina* based in Mannheim, amongst other things in the distribution of thermometers (Kington 1972). Several observers seem to have suggested to the societies the need for meteorological networks; however, although observations were submitted to the societies, official networks were not created immediately and throughout the eighteenth century observations continued, but not systematically. Some examples are described below.

In Copenhagen, the astronomer Peder Horrebow (1728–1812) began conducting regular observations at the Rundetaarn Observatory in 1751. However, he kept his thermometer inside the observatory, and it was not placed outside the observatory until as late as 1767. In 1768 Thomas Bugge (1740–1815) of the Danish Scientific Society continued the Copenhagen records begun by Horrebow but now with the thermometer placed outside. In Lund, the astronomer Nils Schenmark (1720–1788) began making meteorological observations in October 1752 and there is a record of observations made in Iceland between 1752 and 1757, but these have not been found (Jónsson and Garðarsson 2001). Some of Pehr Kalm's (1716–1779) observations, made in the 1750s after returning to Turku from his travels in North America, are also lost. In general, it is not uncommon to find references to meteorological records that have been lost, even though the observations were made by well-known and learned individuals. An important series that has been preserved is that for Stockholm, kept by the astronomer and Secretary General of the Royal Swedish Academy of Sciences, Pehr Wargentin (1717–1783) who began making observations in 1754. This series is noteworthy because it has been continued to the present day and is thus one of the longest temperature series in the world. In 1762, Johan Daniel Berlin began observations in Trondheim, Norway. After his death, this record was continued by Diderik Christian Fester (1732–1811). Other observations from Norway include those made by the minister and zoologist Hans Strøm, who made two long records with a Fahrenheit thermometer. The first from Borgsund, Sunnmøre, covers twelve years and, after he was relocated to Eiker in 1779, he published observations made between 1782 and 1791 (Ström 1798). The Danish-born priest Jacob Nicolai Wilse (1735–1801) who was affiliated with the Societas Meteorologica Palatina stations in Mannheim (Winkler 2023) conducted meteorological observations in Spydeberg and Eidsberg from 1783. Wilse emphasised the expression 'one deduces the future from the past' and the need to be able to predict future weather (Federhofer 2002).

The Development of Meteorological Institutions

In Iceland, early observations were first made by enthusiastic individuals such as the naturalist and doctor Sveinn Pálsson (1762–1840) and Sheriff Magnús Ketilsson (1732–1803) (Jónsson and Garðarsson 2001). Observations were also made in cooperation with the Danish Scientific Society as when Rasmus Lievog (1738–1811) was appointed Royal Astronomical Observer in Iceland in 1779. Lievog used Reaumur's thermometer and made observations from that year possibly to 1789, but many are lost, and their provenance is complex (Jónsson and Garðarsson 2001). As regards Greenland, the first meteorological measurements were associated with the activities of the Moravian missionaries and carried out at Neu-Herrnhut, near modern-day Nuuk (formerly Godthåb) from September 1767 to July 1768 (Demarée and Ogilvie 2020, 2021). Observations were also made by Andreas Ginges (1754–1812). His records are not continuous but include sporadic periods of observations between 1784 and 1792 (Przybylak et al. 2009, 2024).

Table 1. Overview of the 5 earliest meteorological temperature stations in the Nordic countries. Source: see Data section in Bibliography.

Start	Country	Station	º North	º East	End
1722	Sweden	Uppsala	59.861	17.6322	
1737	Finland	Tornio (Torneå)	65.5	24.0847	1749
1748	Finland	Turku (Åbo)	60.45	22.2831	1822
1752	Sweden	Lund	55.4221	13.1135	
1754	Sweden	Stockholm	59.19	18.3536	
1758	Finland	Loviisa (Lovisa)	60.45	26.228	1765
1761	Finland	Pirkkala (Birkkala)	61.27	23.3835	1770
1762	Norway	Trondheim-Berlin	63.4106	10.4533	1787
1767	Denmark	Copenhagen	55.67	12.5683	
1776	Finland	Oulu (Uleåborg)	64.9375	25.3756	1792
1781	Denmark	Faeroe Islands	62	7	
1782	Denmark	Landbohøjskolen	55.68234	12.5404	
1783	Norway	Spydeberg/Eidsberg*	59.62	11.0778	1791
1796	Sweden	Umeå	63.82	20.263	
1800	Norway	Trondheim-Fester	63.4333	10.4167	1802
1816	Norway	Bergen-Bohr	60.6388	5.5833	1826
1816	Norway	Oslo-Øvre-Vollgate	59.9119	10.7457	1838
1820	Iceland	Reykjavik (Nes)	64.12	21.8277	
1823	Sweden	Kreuzburg	60	18.2	1849

Start	Country	Station	° North	° East	End
1823	Denmark	Aabenraa	55.04	9.418	
1845	Iceland	Stykkishólmur	65.06	22.7333	
1861	Denmark	Tarm	55.9	8.5304	
1873	Iceland	Teigarhorn	64.4	14.2066	
1874	Iceland	Grímsey	66.33	18	
1878	Iceland	Vestmannaeyar	63.44	20.2734	

*One of the *Societas Meteorologica Palatina* stations.

Networks and Institutions: Developments in the 1800s

Globally, weather services began to be established in the mid-nineteenth century, with European meteorological institutions starting in the 1850s and 1860s, largely to aid seafarers. The world's first meteorological institute, the Royal Netherlands Meteorological Institute (KNMI), was established by King William III (1849–1890) on 21 January 1854. Later that year, the Meteorological Office of the United Kingdom (now the UK Met Office) was founded in London under Admiral Robert FitzRoy (1805–1865) with a focus on marine safety through storm warnings and shipping forecasts. Safety at sea was also a key issue in the Nordic countries and meteorological observations were initiated at a number of lighthouses.

Meteorological observations from the first half of the 1800s have in general received less attention than those from the 1700s. Nonetheless, meteorological observations were still made by ordinary citizens, possibly to an even higher degree than in the 1700s. This was certainly the case in Iceland. However, the very valuable observations begun in Reykjavík in 1820, and continuing for over thirty years, made by the *Landlæknir* (chief physician of Iceland) Jón Thorsteinsson (1794–1855) were undertaken in cooperation with the Danish Scientific Society (Jónsson 1989). A major effort was also initiated in 1840 by the Icelandic Society of Letters at the urging of the poet and natural scientist Jónas Hallgrímsson (1807–1845). The Danish Scientific Society subsequently donated 45 thermometers to this cause, and these were sent to parish priests around the country. One of the thermometers was acquired by a merchant in Stykkishólmur in the west of Iceland, Árni Thorlacius (1802–1891). He began a major undertaking, making observations many times a day with a set of thermometers and barometers and a precipitation gauge (Berþórsson 1957). These observations have continued from 1845 to the present day (Sigfúsdóttir 1845; Jónsson and Garðarson 2001: 170–171; Hanna et al. 2004). For Greenland, records from the nineteenth century are from Ilulissat (formerly Jakobshavn)

The Development of Meteorological Institutions

and Qaqortoq (formerly Julianehåb), both begun in 1807. Continuous series started by the Danish Meteorological Institute as the first official stations became operational in Ilulissat and Upernavik and Nuuk (formerly Godthåb) were from 1873 and Ivigtut/Narsarsuaq in 1875 (Vinther et al. 2006a, 2006b). Combining observations from southwestern Greenland has facilitated the creation of a record extending back to 1784 (Vinther et al. 2006a).

Norway was the first country in the Nordic region to establish a national meteorological institute. The Norwegian Meteorological Institute, now known internationally as MET Norway, was founded in 1866 at the University of Christiania (Oslo). Henrik Mohn (1835–1916) was appointed as its inaugural director. Under his leadership, weather telegrams to individual ports were initiated in the same year, marking the beginning of a comprehensive network of weather stations designed to map Norway's climate, issue storm warnings and advance atmospheric research. The Norwegian Meteorological Institute is still a government agency (as in all the Nordic countries) subordinate to the Ministry of Climate and Environment since 2014 (formerly under the Ministry of Education), with responsibility for the public meteorological service for both civil and military purposes.

The Danish Meteorological Institute (*Danmarks Meteorologiske Institut* – DMI) was established in 1872 under the Ministry of the Navy. One year after establishing the Institute, the number of contributing stations had already grown to 140. The Meteorological Institute was fused with the Aeronautical Meteorological Service (founded 1926) and the Danish Defence Weather Service (founded 1953) in 1990, leading to the current version of the DMI in 1990. Today, the Institute is situated under the Ministry of Climate and Energy. In addition to the mainland, the Institute also serves the inhabitants of the Faroe Islands and Greenland, while forecasts are undertaken as part of an agreement, in Iceland.

The network started by the Academy of Sciences was transferred to the Meteorological Central Facility (*Statens Meteorologiska Anstalt*), which was originally founded in 1873 and, in 1881, the network gathered observations on temperature and precipitation from circa 400 stations. In 1919 the Facility was joined with the Hydrological Institute (*Hydrologiska Byrån*) which led to the founding of the Swedish Meteorological and Hydrological Facility (SMHA), which was renamed the Swedish Meteorological and Hydrological Institute (SMHI) in 1945. At the present time, SMHI is an agency under the Ministry of Climate and Enterprise. As regards Denmark, in 1820, Hans. C. Ørsted (1777–1851) suggested a systematic collection of meteorological observations across the country. Subsequently, a Meteorological Committee was created for this purpose in Copenhagen in 1827.

Elin Lundstad, Stefan Norrgård and A.E.J. Ogilvie

The development in Finland (during the 1800s a part of Russia) was somewhat different from that in other Nordic countries. The Finnish Economic Society (*Finska Hushållningssällskapet*) established in 1797, attempted to create a meteorological network in 1815. It was not a success, but the foundation was laid for the Finnish Society of Sciences and Letters (*Finska Vetenskaps-Societeten*) established in 1838, which began a network in the winter of 1845. In this same year, 1881, the Helsinki Magnetic Observatory was renamed the Central Meteorological Institute (later the Finnish Meteorological Institute) and the focus shifted from magnetism towards meteorology. During the late nineteenth century, the Meteorological Institute oversaw nationwide meteorological, hydrological and marine observations and research. After Finland gained independence in 1917, following the Russian Revolution, these were reorganised to become separate governmental bodies. In 1918, the Institute was further reorganised as a state research institute and, in 1919, the Meteorological Institute was transferred from the Society of Science and Letters to the Ministry of Agriculture, then, in 1968, to the Ministry of Transportation. At the present time (2025) it is under the Ministry of Transportation and Communication (Nevanlinna 2014; Seppinen 1988). In Sweden, the Academy of Sciences initiated a meteorological network in the 1780s. It was not a success, even though observations continued at some stations well into the mid-1860s. As in Finland, meteorological observations were also initiated at lighthouses from the mid-1800s. The longest series, from 1848, is from the Grönskär lighthouse in the Stockholm archipelago. The Academy attempted to create a network again in 1858. At first, the network included eleven different sites, but this later grew to twenty, and several are still active today. One of the measuring sites that began operating in 1859 was in Haparanda, the sister city to Tornio in Finland, which is currently one of the longest temperature records from subarctic regions (Dienst et al. 2017). The Icelandic Meteorological Office (IMO), *Veðurstofa Íslands*, was established on 1 January 1920 as a purely meteorological institute (Garðarson 2000). It is now a government agency under the Ministry of Environment and Natural Resources. As well as weather forecasting, it is active in volcano monitoring and seismology. It has a research emphasis on atmospheric processes, glacier and avalanche studies, hydrological systems, earthquake and volcanic processes and geohazards. Table 2 shows the year each Nordic country first established its national meteorological institute.

The Development of Meteorological Institutions

Table 2. Establishment of the National Meteorological Institutes in the Nordic countries. Source: see Data section of Bibliography.

Country	Name	Abbreviation	Web address	Established year
Norway	Meteorological institute	MetNorway	https://www.met.no/	1866
Denmark	Danmarks Meteorologiske Institut	DMI	https://www.dmi.dk/	1872
Sweden	Sveriges meteorologiska och hydrologiska institut	SMHI	https://www.smhi.se	1873
Finland	Ilmatieteen laitos	FMI	https://en.ilmatieteenlaitos.fi/	1881
Iceland	Veðurstofa Íslands	IMO	https://www.vedur.is/	1920

The Longest Records from the Nordic Countries

Several of the longest Nordic temperature records are well known by climate historians and climatologists as they have been analysed in research on anthropogenic climate warming (IPCC 2014; Nicholls 2010). The longest record from the Nordic countries is the Uppsala record (see Figure 1) begun by Anders Celsius in 1722 and thus having its 300th year anniversary in 2022. This is followed by the Stockholm record (see Figure 4), which was begun by Wargentin in 1754; and the Norwegian Trondheim record which was initiated by Berlin in 1762 (see Figure 1). The longest record from Denmark is that begun by Bugge in Copenhagen in 1768 (see Figure 4). All the records have some gaps and cannot be considered complete. For the other Nordic countries, the longest records start in the 1800s.

In the case of Finland, the record that was begun in 1748 in Turku was curtailed after the great city fire in 1827. The fire devastated a large part of the city, but it was the political decision of the Russian Emperor to move the University to Helsinki that meant the end of the Turku temperature record. After this move, the physics professor G.G. Hällström (1775–1844) began a record of observations in 1828 and these continued until his death in 1844. Meanwhile, in 1838, conducting meteorological observations became part of the practice of the newly founded Helsinki magnetic observatory and this laid the foundation for the current temperature records from Helsinki (see Figure 4). In 1846, five stations were equipped with thermometers. The most successful stations were those created in Oulu, Kajaana and Kuopio in 1846. Other stations were added in the1850s and 1860s but most were in use for only a few years before being discontinued (Heino 1994). In Turku, observations continued until the 1820s but ceased after the great city fire in 1827 and the relocation of the University to Helsinki in 1828 (by order of the Russian Emperor). In a venture separate from the Finnish Society of Sciences, the Finnish Pilot and Lighthouse Department established meteorological stations from the 1850s onwards. In the latter half of the 1800s, a number of other meteorological observations were begun at locations including the southernmost city of mainland Finland, Hanko, in 1869 and in Mariehamn, in the Åland islands between Finland and Sweden. Observations started up again in Turku in 1873 and at the Utö islands in the Baltic Sea in 1881.

For Greenland, combining observations from southwestern Greenland facilitated the creation of a record extending back to 1784 (Vinther et al. 2006a). For Iceland, the longest record is that of the combined Reykjavik/Stykkisholmur series from 1820 to the present (Jónsson and Garðarson 2001) as noted above (see Figure 4).

In Norway, subsequent to observations from Oslo in 1727, a record of meteorological observations was begun by Jens Esmark (1763–1839) in 1816, and he continued with his observations until 1839 (Hestmark and Nordli 2016). Christoffer Hansteen (1784–1873) took over the observations at the Oslo Observatory in 1837 until his death. After the Meteorological Institute was established in 1866, weather observations were conducted outside its building in the city centre (initially in Kirkegata, later moved to Nordahl Brunsgate) on the initiative of Henrik Mohn, the first director of Met Norway. Temperature measurements in Oslo were officially moved to the Oslo Observatory in 1876, which was still located in the city centre. In 1937, the measuring station was relocated four kilometres from the city centre to Blindern. Regarding the establishment of meteorological stations in Norway, the first group beyond the main stations in Oslo, Bergen, and Trondheim was set up in the north, in Finnmark: Vardø Radio (1829), Hammerfest (1848) and Lebesby (1858) (Barlaup 1966). The next group was linked to the telegraph system, established in 1860–1861, with stations along the coast, including Kristiansund, Ålesund, Skudeneshavn, Sandøsund and Mandal. From 1867, additional stations were installed at lighthouses such as Utsira, Andenes, Oksøy, Brønnøysund and others (Nilsen 2016). This happened at the same time as the institute was established.

Figure 3 shows pairwise scatter plots of annual temperature time records for the five Nordic capitals. The strong correlations among Stockholm, Copenhagen, Helsinki and Oslo reflect their geographic proximity and shared climatic influences. Reykjavik's lower correlations are a result of its maritime climate and

The Development of Meteorological Institutions

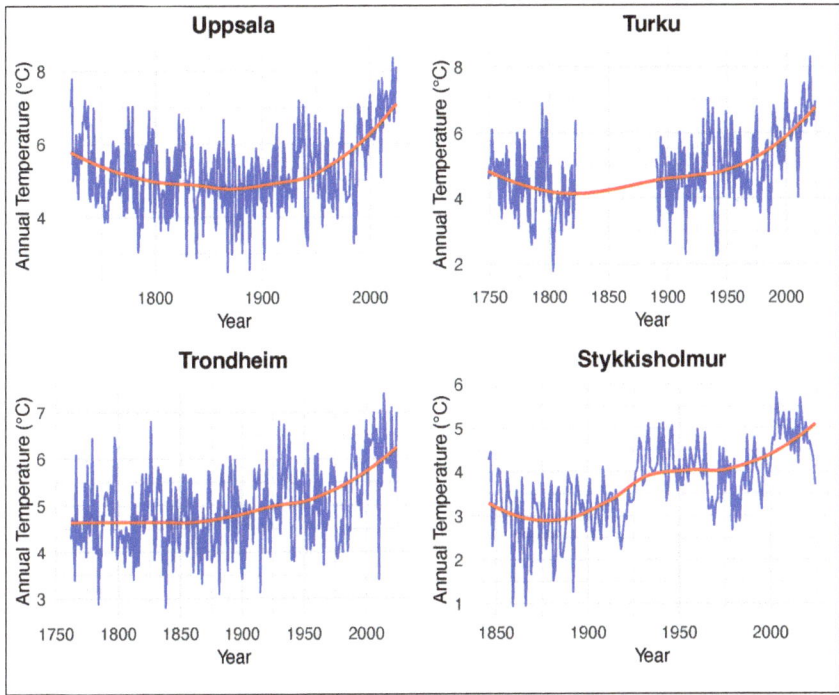

Figure 1. The earliest and longest annual temperature trends from four Nordic locations showing temperature fluctuations (blue) and long-term trends (red). The data suggest a general warming pattern, with temperatures increasing significantly in recent decades. Data sources: see 'Data' section of the Bibliography.

the influence of oceanic weather patterns, differentiating its temperature trends from those of mainland cities.

Figure 1 presents the annual temperature time series from four Nordic locations with the longest records: Uppsala (from 1722), Turku (from 1749), Trondheim (from 1762) and Stykkishólmur (from 1845). Each panel illustrates temperature variations over several centuries, with blue lines representing annual temperature fluctuations and red lines indicating smoothed trends over time. All four locations exhibit long-term temperature variations, with a notable warming trend in recent decades. Early records show significant variability, followed by a general cooling period in the late eighteenth and early nineteenth centuries. A clear warming trend emerges in the twentieth century, accelerating toward the present day. Uppsala and Turku display a U-shaped temperature trend, with a cooler period in the early nineteenth century followed by significant warming.

Elin Lundstad, Stefan Norrgård and A.E.J. Ogilvie

The warming trend in Uppsala appears particularly pronounced in recent decades. Trondheim shows relatively stable temperatures until the late nineteenth century, followed by a gradual warming trend. The increase in temperature appears slightly less abrupt compared to Uppsala and Turku. Stykkishólmur exhibits more pronounced fluctuations but follows a similar overall pattern of initial cooling, stabilisation and subsequent warming. A mid-twentieth-century warm period is evident, followed by continued warming.

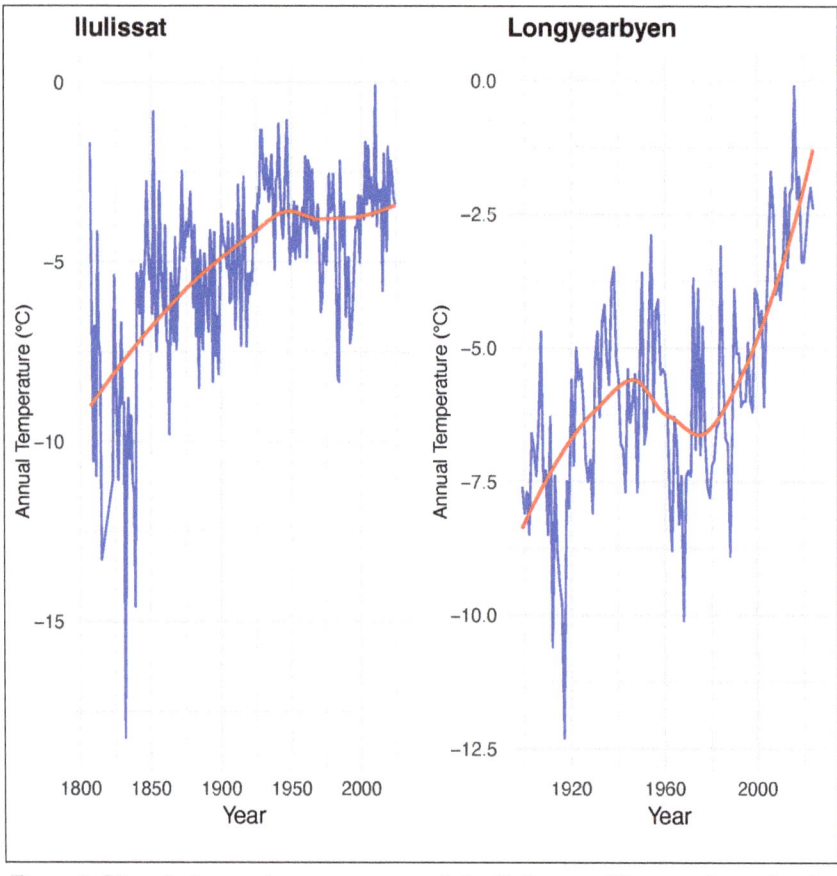

Figure 2. Historical annual temperature trends for Ilulissat and Longyearbyen, showing temperature variations (blue) and long-term trends (red). Both locations exhibit significant warming, particularly in recent decades, highlighting the pronounced impact of climate change in Arctic regions. Data sources: see 'Data' section of the Bibliography.

The Development of Meteorological Institutions

Figure 2 displays historical annual temperature variations for Ilulissat, Greenland and Longyearbyen, Svalbard. Both locations are in the Arctic, where climate change has a particularly strong influence. For Ilulissat (left graph), the data begin in the early nineteenth century, showing initially very cold temperatures. A strong warming trend is visible from the early nineteenth century until around the mid-twentieth century, after which the trend stabilises somewhat. The late twentieth and early twentieth centuries show another sharp increase in temperature, aligning with global warming trends. The overall warming pattern suggests a significant shift in the local climate, which could have major implications for Greenland's ice sheet and sea-level rise. For Longyearbyen (right graph), this dataset starts later, around 1900, but already shows a clear warming trend. A temporary cooling phase appears around the mid-twentieth century, possibly linked to regional climate variations. The most dramatic warming occurs from around 1980 onward, with temperatures rising rapidly. This pronounced warming in Svalbard aligns with other Arctic observations, where temperatures are increasing at more than twice the global average. To summarise, both Ilulissat and Longyearbyen exhibit strong warming trends, particularly in recent decades. The Arctic is one of the fastest-warming regions on Earth, and these graphs reflect that trend. The rapid rise in temperatures has significant consequences, including melting ice sheets and glaciers (Greenland's contribution to sea level rise); thawing permafrost, which releases greenhouse gases; and ecosystem disruptions that affect both Arctic wildlife and human settlements.

Data and Methods

Below, we present temperature time series and investigate the correlation between the series from each capital. We used the Mann-Kendall test (Mann 1945; Kendall 1948) to identify trends. The Mann-Kendall test identifies the presence of a monotonic trend in a time series of a variable. This nonparametric method does not assume any specific distribution for the data and its rank-based measure is robust to extreme values. The method provides three key types of information: Kendall's Tau (τ): This rank correlation coefficient quantifies the monotonicity of the trend. It ranges from –1 to 1, with positive values indicating an increasing trend and negative values indicating a decreasing trend. Sen's Slope represents the overall slope of the time series and is calculated as the median of slopes between all pairs of points in the series.

Elin Lundstad, Stefan Norrgård and A.E.J. Ogilvie

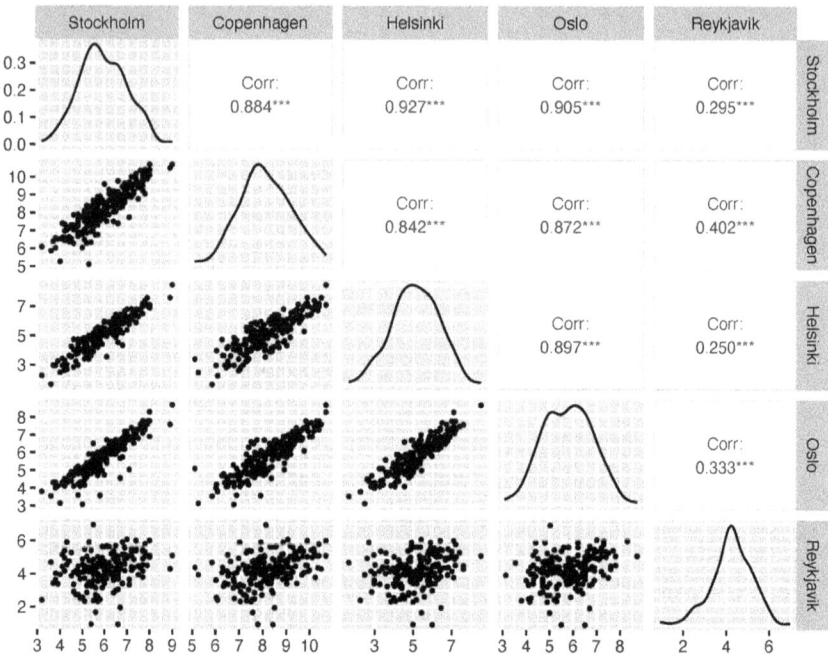

Figure 3. Pairwise scatter plots of annual temperature time records for five Nordic capitals (Stockholm, Copenhagen, Helsinki, Oslo and Reykjavik). The lower left triangle of panels display scatter plots, while the diagonal panels show density plots of individual time records. The upper right triangle of panels present Pearson correlation coefficients, with significance levels indicated. The results suggest strong correlations among mainland cities, while Reykjavik exhibits weaker correlations with the others. Data sources: see 'Data' section of the Bibliography.

Temperature Differences in Nordic Capitals

Figure 3 displays a pairwise scatter plot matrix (also known as a correlation plot) of time series data for five cities: Stockholm, Copenhagen, Helsinki, Oslo and Reykjavik. Stockholm, Copenhagen, Helsinki and Oslo exhibit strong positive correlations (> 0.8), suggesting that their time series trends are similar. Reykjavik has weak correlations (around 0.25–0.40) with the other cities, implying different trends or patterns.

Figure 4 presents historical annual temperature trends for the five Nordic capitals – Stockholm, Copenhagen, Helsinki, Reykjavik and Oslo – spanning from the eighteenth century to the present. The blue lines represent annual temperature fluctuations, while the red curves illustrate long-term temperature trends. Each city's dataset spans over a century, revealing distinct warming patterns.

The Development of Meteorological Institutions

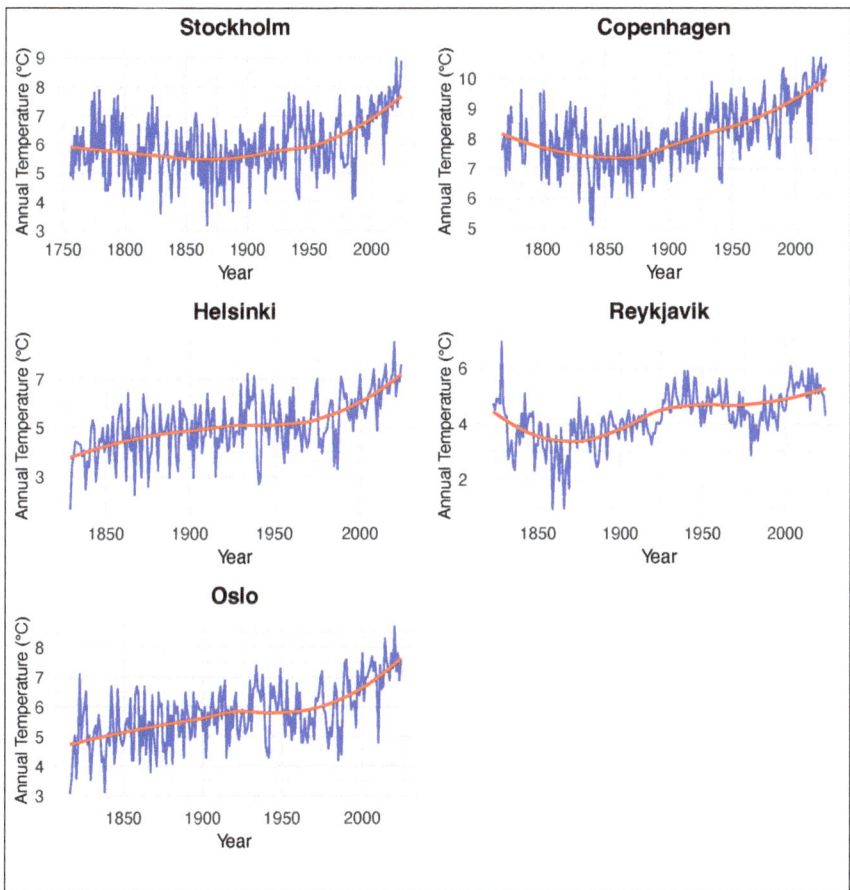

Figure 4. The temperature time series for the five capitals of the Nordic region, with data extending through to 2024: Stockholm, starting in 1756; Copenhagen, starting in 1768; Helsinki, starting in 1829; Reykjavik, starting in 1823; and Oslo, starting in 1816. The blue lines represent annual temperature records, while the red curves show smoothed temperature trends over time. Each panel corresponds to a different city, illustrating long-term warming trends with variations in historical temperature fluctuations. The data highlight a general increase in temperatures, particularly in the twentieth and twenty-first centuries. Data sources: see 'Data' section of the Bibliography.

All five cities exhibit a general increase in annual temperatures, particularly from the late nineteenth century onward. The warming trend is especially pronounced in the latter half of the twentieth century and into the twenty-first century. Reykjavik's temperature trend appears more variable, possibly reflecting

Elin Lundstad, Stefan Norrgård and A.E.J. Ogilvie

its maritime climate influences. The data suggest a strong correlation between rising global temperatures and regional climate changes in the Nordic capitals.

Copenhagen (top right) the southernmost city, is the warmest and shows relatively stable temperatures in earlier centuries, followed by a noticeable upward trend in the twentieth and twenty-first centuries. Reykjavik (middle right), the northernmost city, experiences significant variability, particularly in earlier years, but follows an overall warming pattern. As expected, it is the coolest due to its high latitude and proximity to the ocean in the west. Helsinki (middle left) exhibits a consistent rise in temperatures, with a pronounced increase in recent decades. Oslo (bottom right) demonstrates fluctuations but also a clear warming trend over time. Overall, the figure highlights a general warming trend across all five cities, particularly in the twentieth and twenty-first centuries, aligning with broader global climate change patterns.

Among the Nordic capitals, Stockholm experiences the warmest summers, with an average maximum temperature of 23 degrees Celsius in July, while Copenhagen, Oslo and Helsinki have an average July maximum of 22 degrees Celsius. The Nordic countries each have distinct climates and weather patterns. Norway experiences a greater influence of westerly winds, while Sweden and Finland, the latter also impacted by the continental Eurasian climate, are more sheltered. Denmark, located at a lower latitude compared to the other countries, tends to be warmer as a result. The climate of Iceland is highly variable but is ameliorated by proximity to the Irminger Current. In the past, it was much affected by the sea ice that reached its coasts on the East Greenland Current (Ogilvie and Miles, this volume). Table 3 gives an overview of the maximum and the minimum temperature in each country.

Table 3. Maximum and minimum recorded temperatures in each of the Nordic countries. Source: see Data section of Bibliography.

Country	Coldest	Station	Date	Warmest	Station	Date
Sweden	-52.6 °C	Vuoggatjålme	2 February 1966	38.0 °C	Ultuna / Målilla	9 July 1933 / 29 June 1947
Denmark	-31.2 °C	Thy	8 January 1982	36.4 °C	Holstebro	10 August 1975
Norway	-51.4 °C	Karasjok	1 January 1886	35.6 °C	Nesbyen	20 June 1970
Finland	-51.5 °C	Polla	22 February 1999	37.2 °C	Joensuu airport	29 July 2010
Iceland	-37.9 °C	Grímsstaðir	21 January 1918	30.5 °C	Teigarhorn	22 June 1939

The Development of Meteorological Institutions

Summary of Temperature Trends

The Mann-Kendall test results provide strong statistical evidence of long-term warming in these Nordic cities. The trends are consistent with the effects of climate change, highlighting the need for continued monitoring and climate adaptation strategies in the region. The results confirm a statistically significant warming trend across all five Nordic cities (Stockholm, Copenhagen, Helsinki, Reykjavik and Oslo) over the period 1829–2024. All p-values are extremely low ($p < 0.05$), indicating that the likelihood of these trends occurring by chance is nearly zero. This strongly suggests that temperatures have been rising consistently in all the Nordic capitals. Copenhagen shows the strongest warming trend, with the highest Kendall's tau value ($τ = 0.521$). Stockholm has the weakest trend among the five cities ($τ = 0.344$), though it remains highly significant. Reykjavik ($τ = 0.451$), Helsinki ($τ = 0.386$), and Oslo ($τ = 0.381$) fall in between, all exhibiting statistically significant upward trends. The Mann-Kendall test provides strong statistical evidence of long-term warming in these Nordic cities. The trends are consistent with the effects of climate change, highlighting the need for continued monitoring and climate adaptation strategies in the region.

All four locations exhibit long-term warming trends, with variations in historical cooling and stabilisation periods. However, the most striking feature across all datasets is the rapid temperature increase in the late twentieth and early twenty-first centuries. This aligns with the well-documented global temperature rise attributed to anthropogenic climate change.

Figure 5 demonstrates that Nordic capitals have experienced significant warming over the last two centuries, with an accelerated increase in temperatures in the modern era. The observed patterns align with broader global warming trends, though regional differences in historical cooling and warming phases suggest local climatic influences. The steep rise in recent decades highlights the ongoing impact of climate change in the Nordic region. All five cities exhibit a clear warming trend, particularly in the twentieth and twenty-first centuries. The red curves show that temperatures have increased significantly since the late nineteenth or early twentieth century, consistent with global climate change. All cities show a clear upward trend in annual mean temperature over time, indicating significant warming since the early nineteenth century. The trend lines suggest that warming has been relatively consistent, with some variability in different periods. Copenhagen (red) and Reykjavik (purple) show similarly steep warming curves in the plot (Figure 5). Helsinki (blue) and Oslo (black) display comparable trends but are generally cooler than Copenhagen, while Stockholm (green) shows the lowest warming trend overall. Copenhagen exhibits a strong and steady increase in temperature, especially since the 1980s. This can partly be attributed to the urban heat island effect, which is significant

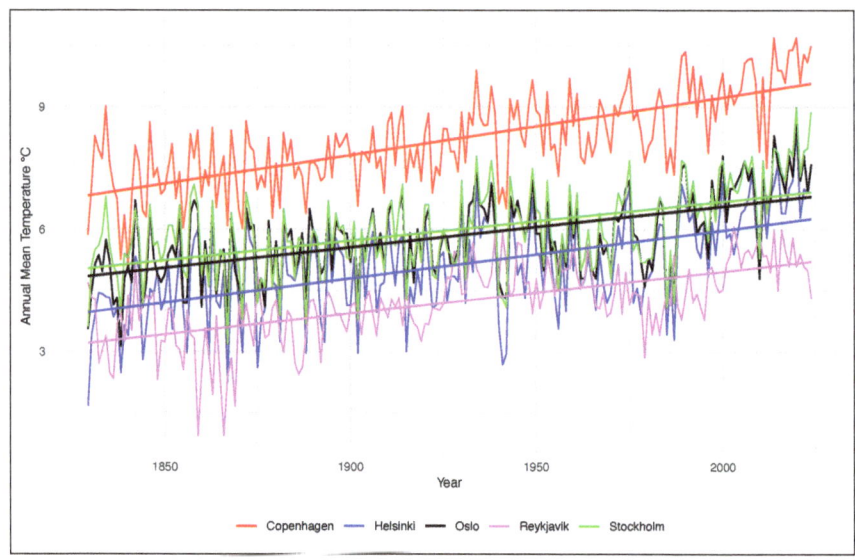

Figure 5. Annual mean temperature trends from 1829 to 2024 for five Nordic capitals: Copenhagen (red), Helsinki (blue), Oslo (black), Reykjavik (purple) and Stockholm (green). The solid lines represent linear trend lines, highlighting long-term warming trends in each city. Among the Nordic capitals, Copenhagen has experienced the most significant temperature change during the common period of observation (1829–2024). The UN's Intergovernmental Panel on Climate Change (IPCC) has highlighted in its reports that observed trends are consistent with the expected impacts of climate change, emphasising the need for continuous monitoring and the development of climate adaptation strategies in various regions. Data sources: see 'Data' section of the Bibliography.

due to the city's dense infrastructure (Alexander, 2021). Additionally, Copenhagen is influenced by both continental and maritime air masses, with warmer southern European influences reaching Denmark more easily. There has also been stronger warming in winter and spring, contributing to a reduction in frost days (DMI Report, 2021). Reykjavik has a maritime subpolar oceanic climate, strongly influenced by the North Atlantic Ocean and the Irminger Current. Temperatures are moderated by the surrounding sea, resulting in less extreme highs and lows. As a result, Reykjavik has lower absolute temperatures due to its maritime climate (Fried et al. 2024). Oslo's continental climate may explain its distinct seasonal variability. Helsinki and Stockholm display nearly parallel trends, suggesting similar regional climatic influences. Both cities lie in the same latitude range, with slight differences arising from their proximity to large bodies of water. Helsinki's station data show a clear acceleration in warming

The Development of Meteorological Institutions

after 1980. To conclude: this figure illustrates a clear warming trend across all five cities, consistent with global climate change. While differences in absolute temperature levels exist due to geographic and climatic factors, the overall pattern strongly suggests that the Nordic region has experienced significant and ongoing warming over the past two centuries.

Summary and Future Work

From the contributions of the many Nordic observers who sent records to the Royal Society in London in the early 1700s, only the Uppsala records survived. While observations were continuously made during the 1700s, and there was a rise in institutional records around the 1750s and 1760s, most non-institutional records were short-lived. In best-case scenarios, they covered ten years of observations, a phenomenon notable in all Nordic countries. The main reason was that making daily observations was a time-consuming task and several records died with the observer. It is therefore easy to understand why Johan Leche, in an article published in 1763, noted that meteorological records, in order to be continuous, should be made at public institutions (Norrgård 2024). While it was popular among enthusiastic individuals to measure the unseeable with the purpose of making meteorological observations at certain hours every day, day after day, week after week, year after year, the results were by no means self-explanatory. It was not understood how to synthesise or process thousands of measurements into useful information. Often, as seen in the journals published by the Royal Societies in Sweden and Norway, a summary of the meteorological diaries, despite having instrumental observations, included qualitative descriptions of the weather for each month. Temperature averages, if extracted, were calculated as the mean from the lowest and highest recorded temperature from each month. This method of presenting information, while occasionally changing, followed James Jurin's suggestions. Jurin was a physician, and he was, as noted in his article, essentially interested in finding out how and where winds were coming from because this caused changes in the weather, which affected people's health. Thus, the idea was never, as argued by Feldmand (1990) to synthesise meteorological observations into what we today would consider *climate*. The pioneers of early meteorological instrumental records could have had no idea that, in centuries to come, their temperature observations would play a vital part in the discussion on anthropogenic global heating. While it is tempting to think that they had a sense of how important their work would be for future scientists, they might be surprised to find out that, in spite of the development of meteorological institutions, it is still a very complex task to predict future weather and, even more so, climate.

Elin Lundstad, Stefan Norrgård and A.E.J. Ogilvie

Acknowledgements

This study is connected to the Research Council of Norway, the University of Oslo and the Centre for Advanced Study (CAS) in Oslo, Norway, which funds and hosts the research project 'The Nordic Little Ice Age' during the 2024/25 academic year. Many thanks to Dominik Collett for his patience and support for finishing this paper. Many thanks also to Fredrik Ljungqvist for his helpful comments on the paper. Elin Lundstad would like to thank the Norwegian Meteorological Institute for their support. The reconstructions and all analyses presented in this data descriptor have been performed based on free and open-source software (R). Stefan Norrgård's research project *Climate, Culture and Society. The Climatic History of Finland during the 18th century* (2023–2026) is funded by The Kone Foundation. Astrid Ogilvie acknowledges support from the National Science Foundation of the USA (current Award 212786) for her work on the past climate of Iceland and also the project *Reconstructions of Climatic and Bioclimatic Conditions in Greenland and Labrador/ Nunatsiavut ca. 1770 to 1939 from Moravian Missionary Observations (MORCLIM)* funded by the National Science Centre of Poland. PI: R. Przybylak. Co-PIs include: A.E.J. Ogilvie, G.R. Demarée.

Contributions

E.L. processed all the data and undertook quality control of the data and also produced the figures. E.L. wrote the first draft, S.N. contributed with the historical connections and enhanced the initial draft, while A.E.J.O. focused on Iceland and Greenland and drew several historical connections. E.L., S.N. and A.E.J.O. contributed to editing and approved the final version of the manuscript. All the authors contributed to reviewing and editing the manuscript.

Bibliography

Alexander, Cici. 2021. 'Influence of the proportion, height and proximity of vegetation and buildings on urban land surface temperature'. *International Journal of Applied Earth Observation and Geoinformation* 95: 102265. https://doi.org/10.1016/j.jag.2020.102265

Barlaup A. 1966. *Det norske meteorologiske institutt 1866-1966* [The Norwegian Meteorological Institute 1866–1966]. Norwegian Meteorological Institute. Oslo.

Bergþórsson, P. 1957. *Loftin blá* [Fair skies]. Reykjavík: Heimskringla.

Demarée, G.R., A.E.J. Ogilvie and P. Mailier. 2020. 'Early meteorological observations in Greenland and Labrador in the 18th century: A contribution of the Moravian Brethren', *Proceedings of the 35th International Symposium on the Okhotsk Sea and Polar Oceans 2020, 16–19 February 2020, Mombetsu, Hokkaido, Japan, Okhotsk Sea and Polar Ocean Research Association (OSPORA)*, C5: 35–38.

The Development of Meteorological Institutions

Demarée, G.R. and A.E.J. Ogilvie. 2021. 'Early meteorological observations in Greenland: The contributions of David Cranz, Christian Gottlieb Kratzenstein and Christopher Brasen'. In F.A. Jensz and C. Petterson (eds), *Legacies of David Cranz's 'Historie von Grönland' (1765)*. Basingstoke: Palgrave Macmillan. pp. 141–164. https://doi.org/10.1007/978-3-030-63998-3.

Derham, W. 1734a. 'An abstract of the meteorological diaries, communicated to the Royal Society, with Remarks upon them, by W. Derham, D.D. Canon of Windsor, F.R.S. [Vide Part III. in Philosophical Trans. N° 433.]'. *Philosophical Transactions of the Royal Society of London* 38 (434): 405–412. https://doi.org/10.1098/rstl.1733.0057.

Derham, W. 1734b. 'An abstract of the meteorological diaries, communicated to the Royal Society, with Remarks upon them, by W. Derham, D.D. Canon of Windsor, F.R.. [Vide Part IV. in Philosophical Trans. N° 434.]'. *Philosophical Transactions of the Royal Society of London* 38 (434): 458–470. https://doi.org/10.1098/rstl.1733.0064.

DMI Report 16-02. 2016. Denmark – DMI Historical Climate Data Collection 1768–2015, ed. by John Cappelen.

Federhofer, M.-T. 2002. 'Værtegn: Om Jacob Nicolaj Wilses (1735–1801) meteorologiske notasjonssystem'. *Nordlit* 11 (March): 91–102. https://doi.org/10.7557/13.2070.

Feldman, T.S. 1990. 'Late Enlightenment meteorology'. In T. Frängsmyr, J.L. Heilbron and R.E. Rider (eds), *The Quantifying Spirit in the Eighteenth Century*. Berkeley: University of California Press. pp. 143–177. https://doi.org/10.1525/9780520321595.

Fried, N., T.C. Biló, W.E. Johns, C.A. Katsman, K.E. Fogaren, M. Yoder et al. 2024. 'Recent freshening of thesubpolar North Atlantic increased thetransport of lighter waters of the Irminger current from 2014 to 2022'. *Journal of Geophysical Research: Oceans*, **129**: e2024JC021184. https://doi.org/10.1029/2024JC021184

Garðarsson, H. 2000. *Saga Veðurstofu Íslands*. Reykjavík: Icelandic Meteorological Office.

Hanna, E., T. Jónsson and J.E. Box. 2004. 'An analysis of Icelandic climate since the nineteenth century'. *International Journal of Climatology* 24 (10): 1193–1210. https://doi.org/10.1002/joc.1051.

Hestmark, G. and Ø. Nordli. 2016. 'Jens Esmark's Christiania (Oslo) meteorological observations 1816–1838: The first long-term continuous temperature record from the Norwegian capital homogenized and analysed'. *Climate of the Past* **12** (11): 2087–2106, https://doi.org/10.5194/cp-12-2087-2016.

Hiebl, J. 2006. *The Early Instrumental Climate Period (1760–1860) in Europe: Evidences from the Alpine Region and Southern Scandinavia*. MA thesis. Wien: Universität Wien. https://www.meteologos.rs/wp-content/uploads/2019/12/THE-EARLY-INSTRUMENTAL-CLIMATE-PERIOD-IN-EUROPE_1760-1860.pdf.

Holopainen J. and H. Vesajoki. 2001. 'Varhainen lämpötilahavaintosarja Torniosta vuosilta 1737–1749'. *Terra* 113 (3): 196–201. https://elektra.lib.helsinki.fi/se/t/0040-3741/113/3/varhaine.pdf.

Holopainen, J. 2004. *Turun varhainen ilmastollinen havaintosarja* [The early climatological observation series from Turku]. Helsinki: Ilmatieteen laitos.

Holopainen, J., S. Helama and S. Holopainen. 2023. 'Kuninkaallisen tiedeakatemian arkistoon sijoitetut Turun vuosien 1748–1800 päivittäiset säähavainnot historiallisena ja elektronisena informaationa' [The observations from Turku 1748–1800, in the Royal Academy of sciences, as an historical and electronic record]. *Auraica: Scripta A Societate Porthan Edita* **14** (1): 39–57. https://doi.org/10.33520/aur.142175.

Huhtamaa, H. and F.C. Ljungqvist. 2021. 'Climate in Nordic historical research – a research review and future perspectives'. *Scandinavian Journal of History* **46** (5): 665–695. https://doi.org/10.1080/03468755.2021.1929455.

Huovila, S. 2001. 'Meteorology'. *Geophysica* **37** (1–2): 287–308.

Intergovernmental Panel on Climate Change (IPCC). 2014. *Climate Change 2014: Synthesis Report. Contribution of Working Groups I, II and III to the Fifth Assessment Report of the Intergovernmental Panel on Climate Change*. Geneva: IPCC.

Jónsson T. 1989. *Afturábak frá Stykkishólmi, Veðurathuganir Jóns Þorsteinssonar landlæknis í Nesi og í Reykjavík* [The observations of Jón Thorsteinsson in Nes and Reykjavík 1820–1854, and their relation to the Stykkishólmur series]. Reykjavík: Icelandic Meteorological Office.

Jónsson, T. and H. Garðarsson. 2001. 'Early instrumental meteorological observations in Iceland'. In A.E.J. Ogilvie and T. Jónsson (eds), *The Iceberg in the Mist: Northern Research in Pursuit of a 'Little Ice Age'*. Dordrecht: Springer. pp. 169–187. https://doi.org/10.1007/978-94-017-3352-6_8.

Kendall, M.G. 1948. *Rank Correlation Methods*. London: Griffin.

Kington, J.A. 1972. 'Meteorological observing in Scandinavia and Iceland during the eighteenth century'. *Weather* **27**: 222–233. https://doi.org/10.1002/j.1477-8696.1972.tb04301.x.

Laakso. L., S. Mikkonen, A. Drebs, A. Karjalainen, P. Pirinen and P. Alenius. 2018. '100 years of atmospheric and marine observations at the Finnish Utö Island in the Baltic Sea'. *Ocean Science* **14** (4): 617–632. https://doi.org/10.5194/os-14-617-2018.

Lamb, H.H. 1977. *Climate: Present, Past and Future Volume 2: Climatic History and the Future*. London: Methuen.

Lundstad, E., Y. Brugnara, D. Pappert et al. 2023. 'The global historical climate database HCLIM'. *Scientific Data* **10**: 44. https://doi.org/10.1038/s41597-022-01919-w.

Lundstad, E. 2004. *Reconstruction of the Climate in Hamar 1749–1835 Based on Farm Diaries*. MA thesis. Bergen: University of Bergen. https://bibsys-almaprimo.hosted.exlibrisgroup.com/permalink/f/8hnp7t/BIBSYS_ILS71481003920002201

Mann, H.B. 'Nonparametric tests against trend. 1945'. *Econometrica: Journal of the Econometric Society* **13**(3): 245–259. https://doi.org/10.2307/1907187.

Moberg, A. and H. Bergström. 1997. 'Homogenization of Swedish temperature data. Part III: The long temperature records from Uppsala and Stockholm'. *International Journal of Climatology* **17** (7): 667–699. https://doi.org/10.1002/(SICI)1097-0088(19970615)17:7<667::AID-JOC115>3.0.CO;2-J.

Nevanlinna, H. 2014. 'On the early history of the Finnish Meteorological Institute'. *History of Geo- and Space Sciences* **5** (1): 75–80. https://doi.org/10.5194/hgss-5-75-2014, 2014.

Nicholls, N. 2010. 'Why do we care about past climates? An editorial essay'. *WIREs Climatic Change* **1** (2): 155–157. https://doi.org/10.1002/wcc.4.

Nilsen, Y. and M. Vollset. 2006. *Vinden dreier: Meteorologiens historie i Norge*. Oslo: Scandinavian Academic Press.

Norrgård, S. 2024. 'En märklig mans liv och arbete. Johan Leche (1704–1764) och meteorologins historia i Åbo, Finland och Sverige under 1700-talet' [A peculiar man's life and work. Johan Leche (1704–1764) and the history of meteorology in Turku, Finland and Sweden during the 18th century]. *Auraica: Scripta A Societate Porthan Edita* **15** (2): 57–86. https://doi.org/10.33520/aur.157036.

Ogilvie, A.E.J. and T. Jónsson (eds). 2001a. *The Iceberg in the Mist: Northern Research in Pursuit of a 'Little Ice Age'*. Dordrecht: Kluwer Academic Publishers.

Ogilvie, A.E.J. and T. Jonsson. 2001b. '"Little Ice Age" research: A perspective from Iceland'. *Climatic Change* **48**: 9–52. https://doi.org/10.1023/A:1005625729889.

Ogilvie, A.E.J. and M.W. Miles. 2025. 'Northern Iceland temperature variations and sea-ice incidence c. AD 1600-1850', this volume. https://doi.org/10.63308/63881023874820.ch08

Pfister, C., R. Brazdil, J. Luterbacher, A.E.J. Ogilvie and S. White. 2018. 'Early modern Europe'. In S. White, C. Pfister and F. Mauelshagen (eds), *The Palgrave Handbook of Climate History*. London: Palgrave Macmillan. pp. 265–295. https://doi.org/10.1057/978-1-137-43020-5_23.

Przybylak, R., G. Singh, P. Wyszyński, A. Araźny and K. Chmist. 2024. 'Air temperature changes in SW Greenland in the second half of the 18th century'. *Climate Past* **20**: 1451–1470. https://doi.org/10.5194/cp-20-1451-2024, 2024.

Przybylak, R., Z. Vizi and P. Wyszyński. 2009. 'Air temperature changes in the Arctic from 1801 to 1920'. *International Journal of Climatology* **30**: 791–812. https://doi.org/10.1002/joc.1918.

Seppinen I. 1988. *Ilmatieteen laitos 1838–1988* [The Meteorological Institute 1938–1988]. Helsinki: Ilmatieteen laitos.

Sigfúsdóttir, A.-B. 1969. 'Hitabreytiingar á Íslandi 1846–1968'. In M.Á. Einarsson (ed.) *Hafísinn*. Reykjavík: Almenna Bókafélagið. pp. 70–79

Skovmand, P.A. 1986a. 'Den meteorologiske aktivitet på Rundetârn i Arrene 1777–1781. (I)' ['The meteorological activity at Rundetaam 1777–1781. (I)']. *Vejret* **28**: 15–18.

Skovmand, P.A. 1986b. 'Den meteorologiske aktivitet på Rundetårn i årene 1777–1781. (II)' ['The meteorological activity at Rundetaam 1777–1781. (II)']. *Vejret* **29**: 11–18.

Ström, H. 1798. 'Udtog af 10 aars meteorologiske iagttagelser paa Eger'. *Det Kongelige Norske Videnskabers Selskabs Skrifter*: 15–42. https://www.ntnu.no/ojs/index.php/DKNVS_skrifter/article/view/850

Elin Lundstad, Stefan Norrgård and A.E.J. Ogilvie

Valler, V., J. Franke, Y. Brugnara et al. 2024. 'ModE-RA: A global monthly paleo-reanalysis of the modern era 1421 to 2008'. *Scientific Data* 11: 36. https://doi.org/10.1038/s41597-023-02733-8.

Vesajoki, H. and J. Holopainen. 2000. 'Lämpötilahavaintoja Oulusta 1700-luvun lopulta' ['Temperature observations from Oulu at the end of the 1700s']. *Terra* 112 (1): 2000.

Vinther, B.M., K.K. Andersen, P.D. Jones, K.R. Briffa and J. Cappelen. 2006a. 'Extending Greenland temperature records into the late eighteenth century'. *Journal of Geophysical Research* 111: D11105. https://doi.org/10.1029/2005JD006810.

Vinther, B.M. et al. 2006b. 'Synchronized dating of three Greenland ice cores throughout the Holocene'. *Journal of Geophysical Research* 111: D13102, https://doi.org/10.1029/2005JD006921.

Winkler, P. 2023. 'The early meteorological network of the Societas Meteorologica Palatina (1781–1792): Foundation, organization, and reception'. *History of Geo- and Space Sciences* 14 (2): 93–120, https://doi.org/10.5194/hgss-14-93-2023.

Webpages

https://www.astro.uu.se/history/Celsius_eng.html [Accessed 12 January 2025].

http://celsius.met.uu.se/default.aspx?pageid=31 [Accessed 12 January 2025].

https://bolin.su.se/data/stockholm-historical-monthly-temperature-3 [Accessed 16 January 2025].

https://wilse.org/Wilse/Jacob_Nicolai_Wilses_liv_og_gjerning_2_4.html [Accessed 17 December 2024].

https://www.smhi.se/polopoly_fs/1.164667!/Hydrologi_81%20Isl%C3%A4ggning%20och%20islossning%20i%20svenska%20sj%C3%B6ar.pdf [Accessed 16 January 2025].

https://www.smhi.se/kunskapsbanken/meteorologi/det-meteorologiska-stationsnatets-historia-1.5248 [Accessed 15 January 2025].

https://munin.uit.no/handle/10037/5584?locale-attribute=en [Accessed 17 December 2024].

Data Sources

DMI. CC BY 4.0. Accessed 14 January 2025. https://opendatadocs.dmi.govcloud.dk/en/Data/Meteorological_Observation_Data.

FMI. CC BY 4.0. Accessed 14 January 2025. https://en.ilmatieteenlaitos.fi/download-observations.

IMO. Accessed 14 January 2025. https://en.vedur.is/Medaltalstoflur-txt/Reykjavik.txt.

MetNorway. CC BY 4.0. Accessed 14 January 2025. https://seklima.met.no/observations/.

SMHI. CC BY 4.0. Accessed 14 January 2025. https://www.smhi.se/data/hitta-data-for-en-plats/ladda-ner-vaderobservationer/airtemperatureInstant

HARD 2.0 – Historical Arctic Database. Accessed 14 January 2025. http://www.hardv2.prac.umk.pl/index.php

Lundstad, E., Y. Brugnara and S. Brönnimann. 2022. Global Early Instrumental Monthly Meteorological Multivariable Database (HCLIM) [dataset bundled publication]. CC BY 4.0, Accessed 9 December 2024. PANGAEA, https://doi.org/10.1594/PANGAEA.940724.

The Authors

Elin Lundstad is a climate researcher at the Norwegian Meteorological Institute. She began studying glaciers in western Norway and climate patterns in 1996. She completed her master's degree in historical climatology in 2004 at the University of Bergen, focusing on an eighteenth-century Norwegian farm diary. Over the years, she has explored various professional paths but found her strongest passion in historical climatology. Currently, she is pursuing a Ph.D. on early instrumental data at the Institute of Geography at the University of Bern, Switzerland.

Stefan Norrgård is a senior researcher and climate historian at the Department of History at Åbo Akademi University in Turku, Finland. Subsequent to reconstructing climate in West Africa during the 1700s, his research interests have centred on riverine ice breakups in Finland. He has reconstructed spring ice breakups for both the Aura River (Turku) and the Kokemäki River (Pori) between the 1700s and 2000s. He has several publications on ice breakups but his research field also covers historical climate adaptation processes and meteorological observations in Finland and Sweden in the 1700s. His ongoing research project, founded by the Kone Foundation, investigates climate, culture and society in Finland in the 1700s.

Astrid E.J. Ogilvie is a Research Professor at the Institute of Arctic and Alpine Research at the University of Colorado and a Senior Associate Scientist at the Stefansson Arctic Institute in Akureyri, Iceland. Her research focuses on the broader issues of climatic change and contemporary Arctic issues, as well as the environmental humanities. Her interdisciplinary, international projects have included leadership of the NordForsk Nordic Centre of Excellence project: *Arctic Climate Predictions: Pathways to Resilient, Sustainable Societies (ARCPATH)*; and *The Natural World in Literary and Historical Sources from Iceland ca. AD 800 to 1800 (ICECHANGE)*. She is currently a Fellow of the project *The Nordic Little Ice Age (1300–1900) Lessons from Past Climate Change (NORLIA)* at the Centre for Advanced Study at the Norwegian Academy of Science and Letters. She is the author of some 100 scientific papers and has three edited books to her credit.

ANCIENT AND MEDIEVAL CLIMATE

Chapter 2

COLD OR CULTURE? EFFECTS OF MID-HOLOCENE TEMPERATURES ON FORAGER AND EARLY FARMER DEMOGRAPHICS IN SOUTHERN NORWAY

Svein Vatsvåg Nielsen

Abstract

Climate has changed considerably throughout the Holocene and humans have continuously adapted to environmental change. However, research is not clear on how, to what extent and in relation to which environmental factors populations have adapted. The mid-Holocene period c. 6200–2200 BCE in Scandinavia involved the transition from purely forager-based (Mesolithic) economies to the establishment of farming-based (Neolithic) economies. In southern Norway there is evidence of a limited introduction of farming and husbandry in the Oslo fjord region in the early fourth millennium BCE, and of foragers experimenting with farming during the Neolithic period. This paper hypothesises that short-term cold events (or Little Ice Age-like events) had negative impact on human demography in southern Norway also during the mid-Holocene. The current record of Little Ice Age-events is compared to a long-term population trajectory based on archaeological data, showing little or no negative impact on human demography during the mid-Holocene.

Introduction

A huge epistemological leap has recently been made within Stone Age studies in Scandinavia. By adopting a palaeodemographic approach, and equipped with recently developed computational methods, several case studies have now convincingly demonstrated a connectivity between fluctuations in demographic proxy records and changes in the archaeological record, e.g., technological evolution, economy, architecture and patterns of mobility (Bergsvik et al. 2021; Damm et al. 2020; Jørgensen 2020; Jørgensen et al. 2020; Lundström 2023; Manninen et al. 2022; Nielsen 2021a; Nielsen et al. 2019; Shennan et al. 2013; Silva and Vander Linden 2017; Solheim and Persson 2018; Timpson et al. 2014). These studies capture an international trend within archaeology focused on assembling large datasets to model global-scale demography throughout prehistory (Bevan et al. 2017; Bird et al. 2022; Manning et al. 2016). More than before, archaeology is now in a position to critically examine the interplay between demographic fluctuations, climate change and social evolution in prehistory.

Recent studies of proxy-based global Holocene temperature reconstruction, from Marcott et al. (2013) and onwards, have enabled a comparative analysis of temperatures and long-term trends in archaeological data. This paper explores whether supra-regional yet – from an archaeological viewpoint – short-term climate events during the mid-Holocene correlate with an archaeological demographic proxy record for a large geographical area, southern Norway. The more recent Little Ice Age (LIA, c. 1300–1800 AD) is one example of such a short-term climate event (Collet and Schuh 2018; Fagan 2000). Little Ice Age-like (or LIA-like) event has become a technical term in recent research and refers to long-lasting cold events at the centennial scale; currently we know of several such events that date to the mid-Holocene. In 2021, Helama et al. reported on two LIA-like events relevant for the northern Hemisphere that pre-dated the Bronze Age (c. 1700 BCE); one at ~5450 BCE and a second at ~3240 BCE (Helama et al. 2021; see also Wanner et al. 2011; Sigl et al. 2022). Then van Dijk et al. identified three additional pre-Bronze Age LIA-like events at ~4535 BCE, ~4378 BCE and ~3941 BCE (van Dijk et al. 2024; see also [Figure 1](#)A in this paper).

These in total five periods of colder temperatures were caused by one or several volcanic eruptions, which have been recognised empirically through documented linear decrease of stratospheric aerosol optical depth (AOD), and in the following atmosphere-ocean-sea ice interaction causing prolonged cooling. Effects of such cooling periods are reduced plant productivity and growing degree days, as well as famine in human populations (Gundersen 2019; Helama et al. 2021; Loftsgarden and Solheim; Arthur et al. 2024). Judging from the impact the Little Ice Age had on societies in northern Europe, there is reason to hypothesise that *LIA-like events in the more distant past also affected the demography of human societies negatively*.

The palaeodemographic turn in archaeology now enables an evaluation of such hypotheses with regard to demographic fluctuations. The mid-Holocene period, c. 6200–2200 BCE, in Scandinavia involves the transition from purely forager-based (Mesolithic) economies to the establishment of farming-based (Neolithic) economies. In southern Norway, it was also a period of significant population growth (Bergsvik et al. 2021; Nielsen 2021a; Solheim and Persson 2018). Although the effects of climate, temperature and changing ecological conditions on prehistoric foragers and early farmers are a central topic of study in archaeology, a general understanding or theory of causal links between humans and the environment has not developed (Binford 2001; Kelly 2013).

The central question has developed beyond the nature-culture divide, for we are not questioning whether culture or nature shape human behavioural patterns, but rather which factors shape human demography and carrying capacity. Following Kelly's (2013: 184) first definition of carrying capacity, anthropologists can construct viable models when information about resource abundance and extractive efficiency is available, but this is rarely the case for studies of prehistoric foragers and early farmers. Further, the complexity of such predictive models increases when other-than-terrestrial food sources are abundant, such as marine and lacustrine food (Binford 2001: 368), which was the case in southern Norway. For mid-Holocene foragers in the northern hemisphere, it could for instance be expected that reduced growing seasons would have had a greater negative impact on early farmers compared to foragers, as the latter would rely less on plant productivity. However, such generalisations cannot be considered reliable hypotheses, as we remain oblivious as to how cold spells would affect the total biotope surrounding the contemporary foragers. Conversely, we would potentially misjudge the strong social ties within early farming groups in the northern Europe, which could have made them more resilient to external pressure.

This leads us to Kelly's (2013: 184) second definition of carrying capacity, namely as a 'density-dependent limit on a population growth rate'. The literature within this field of study goes back to Malthus (1826), who stated that food supply growth was linear within a system of production, while the growth of human population was exponential. The rule was therefore that, without control mechanism, so-called 'checks', populations would eventually grow beyond ecological carrying capacity. Childe (1936) even based his theory of 'revolutionary innovations' in prehistory on observed effects on populations from the Industrial Revolution, which had resulted in a 'kink in the population graph'. It was therefore logical to presume that the emergence of farming and similar innovations have had similar effects on populations in prehistory.

This paper relies on Kelly's second definition of carrying capacity, and the empirical foundation is an archaeological demographic proxy record. However, based on this dataset we are not able to determine for certain *if* the prehistoric populations under study reached carrying capacity at any point in time or *how close* they were to it. As we cannot fulfil Kelly's first definition of carrying capacity and gain knowledge about total population size, it is problematic to discuss effects such as famine and infections. What we can infer from the presented data are long-term trends, as well as – from an archaeological perspective – more short-term fluctuations.

Considering mid-Holocene demographics among foragers and early farmer societies in southern Norway, previous case studies of the impact of annual mean temperatures have at best found weak causal relations (Bunbury et al. 2023; Lundström 2023; Lundström et al. 2024). Rather, it has been suggested that regional specific environmental and demographic change, as well as a myriad of purely 'cultural factors' (e.g., trade, social boundaries, mobility), have played significant roles within the domains of society available for archaeologists to study. Does this hold also when the archaeological data is compared to evidence of more short-term cold period, such as the LIA-like events?

Data and Methods

To compare the evidence of the five relevant mid-Holocene LIA-like events mentioned above with changes in population growth, a long-term population proxy record was constructed using radiocarbon dates from the STAGED dataset and the MCSPD method, or *modeltest*, in RStudio through the 'rcarbon' package (Crema and Bevan 2019; Nielsen 2022; RStudio 2024). The archaeological dataset contained 2,645 anthropogenic radiocarbon dates older than 2999 uncalibrated BP, collected from archaeological sites located in southern and central Norway. The Monte Carlo summed probability density method, or MCSPD (Crema 2022), was used to fit a logistic curve to the empirical summed probability density function (SPD) and simulated 500 SPDs based on random sampling with replacement, a procedure that is similar to bootstrapping.

This approach is based on the assumption that the *number* of specific types of physical traces from past human activity reflects the density of the population within the given area. As the samples used for radiocarbon dating in archaeology have the most exact known age, they are also the most easily applicable material for such statistical analyses. Using an SPD as a population proxy data involves stacking the probability distributions of each calibrated sample together and presenting it on a 2D graph. To judge whether or not variations (ups and downs) in the graph are significant and not simply random variations, or even reflections of the calibration curve, the MCSPD method simulates 'all possible' outcomes under the theory that our empirical dataset reflects a logistic growth pattern.

A number of potential sources of bias can be dealt with, such as variations caused by the curve, although not in their entirety. Overrepresentation by sampling strategy, a typical known bias in radiocarbon datasets, was coped with using a 200-year binning procedure for each site. This means that the effect on the SPD of a high number of samples from one single structure is effectively erased, and such structures are thus weighted more equally in the SPD. More short-term fluctuations caused by the shape of the calibration curve – in this case the IntCal20 curve was applied (Reimer et al. 2020) – were smoothed by adding a fifty-year running mean to the SPDs, both empirical and simulated. Smoothing algorithms cannot erase all impacts from the calibration curve but can remove very prolific variations that we know do not represent real variations in the underlying data.

The 500 simulated SPDs through the MCSPD method are then z-scored. Through z-scoring of the year-by-year values of the simulated SPDs, the probability distributions (95.5 per cent) form the basis of a simulated envelope that works as a threshold for significant positive and negative deviations from the logistic null model. With zero deviations, we are henceforth encouraged to confirm the null model of a logistic population growth pattern. However, with one or several significant deviations from the simulated model we are, conversely, inclined to question the validity of the null-hypothesis, and to formulate an alternative hypothesis.

To investigate in more detail the spatial distribution of periods of significant fluctuations in the population model, a permutation test was used, here applying the *sptest* function in the 'rcarbon' package. Local spatial statistics enables the identification of local *hot* or *cold* spots (Crema et al. 2017), meaning in this case that regions of southern Norway with higher or lower growth rates within periods of marked fluctuations can be exposed spatially. Long-term population proxies for eastern, southernmost, western and central Norway were also made using the standard summed probability functions in 'rcarbon'.

Results

The MCSPD method (Figure 1A) produced a long-term proxy record with a slow increase from the late Glacial period and until the mid-Holocene before a flattening and then a significant increase from around c. 2300 BCE. Although this model captures the broad contour of a logistic development in the forager to fisher-farmer population from the Early Mesolithic and until the full-blown introduction of farming in the Late Neolithic – i.e., after c. 2350 BCE – there are notable deviations from the null model that need to be highlighted. The introduction of farming in the Late Neolithic is here demonstrated by SPDs of dated charred cereals and sites located in agricultural regions within the investigated area (Figure 1C). Trends within the main dataset of radiocarbon dates can also be inspected within the main landscapes of southern Norway in Figure 1D.

Between c. 9300 and 2000 BCE the simulated model identified three main periods of negative departures from the null model. The first is in the segment c. 7400–6700 BCE. Before this negative deviation there is positive deviation (7700–7500 BCE). As argued by Roalkvam and Solheim (2025), the

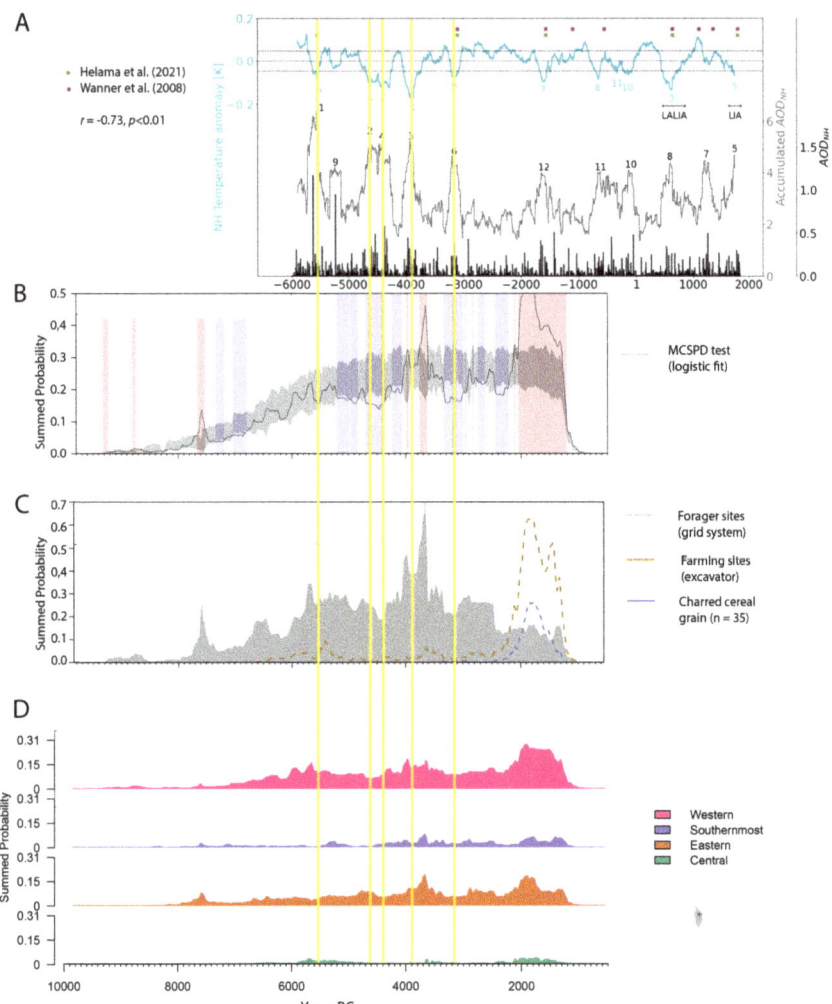

Figure 1. Comparison of pre-Bronze Age cold LIA-like events numbers 1, 2, 4, 6 and 9 (van Dijk et al. 2024), with archaeological demographic proxy data for southern Norway. Panel A: Figure 3 in van Dijk et al. 2024. Panel B: Monte Carlo summed probability distribution with a logistic fit. Panel C: Descriptive SPDs of radiocarbon dates analysed from forager sites (excavated in grid systems), farming sites (excavated in ploughed fields with use of an excavator), and the 135 dated charred cereals in the dataset. The SPD for charred cereals shows limited farming c. 3300–2800 BCE, and the establishment of permanent farming after c. 2350 BCE. Panel D: SPDs stacked according to geographical regions. Note that the SPDs in D were not normalised, meaning that the visual volume of each SPD in this figure reflects the number of samples from the respective region in the archaeological dataset.

Cold or Culture?

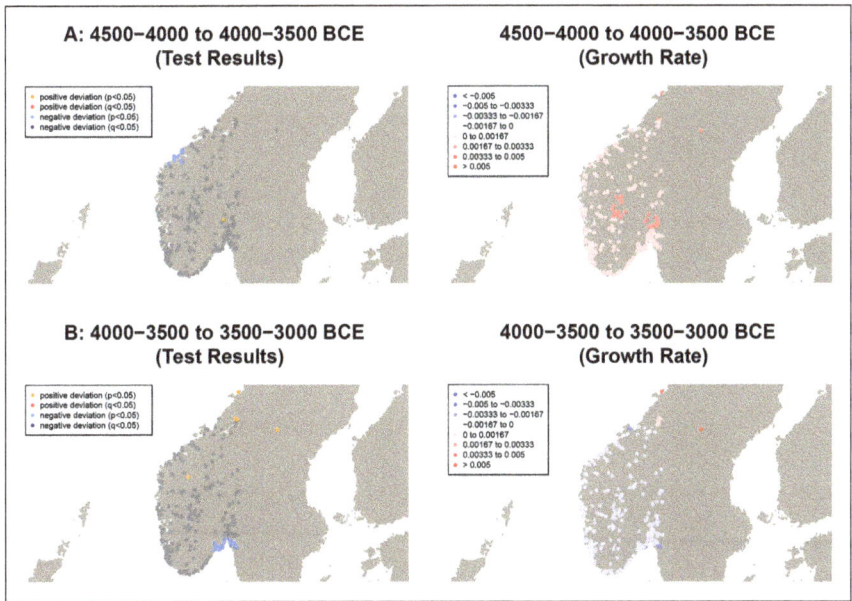

Map 1. Spatial visualisation of test results from the simulation analysis, and growth rates within the archaeological dataset before and after LIA-like periods number 1 (A in figure) and 9 (B in figure). Map by author (Svein Vatsvåg Nielsen).

Middle Mesolithic settlement in eastern Norway followed a logistic growth that plateaued in the centuries after c. 8000 BCE, and the observations here fit this scenario well. The development towards a peak in population after 8000 BCE followed by a decline and levelling during the next centuries supports the relevance of this demographic theory for southern Norway as a whole.

The second negative deviation is a longer period of a levelling or slow decrease in the Late Mesolithic, more precisely in the period c. 5200–4100 BCE. This attests to stagnation in the population, probably as a response to carrying capacity, before what seems to be a period of marked population increase leading into the Neolithic and peaking in the period c. 3800–3600 BCE. This is the period when pottery was introduced along the coast from the Oslo fjord region to the western coast, and when farming was introduced in the former region (Nielsen 2021b). However, then the population development levels out again, leading to a significant negative deviation from the null model in the period c. 3300–2020 BCE. There are shorter phases when the population is within the expected growth curve, meaning that the development within this segment represents a plateau with short-term variations. After this period of levelling and short variations, the growth in population increased drastically after the introduction of farming in the Late Neolithic period. After this, it increased steadily until the Early Iron Age and the AD 536/540 double-eruption event, when another significant decline has been identified using the same methods as applied here (Loftsgarden and Solheim 2023).

To summarise the population trends described above, we can see in southern Norway a development of logistic growth from the Early Mesolithic to around 2200 BCE. The growth pattern reaches a plateau roughly around 5500–5000 BCE. After this the population is relatively stable with more short-term variations except for two periods of unprecedented growth; first in 3800–3600 BCE and secondly from 2100 BCE and leading towards the Bronze Age. These deviations are discussed further below.

When this Mesolithic to Neolithic population model is compared to the relevant LIA-like events in the northern hemisphere, there is little correlation between these cold events and the identified periods of significant population decline and decline. From [Figure 1](#) we can infer that event 6, also called the 7.6 event and referring to four eruptions including Mt. Mazama in Oregon (US), had no visible effect on the contemporary forager population in southern Norway (van Dijk et al. 2024: 4, Table 1). Events number 4 (6.6 event) and 2 (6.4 event) intruded shortly before the mentioned increase in population in the Late Mesolithic, which also indicates little or no negative effect on demography. Further, events number 1 (5.9 event) and 9 (5.1 event) both intruded before periods of population increase in the archaeological data, the first of which is marked as a positive deviation from the simulated null model.

It is still interesting to note that the events number 1 and 9 do correlate with deviations in the archaeological proxy, although not as first expected. Event 1 (3895 BCE) was the strongest multi-centennial cold anomaly reported by van Dijk et al. (2024) in terms of cooling, and here it intrudes shortly before a prominent peak in population within the research area. A spatial analysis of this transition ([Map 1](#)) here studied as the transition in the archaeological data from 4500–4000 BCE to 4000–3500 BCE show a spatial concentration of increase in two specific areas: the inner Oslo fjord, where the conditions for farming in the early fourth millennium BCE were optimal, and on the Hardangervidda mountain plateau, where there is evidence of large game hunting in the Early Neolithic period (Nielsen 2021b; Nielsen et al. 2019; Solheim 2012).

The spatial test for the transition from the Early Neolithic population 'boom' (4000–3500 BCE) to the following decline (3500–3000 BCE) does not identify a regional hot spot but instead we see a picture of general decline in southern Norway ([Map 1](#)). The decline is then in this model interrupted by event number 9 (3152 BCE) and a period of slightly higher population, although the increase in the population data is not as vivid as in the early fourth millennium BCE and is therefore not marked as a significant deviation in the simulated model. To summarise then, although there are interesting patterns to be discussed in the presented data, a persistent pattern of LIA-like events leading to population decline, as hypothesised in this paper, is not identified.

Discussion

Since the first paper by Marcott et al. (2013), several studies of proxy-based global Holocene temperature reconstruction have shown warming after the Younger Dryas culminating around 8,000–6,000 years ago ('Holocene thermal maximum') followed by continuous cooling until nineteenth-century warming (Bova et al. 2021; Kaufman et al. 2020; Osman et al. 2021). However, model simulations of global temperatures based on observations pertaining to orbital insolation, retreating ice sheets and rising atmospheric greenhouse gases have shown a globally averaged warming in the period, i.e. with no 'Holocene thermal maximum' followed by continuous cooling, but rather a long-term warming trend (Hopcroft et al. 2023; Liu et al. 2014, 2009; Rao et al. 2022). This mismatch between proxy-based and simulation-based temperature reconstructions is usually referred to as the 'Holocene temperature conundrum' (Essell et al. 2023). The mismatch between proxy-based and simulation-based reconstructions has not been solved and is probably due to both methodological issues as well as regional variations (Hopcroft et al. 2023). Archaeology has also contributed to this debate through the early anthropogenic hypothesis (EAH), which is yet to be verified but proposes that neolithization, understood as a global phenomenon, added significant CO_2 to the atmosphere by massive deforestation starting around 7,000 years ago (Stephens et al. 2019; Ruddiman et al. 2020).

Perspectives have also changed with regard to the magnitude and effects of repetitive climate fluctuations on forager and early farming-based societies, such as the Bond events; in the case of the 4.2 event, perspectives on northern Europe have changed from disaster scenarios to local and thus often insignificant variations (Bradley and Bakke 2019; McKay et al. 2024). As has been stressed in this paper, studies of the effects on temperature of volcanic activity, such as the Little Ice Age (LIA), have identified a number of LIA-like climatic regimes in northern Europe during the last 8,000 years (Helama et al. 2021; van Dijk et al. 2024). However, this paper found no convincing causal relationship between fluctuations in temperature caused by LIA-like events in the mid-Holocene and contemporary human demography in southern Norway. This begs the question: which known 'cultural factors' were at work during this period, and how could they affect the population proxy data?

Leaving the earliest Holocene short-term deviations in Figure 1 aside as artefacts primarily of the model itself, a prominent feature in the palaeodemographic model is the significant peak in population in the period c. 3800–3600 BCE. As shown in Map 1, this peak is caused primarily by an increase of anthropogenic radiocarbon dates in the Oslo fjord region and on the high-mountain plateau between Oslo and Bergen. We can also infer from Map 2 that the increase in the 14C-data is most pertinent in eastern Norway, when compared to the southernmost, western and central regions. Recent reviews of the physical traces of settlements in the Oslo fjord region found that this time slot (c. 3800–3600 BCE) was precisely when the hinterland and inland became settled by people who probably practised a Neolithic economy (Nielsen 2021b; see also Roalkvam 2023:15). Thus, a certain measure of the surplus of radiocarbon dates from this phase derive from waste pits, cooking pits and postholes from longhouses at sites inhabited by the first farming-based societies in Norway.

This Early Neolithic *hot* spot in the palaeodemographic model correlates both with settlements and the distribution of imported flint axes (Figure 2). It should also be stressed that, starting around 3900 BCE and lasting until c. 3600 BCE, there is also evidence of foragers with Early Neolithic pottery in the Oslo fjord region and along the coast further west (Johansen 2005; Reitan, 2014). However, demographic studies have repeatedly found a steady decrease in forager settlement in the Oslo fjord region after 3900 BCE (Solheim and Persson 2018; Roalkvam and Solheim 2025). There is therefore also in southern Norway evidence of a period of cultural overlap, what has been called a 'latest Mesolithic', or a 'negotiation phase' in the neolithisation process (Furholt 2010; Sørensen 2020; Zvelebil and Rowley-Conwy 1984). In short, the Early Neolithic population increase in southern Norway correlates with other evidence of a migrant population entering the Oslo fjord region. As argued by Nielsen (2021b), there is evidence of farming-based societies in the Oslo fjord region until about 2800 BCE.

The second prominent feature in the palaeodemographic model for southern Norway is the multiple negative deviations from the simulation at c. 3300–2200 BCE. We can see in Figure 1 that there is a marked decline in the 14C-data after about 3400 BCE, and from Map 1 and Map 2 we can infer that this decline is effective all over southern Norway; it is not only caused by the lack of dates from farming-based occupations in the Oslo fjord region but reflects a general decline in dates also from forager sites. Already in 1955 – before the advent of radiocarbon dating – it was hypothesised by the Norwegian archaeologist Erik Hinsch that the first farming-based societies in eastern Norway collapsed towards the end of the fourth millennium BCE, and that they adapted their economy to foraging, and fishing in particular (Hinsch 1955). In the 1950s,

Cold or Culture?

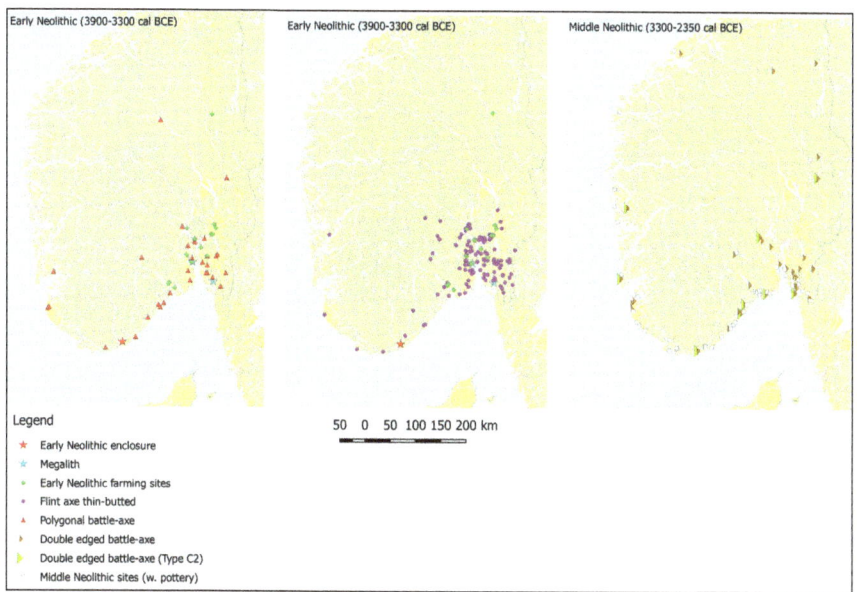

Map 2. *Spatial distribution of Early and Middle Neolithic artefacts and site types. Previously published in Nielsen 2021b. Map by author (Svein Vatsvåg Nielsen).*

the most prominent signature of this collapse was a radical change in the spatial distribution of flint axes and stone battle axes, as Hinsch observed that before c. 3000 BCE most of these finds occurred in the hinterland and inland of eastern Norway, while in the period ca. 3000–2800 BCE (MN III–V of the Middle Neolithic period) they suddenly occurred along the coast.

The spatial pattern identified by Hinsch in 1955 is still visible on updated distribution maps (Map 1 and Map 2). In light of this hypothesis, often referred to as the 'de-neolithisation hypothesis', it is tempting to interpret the negative deviation observed in the archaeological proxy data from around c. 3300 as representing a demographic trend leading towards, and thus partly causing, the radical shift in settlement and artefact distribution around 3000 BCE (Nielsen et al. 2019). However, in contrast to the model presented by Hinsch in 1955 of a full societal collapse around 3000 BCE, there are now indications that farming practices were adapted to the hunter-fisher-gatherer lifestyle that we see re-emerging in eastern and southernmost Norway after c. 3300 BCE (Nielsen and Stokke 2023). It is relevant to mention that the Early Neolithic farming societies in the Oslo fjord region were 'surrounded' by forager groups in southernmost, western and central Norway who persisted until around the transition to the Late Neolithic.

Figure 2. Early Neolithic flint axes, of the polished and thin butted type, found at Krågevoll in Klepp, southwestern Norway. Inventory numbers S9084 and S8070 at the Museum of Archaeology, University of Stavanger. Image is CC BY-NC-ND 4.0, number Sf35690 from Arkeologisk museum, Universitetet i Stavanger (accessed 5 March 2025 at https://digitaltmuseum.no/0210112704752/to-tynn-nakkede-okser-fra-yngre-steinalder).

The second major growth phase in the 14C-data for southern Norway starts around 2200 BCE and continues into the Bronze Age. Judging here based on the considerable data analysis presented recently by Loftsgarden and Solheim (2023: 102; Figure 2), the development in population that started around 2200 continued as a steady growth period until the Early Iron Age and the 536/540 AD double-eruption event, when the population in eastern Norway was reduced considerably. As shown above, the start of this phase corresponds with the (second) introduction of farming in Norway, as evidenced in the charred cereals in the 14C-data. However, this was also a period of significant changes in settlement organisation and architecture (Sand-Eriksen 2025), and we see a gradual decline and eventual disappearance of 'forager sites' in southern Norway (Nielsen et al. 2019). In many aspects, this transition marks the end of the Stone Age in this region of Europe.

The fourth millennium BCE was a turbulent period in southern Scandinavia, including the Oslo fjord region in eastern Norway. A fundamental shift in the primary mode of production from (Mesolithic) foraging to (Neolithic) farming and herding had a profound impact on the locally adapted forager societies. An even more fundamental shift occurred at the transition to the Late Neolithic. It should be stressed that a central factor of the Early Neolithic introduction of Neolithic economies in these regions was driven by mobility and migrations, which limits the weight of the arguments presented here against the impact of the LIA-like events on demographic changes. There is certainly the possibility that these cold spells could have had significant impact on nearby and more densely populated geographical regions, such as northern Germany, Denmark and southern Sweden, where the first farmers of Scandinavia settled (e.g., Iversen et al. 2025). If this was the case, what we observe in the presented data for southern Norway could be but ripples in a much larger scene of human demography. Secondly, there is also the possibility that the demise of one 'cultural group' within a specific area occurred at the expense of increased mobility and migration among neighbouring groups, thus leaving an invisible imprint on proxy-based long-term demographic models despite climate-induced population crash. These are venues that future research should seek to explore further.

Acknowledgements

The author would like to thank the organisers of the conference *Nordic Climate History: Learning from Past Experience* in Oslo in early May 2024, and for inviting me to contribute to this edited volume. Thanks is also due to the internal and the anonymous external reviewers, for critical comments and thoughtful suggestions on a previous draft of this chapter.

References

Arthur, F., K. Hatlestad, K.-J. Lindholm et al. 2024. 'The impact of volcanism in Scandinavian climate and human societies during the Holocene: Insights into the Fimbulwinter eruptions (536/540 AD)'. *The Holocene* 34 (5): 619–633. https://doi.org/10.1177/09596836231225718.

Bergsvik, K.A., K. Darmark, K.L. Hjelle, J. Aksdal and L.I. Åstveit. 2021. 'Demographic developments in Stone Age coastal western Norway by proxy of radiocarbon dates, stray finds and palynological data'. *Quaternary Science Reviews* 259: 106898. https://doi.org/10.1016/j.quascirev.2021.106898.

Bevan, A., S. Colledge, D. Fuller, R. Fyfe, S. Shennan and C. Stevens. 2017. 'Holocene fluctuations in human population demonstrate repeated links to food production and climate'. *Proceedings of the National Academy of Sciences*: 201709190. https://doi.org/10.1073/pnas.1709190114.

Binford, L.R. 2001. *Constructing Frames of Reference: An Analytical Method for Archaeological Theory Building Using Ethnographic and Environmental Data Sets*. California: University of California Press.

Bird, D., L. Miranda, M.V. Linden et al. 2022. 'A synthetic global database of archaeological radiocarbon dates'. *Nature Scientific Data* 9: 1–19. https://doi.org/10.1038/s41597-022-01118-7.

Bova, S., Y. Rosenthal, Z. Liu, S.P. Godad and M. Yan. 2021. 'Seasonal origin of the thermal maxima at the Holocene and the last interglacial'. *Nature* 589: 548–553. https://doi.org/10.1038/s41586-020-03155-x.

Bradley, R.S. and J. Bakke. 2019. 'Is there evidence for a 4.2 ka BP event in the northern North Atlantic region?' *Climate of the Past* 15: 1665–1676. https://doi.org/10.5194/cp-15-1665-2019.

Bunbury, M., K. Ivar, E. Kirkeng, S. Vatsvåg, J. Kneisel and M. Weinelt. 2023. 'Understanding climate resilience in Scandinavia during the Neolithic and Early Bronze Age'. *Quaternary Science Reviews* 322. https://doi.org/10.1016/j.quascirev.2023.108391.

Childe, V.G. 1936. *Man Makes Himself*. New York: New American Library.

Collet, D. and M. Schuh (eds). 2018. *Famines During the 'Little Ice Age' (1300–1800)*. New York: Springer. https://doi.org/10.1007/978-3-319-54337-6.

Crema, E. and A. Bevan. 2019. 'Analysing radiocarbon dates using the rcarbon package'. https://cran.r-project.org/web/packages/rcarbon/vignettes/rcarbon.html.

Crema, E. 2022. 'Statistical inference of prehistoric demography from frequency distributions of radiocarbon dates: A review and a guide for the perplexed'. *Journal of Archaeological Method and Theory*. https://doi.org/10.1007/s10816-022-09559-5.

Crema, E., A. Bevan and S. Shennan. 2017. 'Spatio-temporal approaches to archaeological radiocarbon dates'. *Journal of Archaeological Science* 87: 1–9. https://doi.org/10.1016/j.jas.2017.09.007.

Damm, C.B., M. Skandfer, E.K. Jørgensen et al. 2020. 'Investigating long-term human ecodynamics in the European Arctic: Towards an integrated multi-scalar analysis of early and mid Holocene cultural, environmental and palaeodemographic sequences in Finnmark County, Northern Norway'. *Quaternary International* 549: 52–64. https://doi.org/10.1016/j.quaint.2019.02.032.

Essell, H., P.J. Krusic, J. Esper et al. 2023. 'A frequency-optimised temperature record for the Holocene'. *Environmental Research Letters* 18. https://doi.org/10.1088/1748-9326/ad0065.

Fagan, B.M. 2000. *The Little Ice Age*. New York: Basic Books.

Furholt, M. 2010. 'A virtual and a practiced Neolithic? Material culture symbolism, monumentality and identities in the Western Baltic Region'. *Journal of Neolithic Archaeology* 2010: 1–16.

Gundersen, I.M. 2019. 'The Fimbulwinter theory and the 6th century crisis in the light of Norwegian archaeology: Towards a human-environmental approach'. *Primitive Tider* 21: 101–120.

Helama, S., M. Stoffel, R.J. Hall et al. 2021. 'Recurrent transitions to Little Ice Age-like climatic regimes over the Holocene'. *Climate Dynamics* 56: 3817–3833. https://doi.org/10.1007/s00382-021-05669-0.

Hinsch, E. 1955. 'Traktbegerkultur–megalitkultur. En studie av Øst-Norges eldste, neolitiske gruppe'. *Universitetets Oldsaksamlingen Årbok*, 1951–1953: 10–177.

Hopcroft, P.O., P.J. Valdes, B.N. Shuman, M. Toohey and M. Sigl. 2023. 'Relative importance of forcings and feedbacks in the Holocene temperature conundrum'. *Quaternary Science Reviews* 319. https://doi.org/10.1016/j.quascirev.2023.108322.

Iversen, R., P.O. Nielsen, L.V. Sørensen et al. 2025. 'Sun stones and the darkened sun: Neolithic miniature art from the island of Bornholm, Denmark'. *Antiquity* 1–7. https://doi.org/10.15184/aqy.2024.217

Johansen, K.B. 2005. 'Vestgård 3 og Vestgård 6, to tidligneolittiske boplasser'. *In Situ* 6: 103–112.

Jørgensen, E.K. 2020. 'The palaeodemographic and environmental dynamics of prehistoric Arctic Norway: An overview of human-climate covariation'. *Quaternary International* 549: 36–51. https://doi.org/10.1016/j.quaint.2018.05.014.

Jørgensen, E.K., P. Pesonen and M. Tallavaara. 2020. 'Climatic changes cause synchronous population dynamics and adaptive strategies among coastal hunter-gatherers in Holocene northern Europe'. *Quaternary Research* 108: 1–16. https://doi.org/10.1017/qua.2019.86.

Kaufman, D., N. McKay, C. Routson et al. 2020. 'Holocene global mean surface temperature, a multi-method reconstruction approach'. *Nature Scientific Data* 7: 1–13. https://doi.org/10.1038/s41597-020-0530-7.

Kelly, R.L. 2013. *The Lifeways of Hunter-Gatherers. The Foraging Spectrum*. Cambridge: Cambridge University Press.

Liu, Z., B.L. Otto-Bliesner, F. He et al. 2009. 'Transient simulation of last deglaciation with a new mechanism for bolling-allerod warming'. *Science* 325: 310–314. https://doi.org/10.1126/science.1171041.

Liu, Z., J. Zhu, Y. Rosenthal, X. Zhang, B.L. Otto-Bliesner, A. Timmermann, R.S. Smith, G. Lohmann, W. Zheng and O.E. Timm. 2014. 'The Holocene temperature conundrum'. *Proceedings of the Natural Academy of Science* 111. https://doi.org/10.1073/pnas.1407229111.

Loftsgarden, K. and S. Solheim. 2023. 'Uncovering population dynamics in Southeast Norway from 1300 BC to AD 800 using summed probability distributions'. In M. Ødegaard, and I. Ystgaard (eds), *Complexity and Dynamics. Settlement and Landscape from the Bronze Age to the Renaissance in the Nordic Countries (1700 BC–AD 1600)*. Leiden: Sidestone Press. pp. 99–112.

Lundström, V. 2023. 'Living through changing climates: Temperature and seasonality correlate with population fluctuations among Holocene hunter-fisher-gatherers on the west coast of Norway'. *Holocene* 33: 1376–1388. https://doi.org/10.1177/09596836231185839.

Lundström, V., D. Simpson and P. Yaworsky. 2024. '"Here by the sea and sand": Uninterrupted hunter-fisher-gatherer coastal habitation despite considerable population growth'. *Open Quaternary* 10: 1–16. https://doi.org/10.5334/oq.129-.

Malthus, T. 1826. *An Essay on the Principle of Population*. London: John Murray.

Manninen, M.A., G. Fossum, T. Ekholm and P. Persson. 2022. 'Early postglacial hunter-gatherers show "false logistic" growth in a low productivity environment'. *Journal of Anthropological Archaeology* 70: 101497. https://doi.org/10.1016/j.jaa.2023.101497.

Manning, K., S. Colledge, E. Crema, S. Shennan and A. Timpson. 2016. 'The cultural evolution of Neolithic Europe. EUROEVOL Dataset 1: Sites, phases and radiocarbon data'. *Journal of Open Archaeology Data* 5: 1–5. https://doi.org/10.5334/joad.40.

Marcott, S.A., J.D. Shakun, P.U. Clark and A.C. Mix. 2013. 'A reconstruction of regional and global temperature for the past 11,300 years'. *Science* 339: 1198–1201. https://doi.org/10.1126/science.1228026.

McKay, N.P., D.S. Kaufman, S.H. Arcusa et al. 2024. 'The 4.2 ka event is not remarkable in the context of Holocene climate variability'. *Nature Communications* 15: 6555. https://doi.org/10.1038/s41467-024-50886-w.

Nielsen, S.V. 2022. *In the Wake of Farming: Studies of Demographic and Economic Variability in Southern Norway in the Period 4500–1700 BCE*. Ph.D. thesis. Oslo: Department of archaeology, Museum of Cultural History.

Nielsen, S.V. 2021a. 'A Late Mesolithic forager dispersal caused pre-agricultural demographic transition in Norway'. *Oxford Journal of Archaeology* 40: 153–175. https://doi.org/10.1111/ojoa.12218.

Nielsen, S.V. 2021b. 'Early farming in Southeastern Norway: New evidence and interpretations'. *Journal of Neolithic Archaeology* 23: 1–31. https://doi.org/10.12766/jna.2021.4.

Nielsen, S.V., P. Persson and S. Solheim. 2019. 'De-Neolithisation in southern Norway inferred from statistical modelling of radiocarbon dates'. *Journal of Anthropological Archaeology* 53: 82–91. https://doi.org/10.1016/j.jaa.2018.11.004.

Nielsen, S.V. and J.S. Stokke. 2023. 'Neolithic farming in forager-resource systems: A case from southern Norway'. In D. Groß and M. Rothstein (eds), *Changing Identities in a Changing World: Current Studies on the Stone Age around 4000 BCE*. Leiden: Sidestone Press, pp. 189–198.

Osman, M.B., J.E. Tierney, J. Zhu et al. 2021. 'Globally resolved surface temperatures since the Last Glacial Maximum'. *Nature* 599: 239–244. https://doi.org/10.1038/s41586-021-03984-4.

Rao, Z., Y. Tian, K. Guang et al. 2022. 'Pollen data as a temperaturei in the late Holocene: A review of results on regional, continental and global scales'. *Frontiers in Earth Science* 10. https://doi.org/10.3389/feart.2022.845650.

Reimer, P.J., W.E.N. Austin, E. Bard et al. 'The IntCal20 nothern hemisphere radiocarbon age calibration curve (0-55 cal kBP)'. *Radiocarbon* 62: 725–757. https://doi.org/10.1017/RDC.2020.41.

Reitan, G. 2014. 'Langangen Vestgård 6. En strandbundet boplass med keramikk fra tidligneolitikum'. In G. Reitan and P. Persson (eds), *Vestfoldbaneprosjektet. Arkeologiske Undersøkelser i Forbindelse med ny Jernbane Mellom Larvik og Porsgrunn. Bind II. Seinmesolittiske, Neolittiske og Yngre Lokaliteter i Vestfold og Telemark*. Oslo: Portal forlag og Kulturhistorisk Musem. pp. 171–220.

Roalkvam, I. 2023. 'A simulation-based assessment of the relation between Stone Age sites and relative sea-level change along the Norwegian Skagerrak coast'. *Quaternary Science Reviews* 299: 1–20.

Roalkvam, I. and S. Solheim. 2025. 'Comparing summed probability distributions of shoreline and radiocarbon dates from the Mesolithic Skagerrak coast of Norway'. *Journal of Archaeological Method and Theory* 32: 1–29.

RStudio. 2024. *RStudio: Integrated Development for R. RStudio*. PBC, Boston, MA. http://www.rstudio.com/.

Ruddiman, W.F., F. He, S.J. Vavrus and J.E. Kutzbach. 2020. 'The early anthropogenic hypothesis: A review'. *Quaternary Science Review* 240: 106386. https://doi.org/10.1016/j.quascirev.2020.106386.

Sand-Eriksen, A. 2025. 'Building small, living large: A corpus of south-eastern Norwegian settlement evidence, 2350–500 BC'. *European Journal of Archaeology* 2025: 1–24. https://doi.org/10.1017/eaa.2024.50.

Shennan, S., S.S. Downey, A. Timpson, K. Edinborough, S. Colledge, T. Kerig, K. Manning, and M.G. Thomas. 2013. 'Regional population collapse followed initial agriculture booms in mid-Holocene Europe'. *Nature Communications* 4: 2486. https://doi.org/10.1038/ncomms3486.

Sigl, M., M. Toohey, J.R. McConnell, J. Cole-Dai and M. Severi. 2022. 'Volcanic stratospheric sulfur injections and aerosol optical depth during the Holocene (past 11500 years) from a bipolar ice-core array'. *Earth System Science Data* 14 (7): 3167–3196. https://doi.org/10.5194/essd-14-3167-2022.

Silva, F. and M. Vander Linden. 2017. 'Amplitude of travelling front as inferred from 14C predicts levels of genetic admixture among European early farmers'. *Nature Scientific Reports* 7: 31–34. https://doi.org/10.1038/s41598-017-12318-2.

Solheim, S. 2012. 'Mobility, points and people. Technological and social changes towards the Neolithic of Southern Norway'. In R. Berge, M.E. Jasinski and K. Sognnes (eds), *N-TAG TEN. Proceedings of the 10th Nordic TAG Conference at Stiklestad, Norway 2009*. Oxford: Archaeopress. pp. 205–215.

Solheim, S. and P. Persson. 2018. 'Early and mid-Holocene coastal settlement and demography in southeastern Norway: Comparing distribution of radiocarbon dates and shoreline-dated sites, 8500–2000 Cal. BC'. *Journal of Archaeological Science Reports* 19: 334–343. https://doi.org/10.1016/j.jasrep.2018.03.007.

Sørensen, L. 2020. 'Biased data or hard facts? Interpretations of the earliest evidence of agrarian activity in southern Scandinavia from 6000 to 4000 cal BC in a theoretical discourse random down-the-line exchanges and structured migrations'. In K.J. Gron, L. Sørensen and P. Rowley-Conwy (eds), *Farmers at the Frontier. A Pan-European Perspective on Neolithisation*. Oxford: Oxbow Books. pp. 289–316.

Stephens, L., D. Fuller, N. Boivin et al. 2019. 'Archaeological assessment reveals Earth's early transformation through land use'. *Science* 365: 897–902. https://doi.org/10.1126/science.aax1192.

Timpson, A., S. Colledge, E. Crema et al. 2014. 'Reconstructing regional population fluctuations in the European Neolithic using radiocarbon dates: A new case-study using an improved method'. *Journal of Archaeological Science* 52: 549–557. https://doi.org/10.1016/j.jas.2014.08.011.

van Dijk, E.J.C., J. Jungclaus, M. Sigl, C. Timmreck and K. Krüger. 2024. 'High-frequency climate forcing causes prolonged cold periods in the Holocene'. *Nature Communications Earth and Environment* 5: 242. https://doi.org/10.1038/s43247-024-01380-0.

Wanner, H., O. Solomina, M. Grosjean, S.P. Ritz and M. Jetel. 2011. 'Structure and origin of Holocene cold events'. *Quaternary Science Reviews* 30: 3109–3123. https://doi.org/10.1016/j.quascirev.2011.07.010.

Wanner, H., J. Beer, J. Bütikofer, T.J. Crowley, U. Cubash, J. Flückinger, H. Goosse, F. Joos, J.O. Kaplan, M. Küttel, S.A. Müller, I.C. Prentice, O. Solomina, T.F. Stocker, P. Tarasov, M. Wagner and M. Widmann. 2008. 'Mid- to Late Holocene climate change: An overview', *Quaternary Science Reviews* **27** (19–20): 1791–1828. https://doi.org/10.1016/j.quascirev.2008.06.013.

Zvelebil, M. and P. Rowley-Conwy. 1984. 'Transition to farming in Northern Europe: A hunter-gatherer perspective'. *Norwegian Archaeological Review* 17: 104–128. https://doi.org/10.1080/00293652.1984.9965402.

The Author

Svein Vatsvåg Nielsen is an archaeologist at the Stavanger Maritime Museum in Norway. He has a Ph.D. in archaeology from the University of Oslo where his thesis focused on demographic transitions in northern Europe during the Late Mesolithic and Neolithic periods. Nielsen specialises in economic and demographic theories, and the use of quantitative and statistical methods in archaeological research. He is a frequent practitioner of research-driven field archaeology. During recent years he has focused on excavation and the sampling of wetland areas in southern Norway. He is also trained in underwater archaeology.

Chapter 3

A SERIES OF UNFORTUNATE EVENTS: TWO CENTRAL NORWEGIAN SETTLEMENTS FACING THE CLIMATIC DOWNTURN AFTER A.D. 536–540

Ingrid Ystgaard and Raymond Sauvage

Abstract

This chapter explores the impact of the mid-sixth century climatic downturn, triggered by volcanic eruptions in AD 536 and 540, on two settlements in central Norway: Vik in Ørland and Vinjeøra in Vinjefjord. The study examines how these communities, with differing geographical and cultural contexts, responded to the cooling event. Vik, situated on the outer coast, experienced a decline due to its vulnerable position and the retraction of its harbour. In contrast, Vinjeøra, located in a fjord, showed resilience and adaptability, quickly re-establishing itself after a brief period of abandonment. The analysis highlights the concepts of resilience, vulnerability and adaptation, demonstrating how local geographical conditions and pre-existing social structures influenced the communities' ability to cope with climatic stress. The findings underscore the importance of multi-scalar approaches in understanding the varied human responses to global climatic events.

Introduction

Global crises are experienced locally (Eriksen 2016). Effects of a global climatic incident in prehistory may be observed on a global scale from the point of view of the modern-day observer. However, the outcome of such incidents varies greatly between regions and communities. The increasing amount and resolution of proxy data on the mid-sixth century volcanic eruptions and the following cooling event highlights these differences. In addition to geographical variations caused by weather systems, latitudes, height above sea level and precipitation and soil conditions, cultural circumstances would also contribute

doi: 10.63308/63881023874820.ch03

to considerable variations of the effects the cooling event had on local communities (Gundersen 2019, 2021).

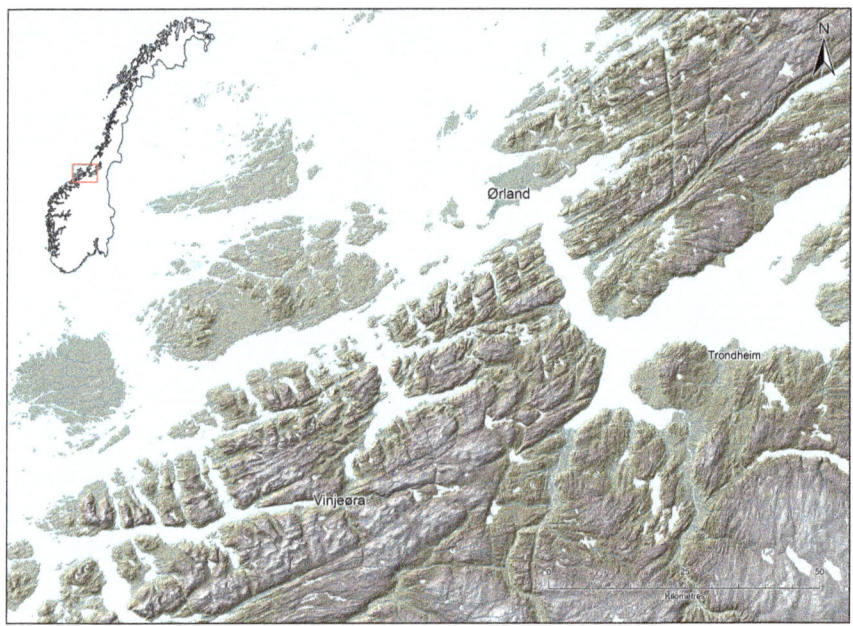

Map 1. The location of Vik and Vinjeøra, central Norway. Source: Magnar Mojaren Gran, NTNU University Museum (CC BY-SA 4.0).

In this chapter, we will explore the interplay between the mid-sixth century climatic events and two settlements in central Norway, at Vik in Ørland on the outer, open coast, and Vinjeøra in Vinjefjord in the inner part of a deep fjord (Ystgaard 2019; Sauvage 2024; Map 1). Through an analysis of the two settlements, we aim to highlight how they were differently equipped to face the climatic downturn of the mid-sixth century, and how the local experience of this global crisis varied within a regional landscape. In doing so, we will work with the concepts of resilience, vulnerability and adaptation (Redman 2005), focusing on how these concepts can direct attention to multiple timescales (Torrence 2018; Gundersen 2021: 141–142). At the same time, we aim to explore how resilience, vulnerability and adaptation in Vik and Vinjeøra may also be understood in multiple spatial frameworks (Løvschal 2022: 202).

A Series of Unfortunate Events

The Cooling Event of the Sixth Century AD

The climatic downturn following the two volcanic eruptions in the years AD 536 and 540 severely affected the Northern Hemisphere (van Dijk et al. 2022). Aerosols from the eruptions entering the Earth's atmosphere partially blocked incoming sunrays, leading to darkened skies, as indicated by sulphate peaks in ice cores as well as entries in contemporary written sources that described how the sun was covered by a dust veil and the rays of the sun lost their effect (Arjava 2005; Sigl et al. 2015). Large volcanic eruptions are major drivers of natural climate variability, and climate simulations indicate that volcanic eruptions in the sixth and seventh centuries set ocean-sea ice interactions in motion. Following the eruptions, a cooling of the Arctic in winter led to an increase in the Arctic Sea ice. This, in turn, has been related to a decrease in the northward ocean heat transport and an increase in the Atlantic meridional overturning circulation, which together caused the cooling to last longer than the direct effect of the volcanic aerosols (Toohey et al. 2016; van Dijk et al. 2022). Tree-ring data indicate that the cooling following the AD 536 and 540 events lasted for some time – in some places perhaps until AD 660 – but that the duration of the cooling effect varied between geographical regions (Büntgen et al. 2016; Helama et al. 2017). In Scandinavia, the cooling effect seems to have lasted only for a couple of decades, with varying effects on regional climate (van Dijk 2022).

Modelling of the impacts of lower temperatures in Norway on growing degree days necessary for crops to ripen indicates that food production in climatic zones with marginal conditions was severely affected, and that famine likely occurred in several areas (Stamnes 2016; Toohey et al. 2016; Gundersen 2021; van Dijk et al. 2023). Crop failure and hunger led to lower resistance to epidemics, and the outbreak of the Justinian plague has been linked to the mid-sixth century cooling event (Little 2007). Even though the *Yersinia pestis* bacteria which caused the plague has not been discovered in Scandinavian cemeteries from the sixth or seventh centuries, authors find it plausible that the plague severely affected population numbers here, too (Gräslund and Price 2012; Wagner et al. 2014; Benedictow 2016; Iversen 2016). Geochemical and palynological analyses of sediments in bogs, lakes and archaeological sites indicate that communities adjusted to the cooling climate by taking up hardier crop species and by shifting their focus to animal husbandry (Bajard et al. 2022; Westling et al. 2022). The dust veil, the cooling temperatures and the depopulation have been discussed as inspirations for the tales of the mythological Fimbulwinter in the Eddic poetry (Gräslund and Price 2012).

Moving forward, a more nuanced understanding of the effects of climate change on the interaction between humans and their environments requires studies across different spatial and temporal scales, from the global to the local and from long-term surveys to fine-grained chronological resolution.

Vulnerability, Resilience and Adaptation: A Multi-Scalar Approach

The response of communities to climatic events has been described in terms of their vulnerability, resilience and adaptation (Redman 2005; Torrence 2018). Vulnerability studies have focused on the immediate preface, presence and consequences of disaster development (Torrence 2018: 259; Gundersen 2021: 141). In contrast, resilience studies examine the interaction between society and environment in larger and multiple timeframes. Resilience theory, introduced from ecology studies to archaeology, understands the interaction between societies and their environments in adaptive cycles. Adaptive cycles are described as patterns of change in four phases, depicted as a loop moving through growth, conservation, release and reorganisation. Multiple adaptive cycles of different durations could be nested together in panarchies (Redman 2005). The concept of resilience, then, implies that the speed of change varies across scales, and that rapid change is linked to long-term processes (Løvschal 2022: 198). Adaptation is vital to a society's restoring and reorganising after incidents of rapid change, and therefore at the core of a society's development of resilience (Torrence 2018: 259).

For this analysis, we will apply a somewhat simplified understanding of the aspect of multiple time scales embedded in the concepts of vulnerability, resilience and adaptation. Vulnerability has to do with the community's immediate response in the presence of a crisis. Resilience relates to how communities work before the event of a crisis, and thus how they are equipped to encounter it, both through their interaction with their local environment and with other communities in political, economic and cultural networks. Adaptation relates to how communities adapt to their changing situation after the crisis has occurred (cf. Torrence 2018; Gundersen 2021: 141–142). Crossing temporal and spatial scales provides a more detailed account of experiences in the disruptive decades following the AD 536–540 events in central Norway, highlighting how resilience was built on previous generations' interactions with their local environment and with society both on a regional and inter-regional scale. The following generations after climatic disruptions would, for their part, act upon their social memory as well as on memories carried in the landscape itself, in their adaptation to the consequences of the climatic events.

A Series of Unfortunate Events

Local Communities and Global Climatic Events on Multiple Scales

The sites at Vik and Vinjeøra have been chosen for analysis because of the sampling strategy of the excavations (Ystgaard et al. 2019: 27; Sauvage 2024: 16). The high number of dated radiocarbon samples from both sites provides fine-grained chronological sequences around the time of the AD 536–540 events. Statistical modelling of these dates (Bronk Ramsey 2017) provides detailed insights into both communities in the period leading up to and following the AD 536–540 events. At Vik, the longevity of the settlement provides data all the way back to the last part of the Bronze Age and until the High Medieval period (Ystgaard et al. 2019). At Vinjeøra, chronological data covers the period from the beginning of our era until the Late Medieval period (Sauvage 2024). Good data on the spatial extent and arrangement of the settlements are also required. The excavations at Vik and Vinjeøra provide detailed information regarding the interrelationship between the physical installations of the settlement (buildings, roads, fences, cultivated land, cemeteries) and the immediate surrounding environment, including cultivated land, pastures, hunting and foraging grounds, fishing grounds and communication routes. Vegetation history data modelled into estimations of the composition of prehistoric vegetation cover provides important information on the interaction between humans, animals and vegetation cover (Hjelle et al. 2022; Overland and Hjelle 2024). The relationship between community and landscape is of vital importance. In the Vik case, in particular, a thorough understanding of the transformation of the coastal landscape is vital, due to land upheaval after the melting of the ice sheet towards the end of the Pleistocene (Romundset and Lakeman 2019; Hjelle et al. 2022).

A widened spatial frame is also necessary to understand a community's relationship and interconnectedness with a wider society. Such relations would have been multi-dimensional, encompassing kinship relations, travel connections and goods exchange, as well as the exchange of livestock, seeds, know-how and ideas. However, what we are left with in the archaeological record are items of exchange, such as Rhinish manufactured glass from Migration Period Vik (Figure 1) and an insular bronze fitting from early Viking age Vinjeøra (Figure 2). Such hints are important for how we understand the interconnectedness of societies over larger distances.

Figure 1. Fragments of a glass beaker, manufactured in a workshop within the realm of the Western Roman Empire, deposited in a cooking pit and a waste deposit at Area C, Vik, during the fifth century. Source: Åge Hojem, NTNU University Museum (CC BY-SA 4.0).

Figure 2. Insular bronze fitting, placed in Grave 11 at Skeiet 1, Vinjeøra, during the ninth century AD. Source: Åge Hojem, NTNU University Museum (CC BY-SA 4.0).

A Series of Unfortunate Events

Vik, Ørland: Thriving, Then Declining by the Bay

Ørland is a relatively wide and flat peninsula on the central Norwegian outer coast, situated at the mouth of the large Trondheim fjord system, which connects the outer coast and its important communication route with an interior landscape rich in agriculture and possessing wide access to outfield resources through large river systems. The peninsula itself offered rich possibilities for agriculture and animal husbandry as well as fishing grounds in the surrounding sea and foraging grounds in the tidal zone. Quite a few Iron Age settlements and cemeteries have been excavated in the peninsula during the past two decades, adding to our understanding of house construction, spatial organisation, land use and agricultural strategies (Grønnesby 1999; Berglund and Solem 2017; Ystgaard et al. 2018; Ystgaard et al. 2019; Grønnesby et al. 2023; see Map 2). The Iron Age and Medieval Period settlement at Vik, Ørland was excavated during 2015 and 2016 before the enlargement of the Norwegian Main Air Base (Ystgaard et al. 2019).

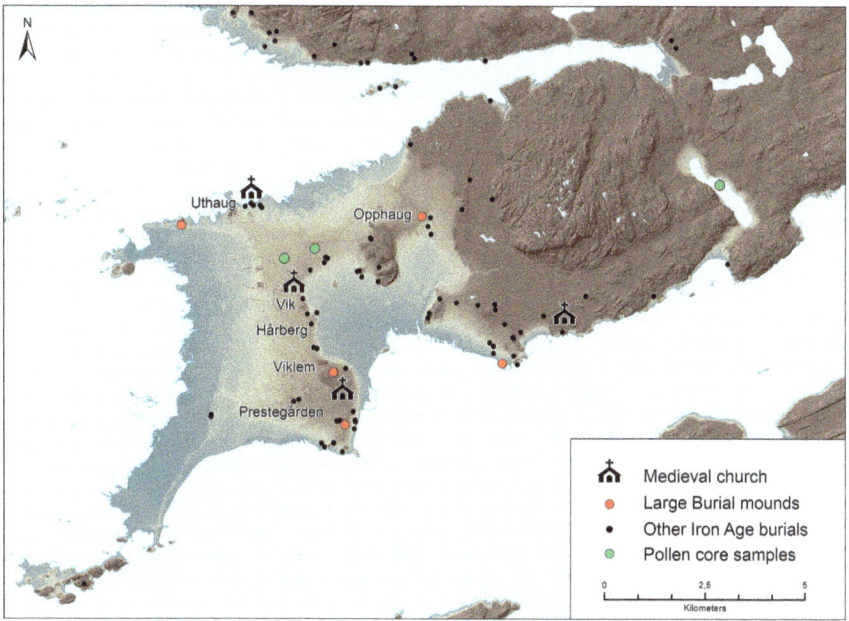

Map 2. The Ørland peninsula showing the excavation site at Vik, sites mentioned in the text, Iron Age burial mounds, and medieval churches. Source: Magnar Mojaren Gran, NTNU University Museum (CC BY-SA 4.0).

The flat character of the Ørland peninsula means that the local landscape has changed considerably since the time when it first appeared from underneath the surface of the ocean. A shoreline displacement curve existed for the area prior to the 2015–2016 excavations (Kjemperud 1986), but additional fieldwork with sampling in lower-lying isolation basins provided a better chronological resolution for the period when the Ørland peninsula first appeared and the shoreline retracted (Romundset and Lakeman 2019). During the period ca. 500 BC–AD 500, a large, shallow bay formed to the west of the peninsula, sheltered from the prevailing south-westerly winds, and reflected in the name of the settlement (Vik meaning bay). However, the new shoreline data suggest that the bay became dry and useless as a harbour during the sixth century AD (Map 3)

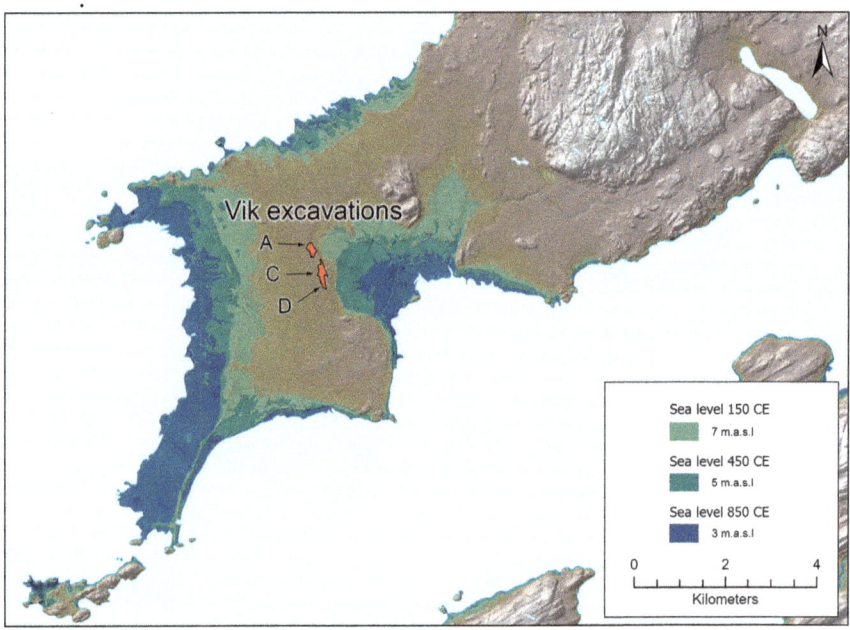

Map 3. *The shoreline at Ørland in AD 150, 450 and 850, with the excavated areas at Vik marked. The three Roman Iron Age farm sites were found in areas A, C and D. Source: Magnar Mojaren Gran, NTNU University Museum (CC BY-SA 4.0).*

The Vik settlement, placed on the highest ridge of the peninsula on dry self-draining land, displays continual occupation from around 800 BC, when the ridge first emerged from the sea and rose to become safe from sea surges and storm floods, down to the mid-sixth century AD. Local settlement, which can be traced through remains of buildings, cultivation layers, cooking pits, wells and waste deposits, stabilised from around 200 BC. Occupation intensity and

A Series of Unfortunate Events

thus population numbers peaked around AD 300, when three farm sites existed simultaneously within the excavation area (Heen-Pettersen and Lorentzen 2019; Ystgaard 2019; Map 3). No cemetery was found within the limits of the excavation area. However, finds of graves on the Ørland peninsula over the years indicate that the spatial organisation of the Roman and Migration Period (ca. AD 1–550) landscape placed the settlements on the top of the ridge, while graves were located between the settlements and the bay to the east (Map 2).

The Vik settlement leaves the impression of a regular, albeit well-off, farming and fishing community. The excavations revealed artefacts that indicate long-distance connections. Shards were found of a Migration Period (ca. AD 400–550) glass beaker, produced in a lower Rhine area workshop (Mokkelbost 2019: 221; Figure 1), and shards of Migration Period pottery which indicate inter-regional connections with the Norwegian Westland coast have also been unearthed (Solvold 2019: 296). Unusually well-preserved waste deposits containing fish and animal bones offer a glimpse into the varied resources on which the communities relied. In the Roman Iron Age and the Migration Periods (ca. AD 1–550), the communities at Vik relied on animal husbandry with sheep/goats and cattle, and fishing, mainly for cod, in inshore waters as well as the open sea. They also consumed shellfish and caught the occasional sea mammal (Storå et al. 2019). Fields surrounding the settlement were fertilised with animal dung, and yielded crops of oats and barley (Linderholm et al. 2019: 115). The landscape in which the settlement thrived consisted of a mosaic of vegetation containing infields and outfields, with animals grazing in the heathlands as well as on the seashores and in wetlands (Hjelle et al. 2022). All in all, it appears that the Vik settlement in the Roman and Migration Periods had a flexible and dynamic subsistence economy, relying on a wide variety of resources such as animal husbandry, fishing, foraging and agriculture, and that it was integrated in regional and supra-regional networks of communication and exchange.

While the Vik community was thriving in the Roman Iron Age (ca. AD 1–400), the settlement saw a decline in activity and population numbers towards the end of the period. One of the former three farm sites, in Area D, was abandoned around AD 350, but occupation continued in Areas A and C. However, occupation declined rapidly during the first half of the sixth century, and particularly from c. AD 536–540. Statistical modelling of the twelve last radiocarbon dates from the site prior to its desertion indicates that people were able to inhabit Vik for a couple of decades after the AD 536–540 events, but that it was most likely left around AD 555 (Figure 4). During the period when people were preparing to abandon, or had just abandoned the site, a six-month-old foal was deposited in a posthole in House 2, Area C. Some years after its inhabitants had left the site, someone returned to bury a calf in a pit fourteen metres north of the ruins of the same house (Storå et al. 2019: 246–247).

After the desertion of the settlement, the surrounding area was reforested. Previous infields turned by successional stages into forests, starting with an increase in shrubs and pioneer trees followed by increase in coniferous forests. Reduced human and animal impacts on the vegetation, combined with the isostatic uplift of the flat peninsula, may have led to poorer drainage and increased peat development. Pre-sixth-century grasslands turned into post-sixth-century heathlands, providing poorer agricultural possibilities (Hjelle et al. 2020: 15). Reoccupation did not occur until ca. AD 950, when a farmstead was established in Area A (Fransson 2019).

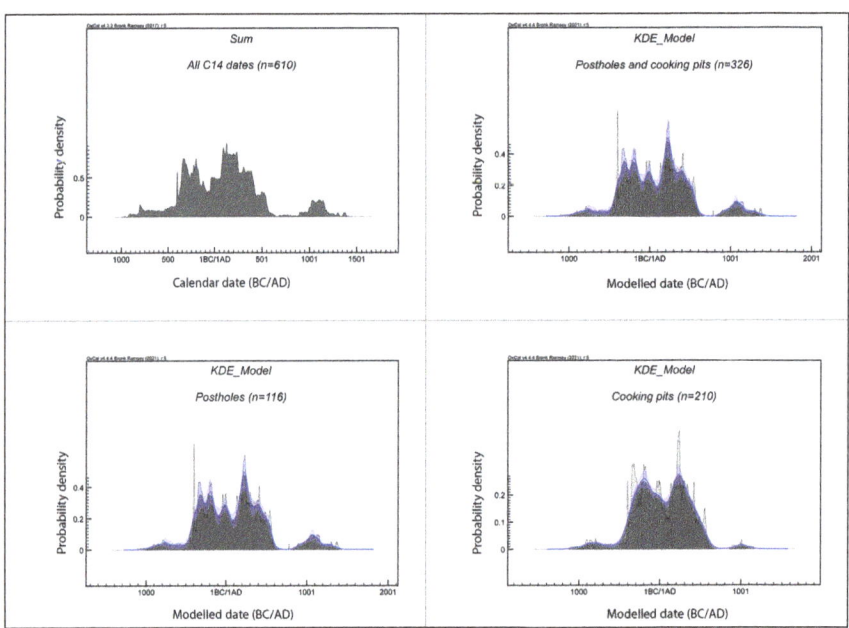

Figure 3. Summed probability distribution (SPD) and Kernel density estimate (KDE) of radiocarbon dates from Vik archaeological excavations. Source: Magnar Mojaren Gran, NTNU University Museum (CC BY-SA 4.0).

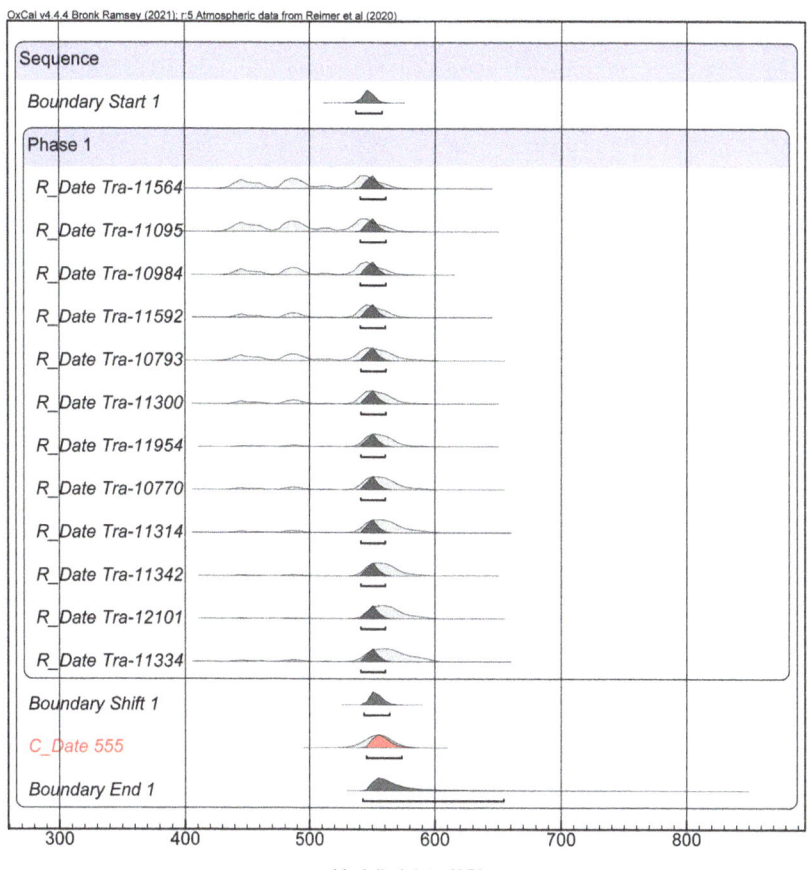

Figure 4. Last dates from Vik, areas A and C, modelled around the hypothetical date of abandonment in AD 555 (in red). Source: Ingrid Ystgaard.

Vinjeøra, Heim: Declining, Then Thriving by the Fjord

During 2019 and 2020, settlement remains and cemeteries dating from the Migration and Merovingian Periods and the Viking Age were excavated in Vinjeøra, prior to the realigning and extension of the interregional main road between Trondheim and Kristiansund (Sauvage 2024). Vinjefjord is a long and narrow waterway, along which communication between the inner fjord areas and the outer coastal areas could be safely conducted. The Vinjeøra settlements are situated in the bottom of the fjord, where communication would transfer from land to sea. An analysis of placenames and archaeological finds in the surrounding area shows how communication on land would be led through natural channels and how communication by sea would also follow natural routes (Maixner and Hellan 2024). In contrast to Vik and Ørland, this area's shoreline is rather steep, and shoreline displacement in the last 2,000 years would not have had a transformative effect on the landscape. A natural harbour would form at the mouth of the river Fjelna, and it would most likely be usable through the landscape alteration that followed land uplift. Therefore, the importance of the Vinjefjord area as a transition between travel at sea and on land would not have changed fundamentally.

Excavations in Vinjeøra in the inner part of the fjord revealed settlement traces in the area where communication transferred between land and sea as well as areas with ample opportunities for agriculture, easy access to resources in the inland and mountains, and rich fisheries in the fjord (Sauvage 2024: 7). Three areas in Vinjeøra were excavated. Skeiet 1 revealed a Merovingian and Viking Period (ca. AD 550–1050) burial ground; and Skeiet 2 contained traces of Migration to Medieval Period (ca. AD 400–1350) settlement; both were located on a well-drained glacial terrace formation overlooking the mouth of the river Fjelna and the fjord. In addition, in an area at Fjelnset, a concentration of Roman and Migration Period (ca. AD 1–550) cooking pits as well as a few burials of unknown, but most likely Late Iron Age, date (ca. AD 550–1050) were examined at a site which was also placed on a glacial terrace, directly above the natural harbour at the Fjelna river estuary (Sauvage 2024: 9–10; [Map 4](#)).

Kernel density estimates of radiocarbon dates from the excavated areas indicate that Vinjeøra first became continuously occupied during the Roman Iron Age (ca. AD 1–400) and in the Migration Period (ca. AD 400–575) (Sauvage 2024: 21–27; [Figure 5](#)). The establishment of cultivated fields at Skeiet and Fjelnset stems from the Roman Iron Age and the Migration Period, as does the use of the cooking pit area at Fjelnset. The cooking pit area was most intensively used between AD 400 and 500. Radiocarbon dates indicate that the intense use of cooking pits ceased between AD 440 and 550 (Sauvage 2024: 22). The cooking pit areas' location at a strategic point in the landscape, easily accessed from both land and sea, along with the find of a rotary quern which could indicate food preparation for larger groups, suggests that the field may have hosted a local gathering (Sauvage 2024: 28). A cooking pit site in nearby Leikvin in Sunndal, as well as other sites in Norway, have been interpreted as similar gathering sites (Narmo 1996; Ødegaard 2019). In the Migration Period, the oldest recorded house (House 2) was built at Skeiet (Bryn 2024: 91). During this period, local pollen data indicate agricultural expansion, with well-established cultural landscapes with cereal cultivation on sandy soils and outfield grazing in more wooded areas (Overland and Hjelle 2024: 73).

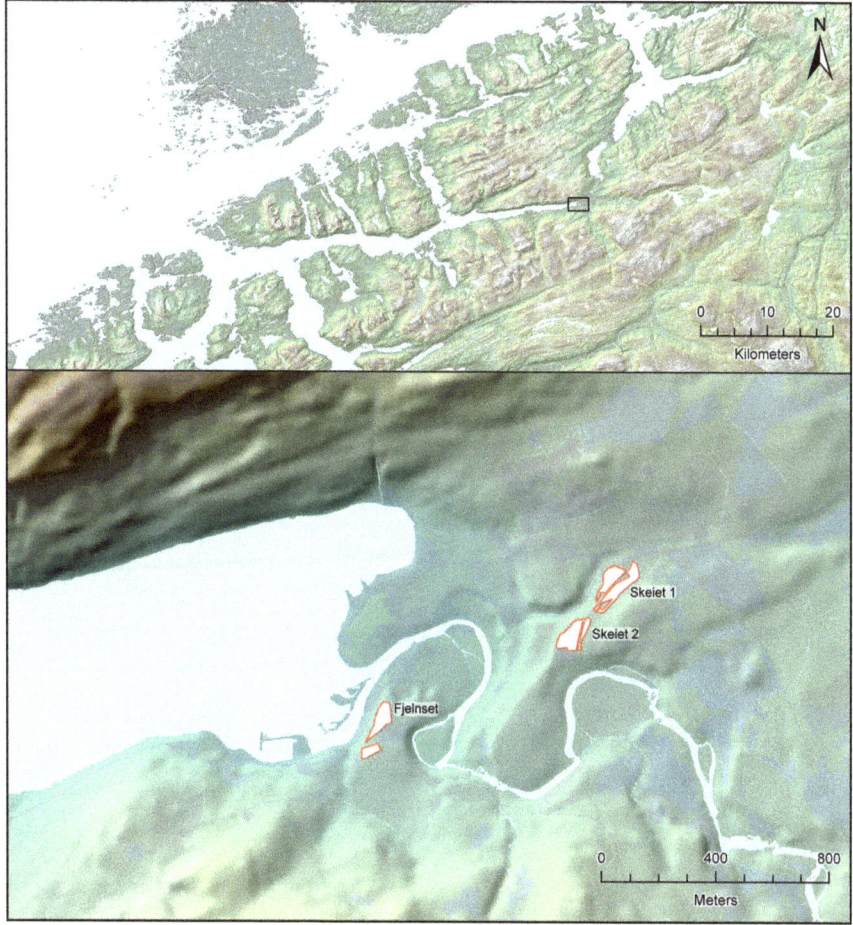

Map 4. *The Vinjefjord area with Vinjeøra enlarged to show sites mentioned in the text. Source: Kristoffer Rantala, NTNU University Museum (CC BY-SA 4.0).*

The dates then demonstrate a marked decline in the use of all three areas starting around AD 500, and the only recorded house in the area, House 2 in Skeiet 2, was abandoned (Bryn 2024: 91; Figure 6).

Statistical modelling of radiocarbon dates from Skeiet 2 around a hypothetical date of abandonment in AD 555, indicates that the inhabitants left Skeiet a couple of decades after the AD 536–540 event, just as in Vik (Sauvage 2024: 21; Figure 6). Palaeobotanical data from nearby bogs indicate reforestation and even the possible establishment of spruce in the region during the period of

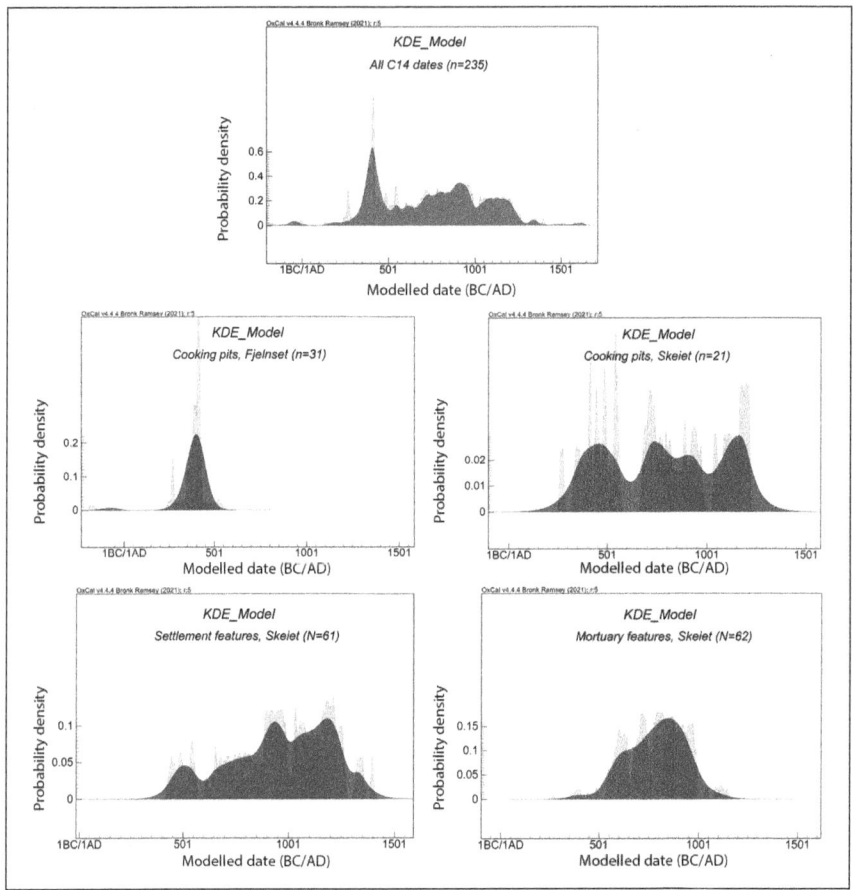

Figure 5. Summed probability distribution (SPD) plots and Kernel density estimates (KDE) of radiocarbon dates from Vinjefjord archaeological excavations. Source: Raymond Sauvage, NTNU University Museum (CC BY-SA 4.0).

abandonment. The rapid expansion of spruce in central Norway in the Merovingian and Viking periods has been related to climatic changes (Overland and Hjelle 2024: 73, with references). However, at the same time, an increased use of outfield resources and an intensification in outfield grazing can be observed (Overland and Hjelle 2024: 73). This possibly indicates a re-organisation of settlement and subsistence as a response to the changed climatic situation.

The settlement at Skeiet 2 was abandoned for only a generation or two. It became re-settled sometime in the first half of the seventh century AD (Bryn 2024: 91–92; Figure 6). Meanwhile, people must still have lived and died in the

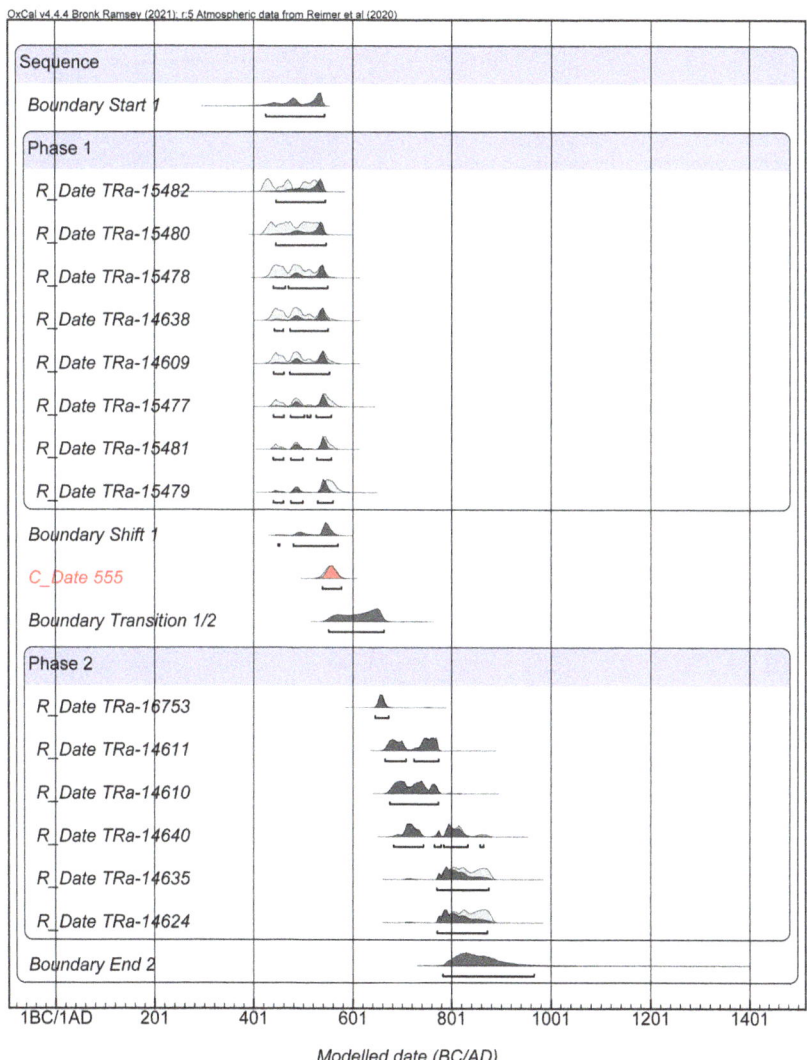

Figure 6. Settlement feature dates from Skeiet 2, modelled around the hypothetical date of abandonment in AD 555 (in red). Source: Ingrid Ystgaard.

local area. During this particularly difficult period in the existence of the local community, a new burial ground was established on formerly cultivated land in Skeiet 1, with the oldest graves dating to as early as AD 575–650 (Sauvage and Lorentzen 2024: 137; Figure 5, below left). This profound re-organisation of landscape use is reflected in nearby Fjelnset, where the former cooking pit area and possible assembly site was replaced by a cemetery during the Merovingian and Viking Periods. Cooking pits are recognised to have been most frequently in use until ca. AD 500 in central Norway and declining from that period onwards (Grønnesby 2016). The establishment of cemeteries on formerly cultivated land and on a former concentration of cooking pits indicates re-organisation of the settlement as well as establishment of new claims to rights and inheritance of the land (Sauvage 2024: 29–30). The Merovingian and Viking Period burials at the cemetery at Skeiet preserved artefacts demonstrating the extensive inter-regional connections of the people who buried their dead there, the most exotic of these being a bronze fitting of insular origin (Sauvage and Lorentzen 2024: 155; Figure 2). The cemeteries were used until Christianisation transformed burial rites, and the nearby settlement site was occupied until the Black Death caused another population decline (Sauvage 2024: 31–33).

Before and During the Crises: Resilience and Vulnerability

Even though volcanic eruptions, such as the AD 536–540 events, have had considerable effects on climate systems, they work alongside many other elements that affect the climate. Warmer and cooler periods alternated throughout the Holocene (Burroughs 2005). Climate can and should be considered as one of many agents – human and non-human – at work in the rise and fall of the Roman empire. The Roman Climatic Optimum lasted from ca. 300 BC until ca. AD 200 (Harper and McCormick 2018) and allowed for larger agricultural yields per annum and for increased tax income, among other effects (Harper 2017: 23–64; Seland and Kleiven 2022: 60). The expansion of the Roman empire on the European continent reached its peak of ambition in AD 9, when the Roman defeat in the battle of the Teutoburger Forest contributed to the withdrawal of the limits of the Roman empire to the Limes area along the Rhine. The presence of Roman military and economic activities in the border zones contributed to an intensified military and economic activity well beyond the limits of the empire (Hedeager and Tvarnø 2001: 11–23; Herschend 2009). This influence extended far into the outlands of the Scandinavian peninsula, which was rich in resources such as iron, furs and antlers (Hennius 2021; Ystgaard 2024: 81–105).

The Roman Empire thus had a profound impact on economic, military and political development in the Scandinavian peninsula. At Vik, a sophisticated farming and fishing community emerged in this period, with a varied subsistence economy, apparent in its three largely independent economic units, which nevertheless co-operated when it came to the daily tasks of producing surplus meat, milk and wool (Storå et al. 2019; Ystgaard 2023). Surplus production

A Series of Unfortunate Events

such as this enabled local communities to save time and resources needed to educate and equip young men to partake in economic and military activities in Northern Europe (Ystgaard 2020). At Vinjeøra, the archaeological material does not yield detailed information on settlement in the Roman Period; however, a cooking-pit field interpreted as a possible assembly site located at an important point of communication indicates how the general demographic and economic expansion in Scandinavia during the Roman Period left its mark even here (Sauvage 2024: 27–28). Both directly, through better conditions for agriculture, and indirectly, through increased economic activity in response to the demands of the Roman Empire and its neighbours, Vik and Vinjeøra benefited from the favourable climatic conditions of the Roman Climatic Optimum.

However, the Roman Warm Period transformed into a cooler and more varied climate from around AD 200 (Harper 2017: 131). Even though climate is not the only mover in the complex trajectories of the rise and fall of the Roman Empire, a general cooling trend and a less stable climate contributed to a change in living conditions both within and beyond the borders of the empire. Drought and reduced growing seasons might have been part of the situation when Asian steppe nomads and regional European groups began to migrate and put increasing pressure on the borders of the empire from ca. AD 300 onwards (Cook 2013). We cannot determine whether a colder climate from around AD 200 had direct impacts on the Vik and Vinjeøra communities, nor what the increasing unrest on the Continent meant for these communities. Nevertheless, we can, on a general level, draw lines between these circumstances and increasing evidence of militarisation and the emergence of ruling elites in Scandinavian societies from ca. AD 200 (Skre 2020). Migration Period archaeological data indicate increasing regional differences in demographic and political development in Norway. Eastern Norway shows signs of stagnation and perhaps even decline in population and economic activity from around AD 400, while western Norway seems to have experienced continued growth into the Migration Period (Myhre 2002: 170–179; Solberg 2003: 160–161; Loftsgarden and Solheim 2023: 103).

At Vik, we can observe a likely decline in population as early as AD 350, with the abandonment of area D (Ystgaard 2019: 388). On a larger scale, this development should be viewed in relation to the increasing unrest in the Continent and to a general economic and demographic stagnation in eastern Norway in the same period. Did the northernmost of our two communities feel the consequences of political and military unrest? In comparison, the data from Vinjeøra suggests solid settlement evidence during the Migration Period, before a more abrupt recession in ca. AD 500–600 (Sauvage 2024: 29). Did the southernmost of our two communities benefit from a closer contact with the still flourishing Migration Period communities in western Norway? Even though there are no clear answers to these questions, large-scale political and economic situations may have contributed to different levels of resiliencies in the two communities.

The regional and local situations are of great importance in understanding a community's resilience. A recent study (van Dijk et al. 2023) has simulated the general cooling during this period, estimated the growing degree days of three areas in north-eastern, south-eastern and south-western Norway, and compared the results to pollen diagrams from bogs within each area and to data from archaeological excavations. With a scenario cooling of three degrees Celsius, crops would have failed to ripen in the north-eastern and south-western areas, while they were more likely to mature in the south-eastern area. The study indicates differing levels of impact on Iron Age societies, focusing on geographical variations. Another study has compared the agricultural strategies of two communities in south-west Norway as they faced the mid-sixth-century downturn (Westling et al. 2022). The Hove-Sørbø community, which enjoyed wider connections, was more centrally located and practised more diverse subsistence strategies survived; however, the Forsandmoen community, which was less centrally located, faced problems with exhausted agricultural soils and depended upon a less varied subsistence basis, ultimately succumbed.

Vik and Vinjeøra do not seem to belong to different zones of climatic impact, as defined in the study of van Dijk et al. (2023). Nevertheless, they do seem to have had significant differences regarding resilience and vulnerability, similar to sites studied by Westling et al. (2022). Both Vik and Vinjeøra were strategically located near important routes of communication. However, the Vik community experienced a significant and increasingly urgent challenge during the sixth century: the uplift of land that made its harbour less usable and that increased the distance between the settlement and the harbour by hundreds of metres (Ystgaard et al. 2019: 36). Vinjeøra's steeper landscape profile may have rendered a more favourable situation, which denotes that the harbour may have stayed at the same place during the same period. Thus, local geographical conditions may have had great importance for the difference between the two communities' ability to adapt to changing circumstances. The Vik community may have become less attractive as early as AD 400 because of the land uplift, causing the decline to start earlier than in Vinjeøra.

After the Crises: Adaptation

The radiocarbon dates from Vik clearly show the effects of its reduced resilience and increased vulnerability. After the community peaked around AD 350, it steadily declined and was finally abandoned by its last inhabitants a few decades after

A Series of Unfortunate Events

AD 536. A combination of the changing inter-regional political and economic situation and the isostatic uplift resulting in the retraction of waters from the local harbour left the community unable to cope with the new climatic situation. This does not mean that they did not try. For example, the burial of a foal and a calf might have been attempts to mitigate the severity of the situation through ritual performances. We do not know whether the last inhabitants died or left; and, if they left, we do not know where they went. They might not have had far to go. Other sites in the Ørland peninsula may have been better equipped to face the new situation – for example, sites with better access to the sea, where people could find replacement for the failed agricultural crops in fishing grounds (cf. Oinonen et al. 2020). However, we still lack material that indicates substantial occupation through the years of crisis in the peninsula. Permanent settlement was not re-established at Vik itself until ca. AD 950 (Fransson 2019). Why did so much time pass? There are several possible answers. One might be that the local population declined so significantly that a considerable amount of time was required for it to revive. More likely, however, was that the increasing distance from the bay caused by isostatic uplift, as well as the swamping of previous grazing lands caused by abandonment and a changing geological situation, made the site less attractive for re-occupation. Other sites closer to the seashore and with better grazing lands might have been preferred.

The general organisation of landscape use in Vik before the sixth century indicates that settlements were concentrated on the well-drained central ridge of the peninsula, while burials were placed in the zone between the central ridge and the lower-lying bay. However, after the sixth century, this pattern changed. Merovingian and Viking Period graves in Ørland are instead found in locations where harbours still existed, such as Uthaug; in inland locations where graves possibly marked land ownership claims or claims to power, such as at Viklem and Opphaug; and along inland roads, such as at Prestegården (Map 2). When Vik was finally re-occupied around AD 950, farmhouses were built according to new architectural principles. The pre-sixth-century longhouses were not re-erected. Several smaller buildings with separate functions were built instead. The new spatial organisation marked a gap between different social groups within the farm (Sørheim 2016). Moreover, the Vik farm itself might have belonged to a lower social stratum, as indicated by re-use of timber in its buildings. Vik might have been in a subordinate relation to nearby Viklem, with its large burial mound and two Viking Period halls (Fransson 2019; Ellingsen and Sauvage 2019).

While the Vik community did not adapt to the new climatic situation at all, the Vinjeøra community certainly did, and in a completely different fashion. Perhaps Vinjeøra was more closely connected to thriving economies in western Norway prior to the crisis, and this contributed to more vibrant contacts and exchange and thus to greater resilience. The nature of the local landscape was not a challenge in the way it was at Vik. Instead, Vinjeøra continued as an important place where communication routes crossed as well as transitioned from land to sea through the sixth century. However, Vinjeøra also experienced a marked decline in the mid-sixth century. The cooking pit field at Fjelnset was abandoned between AD 440 and 550, as was the settlement at Skeiet 2. At the same time, pollen data suggest that the people at Vinjeøra adapted to the cooler climate by increasing their dependence on animal husbandry and outland grazing, perhaps in a similar fashion to the adaptation registered in pollen analysis from Lake Ljøgottjern in south-eastern Norway (Bajard et al. 2022). Even though there was a marked decline in activity during the late 500s, Vinjeøra was re-occupied as early as the end of the century (Sauvage 2024: 29), at a time when social memory of the assembly place with cooking pits at Fjelnset and the settlement area at Skeiet must have been still active.

Even though Vinjeøra was soon re-occupied, there was a marked change in the organisation and use of the landscape. The cooking pit field at Fjelnset and previously cultivated land at Skeiet were turned into cemeteries after the re-occupation. At Fjelnset, this happened at an uncertain date, while at Skeiet the earliest graves can be from around AD 575–600 (Sauvage 2024: 29). Burial mounds in the Migration and Viking Periods have been connected to the expression of land ownership and to new inheritance rights (Zachrisson 2017, Sauvage 2024: 30 with references). Similar transitions have been observed in eastern Norway (Moen 2020: 82–85, 275; Sauvage 2024: 29). In Skeiet 2, the architecture and spatial organisation of the built environment changed. In place of the pre-sixth-century three-aisled multifunctional longhouse, smaller, single-aisled buildings with specified functions were built in a scattered pattern (Bryn 2024). In other parts of central Norway and other parts of Scandinavia, a similar discontinuity of building traditions and settlement organisation has been observed (Göthberg 2000; Löwenborg 2012; Grønnesby 2019; Gjerpe 2023). The communities at Vik and Vinjeøra participated in this overall rearrangement in the architecture and spatial organisation of settlements.

Conclusions

Approaches that cross scales of time and space can yield a more nuanced understanding of how local communities were equipped to meet the climatic events of AD 536–540 and how they adapted to them. When considering interactions between local communities at Vik and Vinjeøra in central Norway, as well as their regional and inter-regional economic and political frameworks and their immediate geographical environments, significant differences emerge in their

A Series of Unfortunate Events

resilience and vulnerability in the face of the AD 536–540 events and in their adaptation to the following cold period. While the Vik community suffered a decline in population as early as AD 350, possibly because of a change in inter-regional networks and possibly because of significant changes in the local landscape, the Vinjeøra community thrived during the Migration Period. A combination of unfortunate events left Vik considerably more vulnerable to the climatic downturn after the AD 536–540 events than Vinjeøra, less resilient in its aftermath and likewise less able to adapt to new situations and re-establish activity during the following century. Vinjeøra, on the other hand, saw renewed activity as early as a few decades after the AD 536–540 events. A re-organisation of settlement and landscape use occurred when new cemeteries were established on the former cooking pit site at Fjelnset and on previously cultivated land at Skeiet. When Vik was finally re-occupied after a 400-year hiatus, the new settlement brought new patterns of building construction and landscape use. The different processes of human/non-human interaction on varying scales in Vik and Vinjeøra contributed to different situations regarding resilience and vulnerability, which were poorer at Vik and better at Vinjeøra. Therefore, people at Vinjeøra were able to adapt to the climatic downturn faster and better than people at Vik.

Acknowledgements

Frands Herschend first modelled the dates from the abandonment phase at Vik. Bente Philippsen gave advice on the models in Figures 7 and 10. Magnar Mojaren Gran produced the maps. Ingar Mørkestøl Gundersen and Samuel A. White most helpfully commented on the text.

References

Arjava, A. 2005. 'The mystery cloud of 536 CE in the Mediterranean sources'. *Dumbarton Oaks Papers* 59: 73–94.

Bajard, M., E. Ballo, H.I. Høeg et al. 2022. 'Climate adaptation of pre-Viking societies'. *Quaternary Science Reviews* 278: 107374. https://doi.org/10.1016/j.quascirev.2022.107374.

Benedictow, O.J. 2016. *The Black Death and Later Plague Epidemics in the Scandinavian Countries: Perspectives and Controversies*. Warsaw: De Gruyter Open. https://doi.org/10.1515/9788376560472.

Berglund, B. and T. Solem. 2017. 'Maktkontinuitet i munningen av Trondheimsfjorden: arkeologi og vegetasjonshistorie 800 BC–1200 AD'. *Heimen* 54 (3): 206–234.

Bronk Ramsey, C. 2017. 'Methods of summarizing radiocarbon datasets'. *Radiocarbon* 59: 1089–1833. https://doi.org/10.1017/RDC.2017.108.

Bryn, H. 2024. 'Building traditions and settlement organization at Skeiet from the Migration Period to the Black Death'. In R. Sauvage (ed.) *From Life to Death in Iron Age and Medieval Vinjefjord. Unveiling Burial Practices, Settlement and Landscape History*. Oslo: Cappelen Damm Akademisk/NOASP. pp. 83–111.

Büntgen, U., V.S. Myglan, F.C. Ljungqvist, M. McCormick, N. Di Cosmo, M. Sigl, J. Jungclaus, S. Wagner, P.J. Krusic, J. Esper, J.O Kaplan, M.A.C. de Vaan, J. Luterbacher, L. Wacker, W. Tegel and A.V. Kirdyanov. 2016. 'Cooling and societal change during the Late Antique Little Ice Age from 536 to around 660 AD'. *Nature Geoscience* 9 (3): 231–236. https://doi.org/10.1038/ngeo2652.

Burroughs, W.J. 2005. *Climate Change in Prehistory: The End of the Reign of Chaos*. Cambridge: Cambridge University Press.

Cook, E.R. 2013. 'Megadroughts, ENSO, and the invasion of Late-Roman Europe by the Huns and Avars'. In W.V. Harris (ed.) *The Ancient Mediterranean Environment Between Science and History*. Leiden: Brill. pp. 89–102.

Ellingsen, E.G. and R. Sauvage. 2019. 'The northern Scandinavian Viking Hall: A case study from Viklem in Ørland, Norway'. In I. Ystgaard (ed.) *Environment and Settlement: Ørland 600 BC–AD 1250*. Oslo: Cappelen Damm Akademisk. pp. 399–426. https://doi.org/10.23865/noasp.89.

Eriksen, T.H. 2016. *Overheating: An Anthropology of Accelerated Change*. London: Pluto Press.

Fransson, U. 2019. 'A farmstead from the late Viking Age and early medieval period. House constructions and social status at Vik, Ørland'. In I. Ystgaard (ed.) *Environment and Settlement: Ørland 600 BC–AD 1250*. Oslo: Cappelen Damm Akademisk. pp. 323–350. https://doi.org/10.23865/noasp.89.

Gjerpe, L.E. 2023. *Effective Houses: Property Rights and Settlement in Iron Age Eastern Norway*. Oslo: Cappelen Damm Akademisk/NOASP.

Göthberg, H. 2000. *Bebyggelse i förändring. Uppland från slutet av yngre bronsålder till tidig medeltid*. Uppsala: Acta Universitatis Upsaliensis.

Gräslund, B. and N. Price. 2012. 'Twilight of the gods? The "dust veil event" of AD 536 in critical perspective'. *Antiquity* 86 (332): 428–443. https://doi.org/10.1017/S0003598X00062852.

Grønnesby, G. 1999. 'Eldre jernalders hus og hall på Hovde i Trøndelag'. *Viking* 62: 69–80.

Grønnesby, G. 2016. 'Hot rocks! Beer brewing on Viking and medieval age farms in Trøndelag'. In F. Iversen and H. Petersson (eds), *The Agrarian Life of the North 2000 BC–AD 1000. Studies in Rural Settlement and Farming in Norway*. Oslo: Cappelen Damm Akademisk. https://doi.org/10.17585/noasp.13.42. pp. 133–149.

Grønnesby, G. 2019. '... en pludselig og stærk omvæltning'? *Eldre jernalder og overgangen til yngre jernalder i Trøndelag. Praksis og overregionale nettverk*. Ph.D. dissertation, NTNU.

Grønnesby, G., G.I. Solvold, E.M.F. Krag and M.M. Gran. 2023. 'The well and rituals in everyday life – three Iron Age wells excavated in Trøndelag, Norway'. *Journal of Wetland Archaeology* 23 (1–2): 62–80.

Gundersen, I.M. 2019. 'The Fimbulwinter theory and the 6th century crisis in the light of Norwegian archaeology: Towards a human-environmental approach'. *Primitive Tider* 21: 101–120.

Gundersen, I.M. 2021. *Iron Age Vulnerability. The Fimbulwinter Hypothesis and the Archaeology of the Inlands of Eastern Norway*. Ph.D. dissertation. University of Oslo.

Harper, K. 2017. *The Fate of Rome. Climate, Disease, & the End of the Roman Empire*. Princeton University Press.

Harper, K. and M. McCormick. 2018. 'Reconstructing Roman climate'. In W. Scheidel (ed.) *The Science of Roman History: Biology, Climate, and the Future of the Past*. Princeton: Princeton University Press.

Hedeager, L. and H. Tvarnø. 2001. *Tusen års europahistorie. Romere, germanere og nordboere*. Oslo: Pax Forlag.

Helama, S., P.D. Jones and K.R. Briffa. 2017. 'Limited Late Antique cooling'. *Nature Geoscience* 10 (4): 242–243.

Hennius, A. 2021. *Outlanders? Resource Colonization, Raw Material Exploitation and Networks in Middle Iron Age Sweden*. Uppsala: Department of Archaeology and Ancient History.

Herschend, F. 2009. *The Early Iron Age in South Scandinavia: Social Order in Settlement and Landscape*. Uppsala: Occasional Papers in Archaeology 45.

Hjelle, K.L., A. Overland, M.M. Gran, A. Romundset and I. Ystgaard. 2022. 'Two thousand years of landscape–human interactions at a coastal peninsula in Norway revealed through pollen analysis, shoreline reconstruction, and radiocarbon dates from archaeological sites'. *Frontiers in Ecology and Evolution* 10: 911780.

Iversen, F. 2016. 'Estate division: Social cohesion in the aftermath of AD 536–7'. In F. Iversen and H. Petersson (eds), *The Agrarian Life of the North 2000 BC–AD 1000. Studies in Rural Settlement and Farming in Norway*. Oslo: Cappelen Damm Akademisk. pp. 41–75. https://doi.org/10.17585/noasp.13.42.

Kjemperud, A. 1986. 'Late Weichselian and Holocene shoreline displacement in the Trondheimsfjord area, central Norway'. *Boreas* 15 (1): 61–82.

Little, L.K. 2007. 'Life and afterlife of the first plague pandemic'. In L.K. Little (ed.) *Plague and the End of Antiquity. The Pandemic of 541–750*. Cambridge: Cambridge University Press. pp. 3–32.

Loftsgarden, K. and S. Solheim. 2022. 'Uncovering population dynamics in Southeast Norway from 1300 BC to AD 800 using summed radiocarbon probability distributions'. In M. Ødegaard and I. Ystgaard (eds), *Complexity and Dynamics. Settlement and Landscape From the Bronze Age to the Renaissance in the Nordic Countries (1700 BC–AD 1600)*. Leiden: Sidestone Press. pp. 99–112.

Løvschal, M. 2022. 'Retranslating resilience theory in archaeology'. *Annual Review of Anthropology* 51: 195–211. https://doi.org/10.1146/annurev-anthro-041320-011705.

Löwenborg, D. 2012. 'An Iron Age shock doctrine: Did the AD 536–7 event trigger large-scale social changes in the Mälaren valley area?' *Journal of Archaeology and Ancient History* 4: 1–29. https://doi.org/10.33063/jaah.vi4.120.

Maixner, B. and T.M. Hellan. 2024. 'Viking period trade and communication in the Vinjefjord and Hemnfjord areas, in light of topography, place names, silver coins and imported beads'. In R. Sauvage (ed.) *From Life to Death in Iron Age and Medieval Vinjefjord. Unveiling Burial Practices, Settlement and Landscape History.* Oslo: Cappelen Damm Akademisk/NOASP. pp. 113–128.

Mokkelbost, M. 2019. 'Roman period waste deposits at Ørland, Norway'. In I. Ystgaard (ed.) *Environment and Settlement: Ørland 600 BC–AD 1250.* Oslo: Cappelen Damm Akademisk. pp. 195–231. https://doi.org/10.23865/noasp.89.

Myhre, B. 2002. 'Landbruk, landskap og samfunn 400 f.Kr.–800 e.Kr.' In N. Myhre and I. Øye (eds), *Norges landbrukshistorie 4000 f.Kr.–1350 e.Kr. Jorda blir levevei. Del I.* Samlaget. pp. 12–214.

Narmo, L.E. 1996. '"Kokekameratene på Leikvin". Kult og kokegroper'. *Viking* 59: 79–100.

Oinonen, M., T. Alenius, L. Arppe, H. Bocherens, H. Etu-Sihvola, S. Helama, H. Huhtamaa, M. Lahtinen, K. Mannermaa, P. Onkamo and J. Palo. 2020. 'Buried in water, burdened by nature – resilience carried the Iron Age people through Fimbulvinter'. *PLoS ONE* 15: e0231787. https://doi.org/10.1371/journal.pone.0231787.

Overland, A. and K.L. Hjelle. 2024. 'Late Holocene vegetation development in Vinjeøra and the Valsøyfjord region'. In R. Sauvage (ed.) *From Life to Death in Iron Age and Medieval Vinjefjord. Unveiling Burial Practices, Settlement and Landscape History.* Oslo: Cappelen Damm Akademisk/NOASP. pp. 45–81.

Pettersen, A.M.H. and A.B. Lorentzen. 2019. 'Roman Iron Age and Migration Period building traditions and settlement organisation at Vik, Ørland'. In I. Ystgaard (ed.) *Environment and Settlement: Ørland 600 BC–AD 1250).* Oslo: Cappelen Damm Akademisk. pp. 167–193. https://doi.org/10.23865/noasp.89.

Redman, C.L. 2005. 'Resilience theory in archaeology'. *American Anthropologist* **107** (1): 70–77. https://doi.org/10.1525/aa.2005.107.1.070.

Romundset, A. and T.R. Lakeman. 2019. 'Shoreline displacement at Ørland since 6000 cal. yr BP'. In I. Ystgaard (ed.) *Environment and Settlement: Ørland 600 BC–AD 1250.* Oslo: Cappelen Damm Akademisk. pp. 51–67. https://doi.org/10.23865/noasp.89.

Sauvage, R. 2024. 'Chonological trajectories at Iron Age and medieval settlements in Vinjefjord based on recent excavations'. In R. Sauvage (ed.) *From Life to Death in Iron Age and Medieval Vinjefjord. Unveiling Burial Practices, Settlement and Landscape History.* Oslo: Cappelen Damm Akademisk/NOASP. pp. 15–43.

Seland, E.H. and K.F. Kleiven. 2023. *En kort introduksjon til klimahistorie. Menneskene og klimaet etter siste istid.* Oslo: Cappelen Damm Akademisk.

Sigl, M., M. Winstrup, J.R. McConnell et al. 2015. 'Timing and climate forcing of volcanic eruptions for the past 2,500 years'. *Nature* **523** (7562): 543–549. https://doi.org/10.1038/nature14565.

Skre, D. 2020. 'Rulership and ruler's sites in 1st–10th-century Scandinavia'. In D. Skre (ed.) *Rulership in 1st to 14th century Scandinavia. Royal Graves and Sites at Avaldsnes and Beyond*. Reallexikon der Germanischen Altertumskunde. Ergänzungsbände 114. Berlin: De Gruyter. pp. 193–243.

Solvold, G.I. 2019. 'The pottery at Vik in the early iron age'. In I. Ystgaard (ed.) *Environment and Settlement: Ørland 600 BC–AD 1250*. Oslo: Cappelen Damm Akademisk. pp. 261–321. https://doi.org/10.23865/noasp.89.

Stamnes, A.A. 2016. 'Effect of temperature change on Iron Age cereal production and settlement patterns in mid-Norway'. In F. Iversen and H. Petersson (eds), *The Agrarian Life of the North 2000 BC–AD 1000. Studies in Rural Settlement and Farming in Norway*. Oslo: Cappelen Damm Akademisk. pp. 27–40. https://doi.org/10.17585/noasp.13.42.

Storå, J., M. Ivarsson-Aalders and I. Ystgaard. 2019. 'Utilization of animal resources in Roman Iron Age Vik: Zooarchaeology at Ørland'. In I. Ystgaard (ed.) *Environment and Settlement: Ørland 600 BC–AD 1250*. Oslo: Cappelen Damm Akademisk. pp. 233–258. https://doi.org/10.23865/noasp.89.

Sørheim, H. 2016. 'House, farmyard and landscape as social arena in a time of transition'. In L.H. Dommasnes, D. Gutsmiedl-Schumann and A.T. Hommedal (eds), *The Farm as a Social Arena*. Münster: Waxmann Verlag. pp. 191–217.

Toohey, M., K. Krüger, M. Sigl, F. Stordal and H. Svensen. 2016. 'Climatic and societal impacts of a volcanic double event at the dawn of the Middle Ages'. *Climatic Change* 136: 401–412. https://doi.org/10.1007/s10584-016-1648-7.

Torrence, R. 2018. 'Social responses to volcanic eruptions: A review of key concepts'. *Quaternary International* 499: 258–265. https://doi.org/10.1016/j.quaint.2018.02.033.

van Dijk, E., J. Jungclaus, S. Lorenz, C. Timmreck and K. Krüger. 2022. 'Was there a volcanic-induced long-lasting cooling over the Northern Hemisphere in the mid-6th–7th century?' *Climate of the Past* 18 (7): 1601–1623. https://doi.org/10.5194/cp-18-1601-2022.

van Dijk, E., I. Mørkestøl Gundersen, A. de Bode et al. 2023. 'Climatic and societal impacts in Scandinavia following the 536 and 540 CE volcanic double event'. *Climate of the Past* 19 (2): 357–398. https://doi.org/10.5194/cp-19-357-2023.

Wagner, D.M., J. Klunk, M. Harbeck et al. 2014. '*Yersinia pestis* and the Plague of Justinian 541–543 AD: A genomic analysis'. *The Lancet Infectious Diseases* 14 (4): 319–326. https://doi.org/10.1016/S1473-3099(13)70323-2.

Westling, S., E.D. Fredh, P. Lagerås and K.A. Oma. 2022. 'Agricultural resilience during the 6th century crisis: Exploring strategies and adaptations using plant-macrofossil data from Hove-Sørbø and Forsandmoen in Southwestern Norway'. *Norwegian Archaeological Review* 55 (1): 38–63.

Ystgaard, I., M.M. Gran, M. Mokkelbost, U. Fransson, A. Heen-Pettersen, A.B. Lorentzen, G.I. Solvold and E.W. Randerz. 2018. 'Arkeologiske utgravninger på Ørland kampflybase 2015–2016'. *NTNU Vitenskapsmuseet arkeologisk rapport*, 27. Trondheim: NTNU University Museum.

Ystgaard, I., M.M. Gran and U. Fransson. 2019. 'Environment and settlement at Vik, Ørland: A phase framework'. In I. Ystgaard (ed.) *Environment and Settlement: Ørland 600 BC-AD 1250: Archaeological Excavations at Vik, Ørland Main Air Base*. Oslo: Cappelen Damm Akademisk/NOASP. pp. 23–48.

Ystgaard, I. 2019. 'Spatial organization of farmsteads at Iron Age and early Medieval Vik (c. 400 BC. AD 1250)'. In I. Ystgaard (ed.) *Environment and Settlement: Ørland 600 BC–AD 1250*. Oslo: Cappelen Damm Akademisk. pp. 373–396. https://doi.org/10.23865/noasp.89.

Ystgaard, I. 2020. 'Warfare and recruitment in Iron Age central Norway'. In H.L. Aannestad, U. Pedersen, M. Moen, E. Naumann and H.L. Berg (eds), *Vikings Across Boundaries. Viking-Age Transformations – Volume II*. London: Routledge. pp. 283–303.

Ystgaard, I. 2023. 'Activities and community organization in Roman Iron Age Vik, Ørland, Central Norway'. In M. Ødegaard and I. Ystgaard (eds), *Complexity and Dynamics. Settlement and Landscape from the Bronze Age to the Renaissance in the Nordic Countries (1700 BC–AD 1600)*. Leiden: Sidestone Press. pp. 143–155.

Ystgaard, I. 2024. *En kort introduksjon til jernalderen i Norge*. Cappelen Damm Akademisk.

Ødegaard, M. 2019. 'Assembling in times of transition – the case of cooking pit sites'. In N. Brady and C. Theune (eds), *Settlement Change Across Medieval Europe. Old Paradigms and New Vistas*. Ruralia XII. Leiden: Sidestone Press. pp. 185–194.

The Authors

Ingrid Ystgaard is Associate Professor in Archaeology in the Department of Historical and Classical Studies at the Norwegian University of Technology and Science in Trondheim, Norway. Her research centres around pre-Viking Age settlement and conflict archaeology in Scandinavia, in addition to environmental archaeology. She managed the archaeological excavations prior to the extension of the Norwegian main Air base at Ørland in 2015–2016, which uncovered extensive settlement traces with faunal, vegetational and geographical data from ca. 500 BC to AD 1050. Her Ph.D. from 2014 explores relations between weapon graves, hill forts, boat houses and military behaviour in a long-time perspective from AD 100–900 in central Norway.

Raymond Sauvage is an archaeologist in the Department of Archaeology and Cultural History at the NTNU University Museum, which is part of the Norwegian University of Science and Technology in Trondheim. His research includes early Medieval and Viking Age studies, with a particular focus on mortuary practices and settlement studies. He has led several large archaeological excavation projects in central Norway and managed the excavations in Vinjefjord during 2019–2020, which uncovered extensive evidence of settlements and pre-Christian cemeteries from ca. AD 350–1350.

Chapter 4

VOLCANIC VULNERABILITY IN MEDIEVAL ICELAND

Carina Damm

Abstract

This article explores the impact of volcanic activity on medieval Icelandic society, focusing on both external shocks and community adaptations. Utilising the vulnerability framework established by Robert Chambers, it examines Iceland's susceptibility to eruptions in pre-modern times, with particular attention to the Eldgjá (c. 939) and Hekla (1104) events. These eruptions had a profound impact on Icelandic society, disrupted settlements, shaped landscapes and influened cultural and religious narratives. While sagas rarely mention volcanic events, annals, legal texts such as *Grágás*, and vow contracts reveal practical coping strategies that included church relocations and community rituals. Archaeological evidence highlights resilience through the reconstruction of settlements on tephra-covered land, and written accounts such as the Norwegian *King's Mirror* portray eruptions as a form of divine judgment by blending spiritual interpretations with pragmatic responses. This interdisciplinary framework underscores medieval Iceland's adaptability to environmental hazards and offers valuable insights into socio-environmental resilience in volatile landscapes that remain relevant for modern disaster management.

Introduction

Iceland serves as a prime example of a volcanically active region, with one of its thirty volcanic systems erupting on average every four to five years (Thordarson and Höskuldsson 2022: 10). This frequent volcanic activity poses significant risks and prompts questions regarding the strategies employed by pre-modern societies to mitigate and develop resilience. Historically, eruptions have played a pivotal role in shaping Icelandic society. For instance, the Eldgjá eruption has been associated with the discourse surrounding the Christianisation of Iceland in the year 1000. Another significant event was the impact of the Hekla eruption

doi: 10.63308/63881023874820.ch04

in 1104 on the nearby Þjórsárdalur valley, which was allegedly left devastated by tephra deposits and thus gained notoriety as the 'Pompeii of the North'.

Using the 'vulnerability approach' by Robert Chambers (1989: 1) – a framework that emphasises that vulnerability has two sides: an external aspect that includes risks such as natural disasters, and an internal aspect characterised by defencelessness or lack of means to cope – this article examines past volcanic eruptions from two perspectives: first, the external shocks faced by medieval Icelanders; and, second, the community adaptations.

Applying this framework to medieval Iceland, I analyse both its external vulnerabilities through recorded historical eruptions between c. 870 and 1500, and coping mechanisms documented in legal texts such as *Grágás* alongside diplomatic 'vow contracts'. A final section analyses the societal impact through the historical record, focusing in particular on the Hekla 1104 event.

The studied medieval period reveals 79 recorded eruptions (Global Volcanism Program, 2025), including major ones at Bárðarbunga (1477), Öræfajökull (1362), Katla (1262) and especially that of Hekla (1104). Fifty-two eruptions are dated from ice cores and tephra analysis, while a further 38 stem directly from Old Icelandic sources – with specific terminology reflecting perceptions of volcanoes, including *brenna* ('burn'), *vikr* ('pumice/ash') and *eldr* ('fire').

Geological Setting and Viking Migrations

In geological terms, Iceland is of recent formation, approximately 25 million years of age (Thordarson and Höskuldsson 2022: 1). It is located in the North Atlantic Ocean within the region where the Eurasian and North American plates are diverging at a rate of two centimetres per year, resulting in the formation of volcanic rift zones. In these zones, the Earth's crust is subjected to a process of fracturing and separation which leads to the creation of a rent that is subsequently filled by the eruption of lava along the sea floor. It is estimated that approximately one-third of Iceland's terrain is covered by volcanoes, which frequently erupt along these rift zones. These events are typically moderate in magnitude, with eruptions ranging from 1 to 3 on the Volcanic Explosivity Index (VEI).[1] However, Icelandic eruptions are characterised by a high degree of diversity, attributable to specific geological and climatological factors. Consequently, Iceland features virtually all the volcano types and eruption styles that are known to occur on Earth (Thordarson and Höskuldsson 2022: 1).

---

1. Invented in 1982, the Volcanic Explosivity Index (VEI) is a tool to measure the relative explosiveness of volcanic eruptions. It indicates how much volcanic material is ejected, the height of the material thrown into the atmosphere, and how long the eruptions last.

Volcanic Vulnerability in Medieval Iceland

What made people settle on such a seemingly uninhabitable island? According to twelfth-century written sources from Iceland, Norwegian settlers began to migrate to Iceland in the 870s, seeking refuge from political turmoil and land scarcity. This period, known in Icelandic as *Landnám* ('land-taking'), gave rise to a distinctive stratigraphic layer known as the *Landnám Tephra Layer* (LTL). This layer is thought to have been deposited by a major eruption in the Törfajökull area and has been identified in soil profiles over a large part of Iceland, just below the earliest traces of human habitation. The layer's distinctive feature is its dual-coloured nature, which sets it apart from other layers. Several attempts have been made to date organic remains found alongside the layer using the radiocarbon (^{14}C) method. Most recently, the layer has been dated to the year 877 ± 1 based on the presence of very fine ash (tephra) in Greenland ice cores (Schmid et al. 2021: 134). This dating is broadly consistent with the first palaeoecological, archaeological, and written evidence of human presence in Iceland. For instance, the twelfth-century *Landnámabók* ('The Book of Settlement') states: 'At the time Iceland was discovered and settled by Norwegians, Pope Adrian (867–72), and after him, Pope John the Fifth (872–82), occupied the Apostolic Seat in Rome'.[2]

The combination of written accounts,[3] such as the *Landnámabók*, as well as a detailed volcanological record in the form of lava flows and tephra layers, allows for a robust chronology of eruptions in Iceland over the past eleven centuries. This data reveals the occurrence of 205 historical eruptions, of which 172 have been verified by natural-scientific proxies such as ^{14}C dating and tephrochronology. The remaining 33 events are documented solely in written accounts. Of the 205 recorded events, 192 (~94%) represent single eruptions, while 13 (~6%) are classified as 'fires', referring to prolonged volcanic episodes that persist for months or years. One of the most notable of these 'fires' occurred during the Laki fissure eruption in 1783/84, and is therefore known in Icelandic as *Skaftáreldar* (= 'Skaftá Fires') after the riverbed of the Skaftá through which the lava flowed (Karlsson 2000: 271). The fissure itself is known as *Lakagígar* (= 'Laki Craters'), and it is this name that has gained the event its widest popular awareness.

The eighteenth-century eruption at the Laki fissure has often been compared to one of the earliest documented 'fires' in Icelandic history, that of

---

2. Pálsson and Edwards 1972 [repr. 2006]: 16: 'At the time Iceland was discovered and settled by Norwegians, Pope Adrian, and after him, Pope John the Fifth, occupied the Apostolic Seat in Rome'.

3. These written accounts of volcanic eruptions date back to the twelfth and thirteenth centuries, and include also references to eruptions extending back to the time of settlement.

the Eldgjá or 'Fire Gorge' fissure in the tenth century, which will be examined in the following in terms of its social impact.

The Eruption That Forced Christianity?

The Eldgjá lava flood is regarded as Iceland's largest historical eruption, having been fed from a magma chamber beneath Katla volcano and occurring along a 75 kilometre-long fissure swarm. In the course of at least sixteen explosive eruptions, 19.6 km^3 of magma were erupted (Thordarson et al. 2001: 35), and ~600 Mt SO_2 were emitted into the atmosphere (Schmidt et al. 2010). This fissure traverses both highland areas and the ice cap of the Mýrdalsjökull glacier, with a substantial part lying beneath the latter. Recent studies of the ejected deposits have revealed that lava flowed southward across the Álftaver district and other areas to the east, producing a total accumulation of 18.3 km^3 of lava (Thordarson et al. 2001: 35). Stratigraphic analysis places the event temporally after the settlement of Iceland in the late ninth century, as it is positioned above the *Landnám Tephra Layer* from 877±1. However, radiocarbon dating of charred wood reveals a more precise temporal window for the Eldgjá eruption, narrowing it down to the first half of the tenth century (Larsen 2000: 11). This temporal discrepancy has led to prolonged debate regarding the precise date. The reconstructed dates for the eruption range from 934±2 to 938±4 (see Hammer et al. 1980; Vinther et al. 2006; and Zielinski et al. 1995, respectively), based on analyses of Greenland ice cores that preserve both sulphate and tephra fallout from volcanic eruptions. The most frequently cited date is 934, based on the correlation of ice-core signals and medieval written sources (see Stothers 1998).

Recent studies suggest a modified chronology of the event. In 2015, Sigl et al. analysed high-resolution glaciochemical records from Greenland ice cores, specifically NEEM-2011-S1 and GISP2 (Sigl et al. 2015). Their data show a pronounced spike in non-sea salt sulphur (NSSS) beginning in 939, which was attributed to the Eldgjá eruption. In 2018, Oppenheimer et al. found that this glaciochemical signature began 7.5 years before the 'millennium eruption' of Changbaishan that extends across the Chinese-North Korean border. This event, in turn, has been accurately dated by several proxies, including radiocarbon dates from charred tree trunks contained in pyroclastic deposits from Changbaishan (Oppenheimer et al. 2017). By correlating the Icelandic eruption with the Changbaishan event of 946, a revised dating of the onset of the Eldgjá eruption has been proposed for spring 939, lasting 1.5 years until autumn 940. Considering this shortened eruption scenario, the effects must have been all the more severe, as I will show below.

Volcanic Vulnerability in Medieval Iceland

Strong eruptions such as that of Eldgjá have long been recognised as a cause of environmental and social change. There are no explicit historical references to the Eldgjá eruption. However, we can infer from medieval Icelandic sources that the effects were immediate on a local scale. The twelfth-century *Landnámabók* states that

> the settler Molda-Gnúpur took possession of land in Álftaver district between the rivers Kúðafjót and Eyjará. At that time a large lake was there and good swan hunting. He sold part of his settlement to many newcomers. The area became populated before it was overrun by *jarðeldur* (an 'earth fire' = fissure eruption), then they fled west to Höfðabrekka and set up a camp at Tjaldavellir. (Pálsson and Edwards 1972, ch. 86)

What we can learn from this episode is that Iceland's early settlers were able to cope with environmental hazards, for example by migrating to other unaffected areas. Apart from migrating as a means of coping, it has been proposed that the severe experience of the Eldgjá stimulated the Christianisation of the Icelanders around the year 1000 through interpreting the eruption 'as signs that the old pagan ways were doomed' (Oppenheimer et al. 2012: 377, 370, 378). It has also been postulated that the eruption of Eldgjá is mirrored in the Eddic poem of *Vǫluspá*. According to Oppenheimer et al., the depiction of *Ragnarök* in *Vǫluspá* represents 'a close literary interpretation of what might have been observed at first hand when Eldgjá erupted in the 930s'. Mathias Nordvig, in his recent book on *Volcanoes in Old Norse Mythology*, also favours the Eldgjá event as the source for the description in *Vǫluspá*, since his argument is based on the claim that myths involving volcanic activity are specifically Icelandic and inspired by what could be observed in the natural environment.

Beyond the realm of speculative geomythological interpretations, insights into shifts in religious practices can be derived from archaeological and tephrochronological data. In Skaftártunga, where extensive lava flows from Eldgjá are present, pagan graves provide evidence of human habitation before the year 1000, when Christianity was adopted in Iceland. However, only one burial at Hrífunes has been found to date to predate the eruption (see Eldjárn 1984; Gestsdóttir et al. 2015; Larsen and Thorarinsson 1984). Skaftártunga is located within a region significantly impacted by tephra deposits, particularly those from the Eldgjá eruption, which resulted in layers up to a metre thick in certain areas. Such substantial tephra deposits typically result in the abandonment of farmland in Iceland. However, archaeological findings reveal that new structures were built directly on these deposits after the eruption. This suggests that, rather than a complete abandonment of the area, it was repopulated following the volcanic event. The construction of settlements directly above Eldgjá tephra provides some of the earliest documented evidence for permanent settlement in

Skaftártunga. Similar patterns have been observed east of the river Skaftá and in the Síða foothills. The development of settlements on these tephra-covered highland margins suggests that significant volcanic eruptions like the one of Eldgjá may have prompted an evacuation from lowlands affected by lava flows towards higher areas. This would however imply that the archaeological data contradicts in that case the written source of the *Landnámabók* which described a redirection of settlements into the opposite direction, from the Skaftá highlands to the lowlands around Höfðabrekka and Kúðafljótsós.

The Silence of the Sources

Given the allegedly drastic effects of the Eldgjá eruption on Icelandic society, it is surprising that the most prominent genre of texts from medieval Iceland, the *Sagas of Icelanders* (e.g., *Egils saga* or *Laxdæla saga*), rarely refer to volcanic events. Instead, when they discuss environmental aspects, they deal with topics such as food production, agriculture, and animal husbandry. However, they do include descriptions of natural events such as eruptions, floods, and bad harvests, which either facilitated or hindered the mostly violent interactions between the sagas' protagonists (Phelpstead et al. 2020: 83). Furthermore, as highlighted by Astrid E.J. Ogilvie and Gísli Pálsson, environmental extremes were employed to establish a specific atmosphere and could function as a backdrop to the prose narratives in which they manifested as metaphors. These extremes could also be utilised as instruments of manipulation in witchcraft and weather magic (Ogilvie and Pálsson 2003: 251–274.).

In 2017, Oren Falk conducted a study of the *Sagas of Icelanders* with a particular focus on their information regarding volcanic events, and drew the following conclusion:

> Few wildfires or famines ravage the Icelandic countryside; harsh winters and disease seldom decimate the population; landslides and floods are mercifully rare; and ravenous polar bears, borne south on breakaway ice floes, almost unheard of … – a meagre haul, in all, which does little to disrupt the impression that saga Iceland must have been a serene paradise, a sort of medieval Tahiti. (Falk 2007: 6)

He concluded that the authors of the sagas 'may … have considered volcanoes too vulgar, too imbued with a self-evident and unavoidable Christian ideology, to be narratively useful'. The claim that volcanic eruptions were conspicuously absent from Icelandic sagas, presumably due to their perceived vulgarity or association with Christian ideology, overlooks however the broader significance of volcanoes in medieval Icelandic society. While genre requirements undoubtedly influenced the content of various literary texts, they do not fully explain why such significant environmental factors were largely absent from saga narratives. This absence is particularly pronounced when juxtaposed with annals, where accounts of natural phenomena, including volcanic eruptions, are prevalent. The contrast underscores the recognition by Icelanders of the significance of these events beyond the confines of genre conventions. This awareness is evident across various Icelandic texts, including annals, sagas, didactic texts and legal codes, which collectively refer to approximately seventy volcanic eruptions.

Many annal entries provide brief descriptions of an eruption, e.g., 1158 *Elldz upkuama i Heklufelli* ('fire-upcoming in Mt Hekla', *Høyersannáll*) or 1211 *Elldr kom vpp ór séa* ('fire came up from the sea', *Annales regii*). Nevertheless, these concise entries can be supplemented by further information derived from saga narratives. For example, when four Icelandic annals merely mention 'fire over the Reykjanes peninsula' (*Elldr fyrir Reykia nesi*) in 1226, the prose narrative of *Íslendinga saga* adds for the winter 1226/27:

> Þessi vetr var kallaðr sandvetr, ok var fellivetr mikill ok dó hundrað nauta fyrir Snorra Sturlusyni út í Svignaskarði. (*Sturlunga saga* 1: 314–15, ch. 60)
>
> That winter was called sand-winter, and it was a very hard winter one hundred of Snorri Sturluson's cattle died out in Svignaskarð.[4]

The here-cited *Íslendinga saga* represents the most substantial contribution to *Sturlunga saga*, a monumental compilation of Icelandic sagas assembled around the year 1300. The entire collection consists of a series of texts by various authors that forms a continuous chronicle of Icelandic history from 1117 to 1264. This period marks the end of the Icelandic Commonwealth and its eventual submission to Norwegian rule in 1262–1264. *Íslendinga saga* was written by Sturla Þórðarson (1214–1284), a nephew of Snorri Sturluson (1179–1241) who features in the quoted episode, with *Sturlunga saga* being the only text in the compilation that can be confidently attributed to an author. The saga begins with the demise of Sturla's paternal grandfather and namesake, Hvamm-Sturla, in 1183, chronicles the history of Iceland and its chieftains throughout the thirteenth century, and culminates in the events that led to Iceland's submission to Norway. Sturla himself was profoundly involved in many of these events and frequently refers to himself in the narrative, though always with restraint and objectivity, employing the third person. The author's portrayals in *Íslendinga saga* are characterised by their vividness and insight into prominent figures such as Snorri Sturluson, who, as Sturla Þórðarson's uncle and a central historical figure, receives considerable attention in these stories.

---

4. Unless otherwise noted, all translations in this article are by the author.

In addition to these personal stories, Icelandic annals also provide detailed accounts of natural phenomena such as volcanic eruptions. *Flateyjarbók*, for example, provides vivid descriptions of such events, including an eruption of Mount Hekla characterised by heavy ashfall that devastated nearby villages and livestock, highlighting the impact of volcanic activity on medieval Icelandic society.

> A fire broke out in Mount Hekla on the Lord's Day after Maundy Thursday, with so much enormity and ashfall that many nearby villages were laid waste, and it was as dark outside on the first day as it is in blackest houses in the dead of winter at night. Rumbling all over the country and ashfall around Borgarfjörður and Skaga, and everywhere in between, so that livestock died. People went to the mountain where the 'casting' [= eruption] was, and they heard that a large rock had been thrown from within the mountain. They appeared to be birds flying in the fire, both small and large, with various sounds. People thought they were souls. There was so much white salt around the opening that a horse could be split, and brimstone.

Flateyjarbók, or GkS 1005 fol., is a comprehensive Icelandic manuscript written by Jón Þórðarson and Magnús Þórhallsson between 1387 and 1390 for Jón Hákonarson, a wealthy Icelandic farmer. The name *Flateyjarbók* originates from its later association with the island of Flatey in Breiðafjörður, where it was kept for some time before being gifted to the Lutheran Bishop of Skálholt, Bishop Brynjólfur Sveinsson, in 1647. The codex contains extensive sagas about Norwegian kings as well as annals documenting historical events such as volcanic eruptions.

The annalistic account in *Flateyjarbók* provides an insight into how people perceived and reacted to volcanic events. The eruption is described with intense imagery, including ash falls so heavy that they caused darkness similar to winter nights, and phenomena interpreted as supernatural, such as 'birds flying in the fire', which were believed to be souls. This mystical understanding reflects a deep sense of awe and fear, with people's responses included approaching the volcano out of curiosity or fascination, despite the destruction it caused. The report highlights the considerable vulnerability caused by the ash falling on livestock and nearby villages, resulting in livestock deaths in regions such as Borgarfjörður and Skaga. Notably, there is no mention of organised mitigation strategies; instead, responses were largely observational or spiritual. In contrast to modern times, where scientific monitoring and evacuation plans during eruptions are common practice (e.g., during the 1970 eruption of Hekla), medieval Icelanders relied more on observation and religious interpretations rather than structured responses to mitigate the effects.

Volcanic Vulnerability in Medieval Iceland

Reasoning Volcanoes

Religious associations to volcanic activity are also shown in the *King's Mirror*. That work, also known as *Konungs skuggsjá*, is an important Norwegian didactic text written around 1250. It was produced during the reign of King Hákon Hákonarson (1204–1263) and was intended for the education of his son, Magnús lagabætir (1238–1280). The work is structured as a dialogue between father and son in which the father imparts wisdom on various topics such as trade, courtly etiquette, military strategy, and legal principles. The text also includes discussions of geography and natural phenomena relevant to navigation.

When questioned by his son about the natural wonders of Iceland, the father tries to explain their origins by drawing a comparison with Sicilian volcanoes and referring to the *Dialogues* of St Gregory. Unlike the fiery eruptions known from Sicily, which are fuelled by organic materials such as wood and earth, Icelandic volcanoes are presented as coming from the depths of hell, as they consume only dead matter such as rock.

> Son: What do you think of the extraordinary fire which rages constantly in that country? Does it rise out of some natural peculiarity of the land, or can it be that it has its origin in the spirit world?
>
> Father: ... concerning the extraordinary fires which burn there, I scarcely know what to say, for they possess a strange nature. I have heard that in Sicily there is an immense fire of unusual power which consumes both earth and wood. I have also heard that Saint Gregory has stated in his Dialogues that there are places of torment in the fires of Sicily. But men are much more inclined to believe that there must be such places of torment in those fires in Iceland ... Now since this fire feeds on dead things only and rejects everything that other fires devour, it must surely be said that it is a dead fire; and it seems most likely that it is the fire of hell, for in hell all things are dead. (Larson 1917 [repr. 1978]: 126–128)

Interestingly, the author of the *King's Mirror* refers to Iceland's volcanic and seismic activity. By stating that 'it rejects what the other fires devour', the text outlines a uniquely Nordic medieval worldview, aware of the barrenness of the Icelandic landscape. This image fits well with the following argument that earthquakes are caused by the movement of air in caverns and voids deep within the earth. The author adds: 'it may be that the great and powerful activity of the air in the foundations of the earth also causes those great fires to be kindled and to appear, which break out in different parts of the land'.

The idea that the earth is spongy and that earthquakes are caused by air currents is not a peculiarly Nordic world view, but has a long historical tradition going back to antiquity. It is found, for example, in the works of Isidore of Seville

(c. 560–636) and Beda Venerabilis (c. 673–735), which go back to the tradition of Aristotle (384–322 BC; see Barnes 1984 [repr. 1995]: 1295) and Pliny the Elder (c. 24–79). The latter seems to have shared similar beliefs when he wrote in his first-century *Natural History*:

> I have no doubts in thinking that winds are the cause [of earthquakes] … [Tremors] also only ever occur after there has been wind, because there are surely blasts secreted in cavities and [earth's] hidden hollows. On earth a tremor is nothing other than a thunderclap in a cloud, nor is a chasm anything other than when lightning bursts out as confined air struggles and strains to free itself and escape. (Turner and Talbert 2022: 78)

The *Etymologiae* of the Spanish church father Isidore of Seville also contain several treatises on natural phenomena in the section *De Natura Rerum*. The origin of earthquakes and volcanic activity is explained as follows:

> Of its motion some say that it is the wind in its hollows that, itself moved, moves the earth. Sallust (Histories 2.28): 'A number of mountains and hills subsided, sundered by the wind rushing through the hollows of the earth'. Others maintain that lifegiving water moves in the earth and simultaneously shakes it, like a vessel, as for instance Lucretius (see *On the Nature of Things* 6.555). Yet others are of the opinion that the earth is σπογγοειδής ('spongy'), and that its mostly hidden, collapsing interior shakes everything placed upon it. Also, an opening in the earth is created through the movement of water in the lower regions, or through repeated thunder, or through winds that erupt from cavities of the earth. (Barney 2006: 375)

Regarding (volcanic) fire, Isidore states that

> Fire also has another variation, for there is one fire that is for human use and another which appears as a part of divine judgment, whether contracting as a lightning-bolt from the sky or bursting forth from the earth through the mountain peaks. Fire (*ignis*) is so named because nothing can be born (*gignere*) from it, for it is an inviolable element, consuming everything that it seizes. (Barney 2006: 375–376)

These passages reveal several aspects. First, the wording of the last section is reminiscent of the Norwegian *Kings' Mirror*, whose author based his knowledge on Isidore's *Etymologiae*. Second, the text seeks to challenge superstitious interpretations of natural phenomena by placing them within a structured worldview that reflects the divine order. What was the motivation for this? Isidore is thought to have written *De Natura Rerum* between 612 and 615 at the request of the Visigoth King Sisebut (r. 612–621). The work may have been prompted by a total solar eclipse on 2 August 612, following a lunar eclipse the previous year, which seems to have provoked superstitious practices and apocalyptic fears that the king and Isidore sought to combat. The intention was to show that anomalous events in the terrestrial or celestial realms, such as seismic activity, volcanic eruptions, solar or lunar eclipses, were neither the work of demons nor harbingers of the end of the world, but rather an intrinsic aspect of the natural order established by God. Isidore's writings can therefore be seen as a Christian interpretation of an enduring theme in ancient cosmological treatises, already exemplified by Lucretius's *De natura rerum*.

Both Isidore's thinking and the title of *De natura rerum* were taken up again by the Anglo-Saxon monk Beda Venerabilis. In chapter 45 of his *De Natura Rerum*, a revised version of Isidore's cosmology, Beda argues that 'earthquakes and volcanoes are caused by air trapped in the earth and struggling to escape to its natural place' (Kendall and Wallis 2011: 161).

Finally, the writings presented, ranging from the Norwegian *King's Mirror* to the Latin works of antiquity, contain significant moral dimensions. The primary aim of cosmology and natural history was to provide ethical guidance. This was achieved by offering rational explanations of the universe and its most awe-inspiring phenomena, including eclipses, earthquakes, thunder and lightning. A second, related purpose was to foster pious admiration for the beauty and order of the world. Isidore of Seville, whose *De natura rerum* was the first Christian non-exegetical treatment of cosmology in Latin, shared the aim of Lucretius (c. 99–55 BC) of demystifying natural phenomena through reason.

In addition to such attempts to rationalise volcanic activity, legal sources provide insights into how medieval societies coped with environmental hazards by offering practical guidance, as will be shown below.

Coping With Volcanoes

The Icelandic law code *Grágás* ('Grey Goose') is a collection of laws of the Icelandic commonwealth predating the year 1271, prior to the implementation of the legal code *Járnsíða* ('Iron side') by the Norwegian King Magnús lagabætir, the recipient of the *King's Mirror*. *Grágás* is a comprehensive legal framework that reflects the social and environmental challenges faced by medieval Icelanders. It contains several provisions that highlight how the community adapted to their harsh natural environment, including threats such as floods, landslides and volcanic eruptions. Notably, *Grágás* includes a provision that allows for the relocation of churches in response to natural hazards such as landslides and snowslides (*skriða*), flooding (*vatnagangr*), fire (*eldsgangr*) or tempest (*ofviðri*):

> Kirkia hver scal standa i sama stað sem vigð er. ef þat ma fyrir skriþvm. eþa vatna gangi. eþa eldz gangi. eþa ofviþri. eþa herod eyþi at. or af dolvm. eþa

út strøndom. þat er rett at føra kirkio ef þeir atbvrþir verþa. þat er rett at føra kirkio ef biskop lofar. (Finsen 1852 [repr. 1974]: 12)

Every church shall remain on the site where it was consecrated if it may on account of landslips or snowslips or flooding or a fire or tempest or unless there is desolation of districts in remote valleys or on ocean coasts. It is lawful to move a church if such things happen. It is lawful to move a church if the bishop permits it. (Dennis, Foote and Perkins 1980: 30)

This passage reflects the awareness of medieval Icelanders of the unpredictability of their environment. The permission to relocate churches shows a practical approach to dealing with environmental hazards. It shows that Icelandic society has been adaptable and willing to make changes necessary for survival in the face of severe natural events. The requirement for the bishop's permission to relocate a church highlights the decisive role of ecclesiastical authority in decision-making processes related to community infrastructure in times of crisis. By allowing relocation under certain conditions (e.g., desolation in remote areas), *Grágás* demonstrates a prioritisation of maintaining community structures essential for social cohesion and religious practice despite environmental challenges. The provision reflects a pragmatic legal system that balances religious tradition with practical needs in times of emergency, and shows how medieval Icelandic society balanced spiritual obligations with survival needs.

Perhaps a combination of the explanations in the *King's Mirror* and the Icelandic legal regulations is represented by another documentary source, the so-called *Heitbréf*. These were public vows that were made in response to natural disasters such as volcanic eruptions and floods, but also plagues. One of the most revealing documents is the so-called *Heitbréf Eyfirðinga* ('Vow Contract of Eyjafjörður'), written by the inhabitants of Grund on 11 March 1477.[5] It invokes many saints such as the Virgin Mary, Archangel Michael, all the angels, John the Baptist, Peter, Paul, John and all the apostles, as well as Stephen, Thomas, Olav and all the martyrs, culminating in an appeal to All Saints (Sigurðsson et al. 1900–1904: 103–107).

a Grund j Eyiafirdi komu saman lærder og leiker … og tauludu þar um þau undur. odæmi og ognaner. sem þa yfer geingu af elldsgangi. sandfalli og öskumyrkrum og ogurligum dunum. af þessum undundrum þreifst fienadr ecki uid. enn þo var sniolaus jörd. (Sigurðsson et al. 1900–1904: 105)

In Grund, in Eyjafjörður, came together learned men and laymen, and there they talked about the wonders, miracles, and horrors that befell them from the fire, the fall of sand, and the darkness of ash and the terrible thunder.

---

5. For the discussion of the *Heitbréf*, see Van Deusen 2019: 41 and Cormack 2020: 133.

Volcanic Vulnerability in Medieval Iceland

Because of all these wonders, the cattle did not thrive. And also the ground was snowless.

These flows of fire, sand and darkness caused by ash certainly refer to a series of eruptions in the Bárðarbunga volcanic system in the early months of 1477, including the VEI-6 eruption of the Veiðivötn craters. These eruptions were so intense that they even triggered a subsequent eruption of another volcano, Torfajökull. The great explosive basalt eruption of 1477 took place in a 65-kilometre long fissure in the southwestern part of the Veiðivötn fissure and deposited about ten cubic kilometres of tephra on land, affecting fifty per cent of the country.

The *Vow Contract* thus shows the will of the inhabitants to alleviate the hardships of the eruptions by promising masses, prayers, and donations to the poor, as the source later states. The gathering in Grund also suggests a common approach to understanding and addressing these challenges. By sharing experiences and discussing impacts, they could collectively assess risks and develop strategies. As already seen in *Grágás*, Icelandic legal codes include provisions for adapting to environmental hazards such as landslides or floods by allowing churches to be relocated. This flexibility suggests an awareness of environmental risks and legal mechanisms to mitigate them. Despite challenges such as poor livestock health due to volcanic ash, Icelandic communities have continued with farming practices adapted to volcanic vulnerability.

Exemplifying Vulnerability? The Hekla 1104 Event

One of the most controversial eruptions in Icelandic history was that of Hekla in 1104, when about twenty farmsteads in the valley of Þjórsárdalur were buried under twenty centimetres of tephra, earning the valley the reputation of being the 'Pompeii of the North' (Bruun 1897: 24). The total volume of tephra produced by the eruption is estimated to be about 2.5 km^3, which corresponds to about 0.5 km^3 of solid rock of the same chemical composition. In terms of the amount of tephra produced, the Hekla 1104 eruption is the second largest in Iceland in historical times (i.e., since the settlement of Iceland in the 870s), the largest being the Öræfajökull eruption of 1362.

The subsequent review of the archaeological material unearthed in the valley will address two main questions: what is known about the first inhabitants of the valley around Hekla, and why did they settle in the vicinity of the volcano despite the danger of an eruption? Out of the approximately twenty Viking-Age farmsteads buried by the 1104 eruption, ten have been excavated. These are Skallakot, Stórólfshlíð, Áslákstunga fremri, Áslákstunga innri, Gjáskógar, Stöng, Skeljastaðir, Sámsstaðir, Sandártunga and Snjáleifartóttir. The most prominent of these is the farmstead at Stöng, which was home to Gaukr Trandilsson, one of the heroes of the settlement period, who is mentioned in both *Landnámabók* and *Njáls saga*. Tephrochronology suggests that medieval settlers such as Gaukr arrived in the valley in the late ninth century. This assertion is corroborated by the analysis of archaeological artefacts, while radiocarbon dating of bones and wood charcoal spans from the settlement period to the thirteenth century (Gísladóttir 2004: 142).

Guðrún Alda Gísladóttir has demonstrated that the archaeological material indicates notable differences in economic conditions between different areas within the valley. The peripheral regions in the south may have emerged as centres for the production of iron that was possibly exchanged for goods that were scarce in the valley, such as grain and maritime products. In contrast, the eastern region of the valley exhibits a distinct pattern, characterised by the presence of iron slag remains, in conjunction with the presence of rare artefacts such as glass beads and pottery fragments at the sites of Skeljastaðir and Stöng. Notably, the eastern part of the valley, despite its proximity to Hekla, is characterised by a particularly fertile soil, a factor that may have attracted settlers during the Viking Age. Furthermore, the presence of churches in these eastern settlements suggests that they may have developed into influential centres.

It is a commonly held belief that the excavated ruins of Gjáskógar and Stöng were abandoned as a result of the 1104 eruption. This assumption has had a significant impact on the field of Icelandic archaeology, leading to the establishment of a predominant narrative. According to this narrative, Þjórsárdalur served as a prime example of how once-flourishing settlements can deteriorate as a consequence of natural disasters (Dugmore et al. 2007: 2). However, the Icelandic archaeologist Vilhjálmur Örn Vilhjálmsson contested this, conducting a re-investigation of Stöng between 1983 and 1993. He proposed that the settlement persisted until the thirteenth century, despite being struck by natural disasters (the next eruption of Hekla followed only 54 years later in 1158). Medieval farmers continued to use the land for summer pastures. The resilience of the local economy is further evidenced by the continuation of coal and iron production, as well as the presence of extensive forests, a phenomenon that persisted until later centuries, as evidenced by forest records of the Bishopric of Skálholt for the years 1555–1587. By the latter year, it is estimated that approximately half of the valley remained wooded or covered in scrub. However, by the late seventeenth century, the condition of these woodlands and scrublands had deteriorated significantly. The eruption of Hekla in 1693 further exacerbated the situation by covering the valley with heavy ash fall, damaging vegetation, and causing the lowlands to swell. Between 1587 and 1708, it is estimated that the wooded areas had decreased by 71 per cent (Gylfadóttir and Guðmundsson 2019: 15).

Volcanic Vulnerability in Medieval Iceland

How can this picture from archaeological sources be supplemented by written sources? The earliest source to mention the eruption of 1104 is *Hungrvaka* ('Appetite-Whetter'), a history of the Christianisation of Iceland and the diocese of Skálholt until the death of Bishop Klængur Þorsteinsson (b. 1102) in 1176. It is generally believed that *Hungrvaka* was written by the same cleric who wrote *Páls saga byskups*, a vita of Bishop Páll Jónsson (1155–1211), in the early thirteenth century. About the eruption of 1104, which happened about one century before *Hungrvaka* was written, the work says that

> during the episcopacy of Bishop Gizurr [Ísleifsson] there were many significant events: the death of King Cnut on Fyn ... the death of the English king William [the Conqueror] ... the translation of Bishop Nicholas the Saint to Bari ... the eruption of Mount Hekla, and many other notable events even if they are not included here. (Andersson 2021: 17)

A similar wording appears in subsequent references to the second and third historical eruptions of Hekla in 1158 and 1206, respectively. The reliability of the account is substantiated by the author's assertion:

> Hefi ek af því þenna boekling saman settan, at eigi falli mér með ǫllu ór minni þat er ek heyrði af þessu máli segja inn fróða mann Gizur Hallsson. (Egilsdóttir 2002: 3)

> I have composed this booklet to prevent losing from memory that which I have heard about these matters from the learned man Gizurr Hallsson and what other notable men have put in story form. (Andersson 2021: 3)

The informant, Gizurr Hallsson, was a chieftain in southern Iceland and law-speaker of the Althing from 1181 to 1202. He resided in Haukadalur for the majority of his life, but spent the final years of his life in Skálholt, where he died in 1206. His position and influence show he was one of the most informed Icelanders of his time, which makes *Hungrvaka* a reliable account of the High Middle Ages in Iceland. Significantly, the eruption of Hekla is the only natural phenomenon included in the list of remarkable events in *Hungrvaka*. Despite the absence of a precise date for the eruption, the author's choice of words suggests that this is the first documented historical eruption of Hekla. The year 1104 is ultimately cited in three Icelandic annals as the specific year of the eruption:

The first coming up of fire in Mount Hekla (*Annales regii* and *Lögmannsannáll*);
Fire in Mount Hekla (*Gottskálksannáll*).

The first annals, dating from the first half of the fourteenth century, also make reference to a 'sandfall winter', that is to say, an ash rain in the year following the eruption.[6] The *Oddaverjaannáll*, a document dating to the late sixteenth century, chronicles two notable events: the first eruption of Mount Hekla in 1106 and a 'sandfall winter' that same year. However, the *Oddaverjaannáll* also records the establishment of the metropolitan see in Denmark for the same year [erroneously for 1106 instead of 1104], which indicates that all these events must have taken place in 1104.

Notwithstanding their customary laconic manner, the Annals offer supplementary insights into the circumstances of the eruption. The *Annales regii* continue

> Var sét blóð flióta vt af brávði. En ioladag var veðr sva grimt at menn máttv eigi komaz til kirkiv. En þeir er við leitaðv. Týnndvz sumir i ánni Fvlda. [1105] Sandfallz vetr. (Storm 1888: 111)
>
> Blood was seen flowing out of bread. On the Christmas day, the weather was so harsh that people could not make it to church.

It is evident that the prevailing weather conditions exerted a significant influence on both the explanation of the eruption and the subsequent population response.

Concluding Remarks

This study of medieval Iceland's volcanic history demonstrated the profound impact of eruptions on the society, culture and environment of the time. Volcanoes shaped the physical landscapes, disrupted settlements, and inspired mythological narratives, as evidenced by the possible link between the Eldgjá eruption and Iceland's Christianisation around 1000. Likewise, the Hekla eruption of 1104 caused widespread devastation in Þjórsárdalur, leading to its subsequent designation as the 'Pompeii of the North'. Despite the dramatic potential of volcanic events, they are conspicuously absent from the *Sagas of Icelanders*, a phenomenon that can be attributed to the genre's conventions and cultural priorities, which placed greater emphasis on human interactions than on environmental catastrophes. However, annals and legal texts such as *Grágás* offer invaluable insights into the coping strategies employed by medieval Icelanders in the face of these challenges.

Utilising Robert Chambers' vulnerability framework, this analysis underscored how communities adapted to external shocks, such as eruptions, through localised responses. These adaptations included the relocation of churches or the organisation of masses (as evidenced by vow contracts like the *Heitbréf Eyfirðinga*), which reflects a pragmatic approach to risk management. Archaeological

---

6. On the dating of the Icelandic annals, see Storm 1888: 46. According to Storm, all annals derive from a single collection written at the end of the thirteenth century.

evidence further reinforces the resilience of these communities, as evidenced by the reconstruction of settlements on tephra-covered land following eruptions like the one at Eldgjá.

Moreover, volcanic activity carried religious and symbolic significance. Texts such as the *King's Mirror* depicted eruptions as manifestations of divine judgment or hellfire, while myths like *Vǫluspá* encoded these events into cultural memory. The selective recording of eruptions in medieval texts underscores broader thematic priorities but also highlights how Icelandic society balanced practical responses with spiritual interpretations.

Ultimately, medieval Iceland's engagement with volcanic hazards demonstrates both remarkable adaptability and strategic resourcefulness within a challenging Nordic environment. By reconciling mythic interpretations of eruptions with practical adaptations, Icelandic society established a model of socio-environmental resilience. This historical precedent continues to provide critical insights for contemporary approaches to disaster management in vulnerable landscapes.

Bibliography

Andersson, T.M. (trans.) 2021. *Bishops in Early Iceland*. Exeter, UK: Viking Society for Northern Research, University College London.

Barnes, J. 1984 [repr. 1995]. *The Complete Works of Aristotle: The Revised Oxford Translation*. Princeton: Princeton University Press.

Barney, S.A. et al. 2006. *The Etymologies of Isidore of Seville*. Cambridge: Cambridge University Press.

Benediktsson, J. (ed.). 1968. *Landnámabók. Íslenzk fornrit 1*. Reykjavík: Hið íslenzka fornritafélag.

Chambers, R. 1989. 'Editorial introduction. Vulnerability, coping, and policy'. *IDS Bulletin* 20: 1–7. https://doi.org/10.1111/j.1759-5436.1989.mp20002001.x.

Cormack, M. 2020. 'How do we know, how did they know? The cult of saints in Iceland in the late Middle Ages'. In K. Kjesrud and M. Males (eds), *Faith and Knowledge in Late Medieval and Early Modern Scandinavia*. Turnhout: Brepols. pp. 123–147. https://doi.org/10.1484/M.KSS-EB.5.117744.

Dennis, A., P. Foote and R. Perkins (trans.) 1980. *Laws of Early Iceland: Grágás I*. Winnipeg: University of Manitoba.

Van Deusen, N.M. 2019. *The Saga of the Sister Saints. The Legend of Martha and Mary Magdalen in Old Norse-Icelandic Translation*. Toronto, ONT: Pontifical Institute of Mediaeval Studies.

Dugmore, A.J. et al. 2007. 'Abandoned farms, volcanic impacts, and woodland management: Revisiting Þórsárdalur, the "Pompeii of Iceland"'. *Arctic Anthropology* **44** (1): 1–11.

Egilsdóttir, Á (ed.). 2002. *Biskupa sögur II: Hungrvaka, Þorláks Saga Byskups in Elzta, Jarteinabók Þorláks Byskups in Forna, Þorláks saga Byskups Yngri, Páls saga Byskups. Íslenzk Fornrit XVI*. Reykjavík: Hið Íslenzka Fornritafélag.

Eldjárn K. 1984. 'Kumlateigur í Hrífunesi í Skaftártungu I'. In I.L. Baldvinsdóttir (ed.) *Árbók Hins Íslenzka Fornleifafélags 1983*. Reykjavík: Hið Íslenska Fornleifafélag. pp. 6–21.

Falk, O. 2007. 'The vanishing volcanoes: Fragments of fourteenth-century Icelandic folklore'. *Folklore* 118: 1–22. https://doi.org/10.1080/00155870601096257.

Finsen, V. (ed.). 1852 [repr. 1974]. *Grágás: Konungsbók*. Odense: Odense Universitetsforlag.

Gestsdóttir, H., U. Ævarsson, G.A. Gísladóttir et al. 2015. 'Kumlateigur í Hrífunesi í Skaftártungu V'. In B. Lárusdóttir, G.S. Sigurðarson, M. Snæsdóttir et al. (eds), *Árbók Hins Íslenzka Fornleifafélags 2014*. Reykjavík: Hið Íslenzka Fornleifafélag. pp. 7–34.

Gísladóttir, G.A. 2004. *Gripir úr Þjórsárdalur*. Unpublished MA thesis. Department of History and Archaeology, University of Iceland.

Global Volcanism Program. 2025. Volcanoes of the World (v. 5.2.6; 5 Feb 2025). Distributed by Smithsonian Institution, compiled by E. Venzke: https://doi.org/10.5479/si.GVP.VOTW5-2024.5.2.

Gylfadóttir, R.G. and G. Guðmundsson. 2019. *Þjórsárdalur: Skráning fornminja úr lofti*. Reykjavík: Fornleifastofnun Íslands.

Hammer, C.U., H.B. Clausen and W. Dansgaard. 1980. 'Greenland ice sheet evidence of post-glacial volcanism and its climatic impact'. *Nature* 288: 230–235. https://doi.org/10.1038/288230a0.

Jackson, T. 2023. 'Icelandic volcanoes in Medieval sagas, chronicles, annals and in modern scholarly literature'. *ISTORIYA* 14 (8): 130. https://doi.org/10.18254/S207987840027707-6.

Karlsson, G. 2000. *Iceland's 1100 Years. History of a Marginal Society*. London: Hurst & Company.

Kendall, C.B. and F. Wallis (trans.) 2011. *Bede: 'On the Nature of Things' and 'On Times'*. Liverpool: Liverpool University Press.

Larsen, G.Þ. 2000. 'Holocene eruptions within the Katla volcanic system, south Iceland: Characteristics and environmental impact'. *Jökull* 49: 1–28. https://jokull.jorfi.is/articles/jokull2000.49/jokull2000.49.001.pdf.

Larsen, G. and S. Thorarinsson. 1984. 'Kumlateigur í Hrífunesi í Skaftártungu IV'. In I.L. Baldvinsdóttir (ed.) *Árbók Hins Íslenzka Fornleifafélags 1983*. Reykjavík: Hið Íslenzka Fornleifafélag. pp. 31–47.

Larson, L.M. 1917. *The King's Mirror (Speculum regale–Konungs skuggsjá)*. New York: American-Scandinavian Foundation.

Nordvig, M. 2021. *Volcanoes in Old Norse Mythology: Myth and Environment in Early Iceland*. Leeds: Arch Humanities Press.

Ogilvie, A.E.J. and G. Pálsson. 2003. 'Mood, magic and metaphor: Allusions to weather and climate in the Sagas of Icelanders'. In S. Strauss and B.S. Orlove (eds), *Weather, Climate, Culture*. London: Routledge. pp. 251–274.

Oppenheimer, C., A. Orchard, M. Stoffel et al. 2018. 'The Eldgjá eruption: Timing, long-range impacts and influence on the christianisation of Iceland'. *Climatic Change* 147: 369–381. https://doi.org/10.1007/s10584-018-2171-9.

Oppenheimer, C. et al. 2017. 'Multi-proxy dating the "Millennium Eruption" of Changbaishan to late 946 CE'. *Quaternary Science Reviews* 158: 164–171. https://doi.org/10.1016/j.quascirev.2016.12.024.

Pálsson, H. and P. Edwards. 1972 [repr. 2006]. *The Book of Settlements: Landnámabók*. Winnipeg: University of Manitoba.

Phelpstead, C. et al. (eds). 2020. *An Introduction to the Sagas of Icelanders*. Gainesville: University Press of Florida.

Schmid, M.M.E., A.J. Dugmore, A.J. Newton and O. Vésteinsson. 2021. 'Multidisciplinary chronological data from Iceland indicate a Viking Age settlement flood, rather than a flow or trickle'. In M.F. Napolitano, J.H. Stone and R.J. DiNapoli (eds), *The Archaeology of Island Colonization*. Gainesville: University of Florida Press. pp. 132–179.

Sigl, M. et al. 2015. 'Timing and climate forcing of volcanic eruptions for the past 2,500 years'. *Nature* 523: 543–549. https://doi.org/10.1038/nature14565.

Storm, G. (ed.). 1888. *Islandske annaler indtil 1578*. Christiania: Grøndahl & Søns Bogtrykkeri.

Stothers, R.B. 1998. 'Far reach of the tenth-century Eldgjá eruption, Iceland'. *Climatic Change* 39: 715–726. https://doi.org/10.1023/A:1005323724072.

Thordarson, Th. and Á Höskuldsson. 2022. *Iceland* (3rd ed.). Edinburgh: Dunedin.

Thordarson, Th. 2011. 'Perception of volcanic eruptions in Iceland'. In I. Martini and W. Chesworth (eds), *Landscapes and Societies*. Dordrecht: Springer. pp. 285–296

Thordarson, Th. and G. Larsen. 2007. 'Volcanism in Iceland in historical time: Volcano types, eruption styles and eruptive history'. *Journal of Geodynamics* 43 (1): 118–152. https://doi.org/10.1016/j.jog.2006.09.005.

Thordarson, Th. et al. 2001. 'New estimates of sulfur degassing and atmospheric mass-loading by the 934 AD Eldgjá eruption'. *Iceland. Journal of Volcanology and Geothermal Research* 108 (1–4): 33–54. https://doi.org/10.1016/S0377-0273(00)00277-8.

Turner, B. and R.J.A. Talbert. 2022. *Pliny the Elder's World, Natural History, Books 2–6*. Cambridge: Cambridge University Press.

Þórarinsson, S. 1967. *The Eruption of Hekla 1947–1948. Volume I. The Eruptions of Hekla in Historical Times. A Tephrochronological Study*. Reykjavík: H.F. Leiftur.

Vinther, B.M. et al. 2006. 'A synchronized dating of three Greenland ice cores throughout the Holocene'. *Journal of Geophysical Research* 111: D13102. https://doi.org/10.1029/2005JD006921.

Zielinski, G.A. et al. 1995. 'Evidence of the Eldgjá (Iceland) eruption in the GISP2 Greenland ice core: relationship to eruption processes and climatic conditions in the tenth century'. *The Holocene* **5** (2): 129–140. https://doi.org/10.1177/0959683695005002.

The Author

Carina Damm is an Assistant Professor and Postdoctoral Researcher at the Centre for Nordic and Old English Studies at the University of Silesia in Katowice. She obtained her MA in History and Scandinavian Studies from the University of Göttingen and her Doctorate from Leipzig University. Her research interests focus on the environmental history of northern Europe, Scandinavian-Slavic interrelations in the Viking Age, and early medieval economies.

Chapter 5

THE MOVING MANORS AND ADAPTATION IN SIXTEENTH-CENTURY DENMARK

Sarah Kerr

Abstract

Early modern Denmark experienced unusual climatic variation, resulting in a catastrophic storm surge in 1593 at Nørre Vosborg manor in the peninsula of Jutland. Nørre Vosborg is a site comprising four late medieval and early modern manor houses, referred to as Vosborgs 1–4. Using the architecture and archaeological remains from surveys and excavations, the four manor houses will be discussed in relation to adaptation and resilience. It is established that Vosborg 1 and 2 were impacted by the 1593 storm surge. Vosborg 1 was replaced by Vosborg 3 approximately 900 metres inland after it sustained devastating flooding and damage, at the turn of the seventeenth century. Vosborg 2 was also damaged by the same environmental event, yet some building material was rescued and reused to create Vosborg 4. It will be suggested that this demonstrates evidence of historic adaptation and resilience to an environmental event.

Introduction

Nørre Vosborg exists today as a multi-period manor house located in Vemb, Denmark, whose earliest range dates from the end of the sixteenth century. While it is mainly known for its architecture, landscape and gastronomic experiences offered to visitors, the history of the manor reveals adaptation and resilience in response to an extreme environmental event: a destructive storm surge in 1593. The extant Nørre Vosborg is one of four manor houses, which all share the same name 'Vosborg', on the western edge of the Jutland peninsula. Two sites are called Nørre Vosborg (North Vosborg) and two are called Sonder Vosborg (South Vosborg). Below, I will describe how the original Vosborg manor was divided into two manor houses owned by two branches of the same family, then both of those were rebuilt further inland after an extreme environmental event. I will suggest that this demonstrates adaptation and resilience in the past.

doi: 10.63308/63881023874820.ch05

Adaptation in archaeological research may be considered as the ways in which humans adjust their cultural systems in response to a changing external environment, while the ability and readiness to adapt are considered central qualities of resilience (Corrigan 2017; Borrero 2014; Redman 2005; Folke et al. 2010). Adaptation to climate changes has emerged in archaeological discourse, providing a balance to the former propensity for studying climate-induced social collapse (Bassett et al. 2013; Tubi et al. 2022). While collapse brought wholesale impacts on all aspects of life, the central narrative in the human past is an ability and willingness to adapt behaviour. Therefore, the more constant reality of the past human-climate relationship included mitigation (Desjardins et al. 2020), adaptation (Jackson et al. 2018) and resilience (Riris et al. 2024).

The Nørre Vosborg case study formed part of the Horizon 2020-funded Project CHICC (Culture, Heritage and Identity: Impacts of Climate Change) carried out in 2021–2023 across a series of archaeological sites threatened by anthropogenic climate change in western Europe (CORDIS 2023). As anthropogenic climate change affects societies across the world, more and larger areas will be impacted and the communities in those areas will have to adapt (Calvin et al. 2023; Fankhauser 2017). It has been established elsewhere that archaeological data may be useful in today's adaptation strategies, despite the varying differences in the events leading to the need for adaptation (Jackson et al. 2018). In this chapter, Nørre Vosborg is discussed as an example of adaptation and resilience. The 1593 storm surge is not equivalent to today's anthropogenic climate change; nor are past and present societies comparable. However, the Nørre Vosborg example may be viewed as a 'completed experiment' (Jackson et al. 2018: 8) of human adaptation and resilience. The CHICC project centred this case study to determine if its retelling in a local-community setting could impact and enhance climate communication and action (Kerr 2023; Kerr et al. 2022). This chapter will focus on analysing the evidence for adaptation and resilience at Nørre Vosborg, in early modern Denmark.

Adaptation and Resilience

Adaptation and resilience are increasingly common themes in academic literature, particularly as understanding how past humans interacted with a changing climate may prove useful in tackling today's anthropogenic climate crisis (Jackson et al. 2018; Redman 2005; Costanza et al. 2007). Yet there are terminological inconsistencies surrounding the usage of adaptation and resilience and little consensus on their definitions (Kerr 2020; Smithers et al. 1997). Similarly, greater academic attention has focused on the concept of vulnerability. The varying definitions and approaches to these concepts are valid, and determining

The Moving Manors and Adaptation in Sixteenth-Century Denmark

the most appropriate is not this chapter's aim – however, clarity on how these terms are used here is important.

The increased use of the term 'adaptation' over the past two decades could be viewed as a revival, as it had been previously somewhat abandoned as a theoretically robust concept (Bassett et al. 2013). This resurgence cannot be separated from today's climate crisis in that it is probably linked to the realisation that useful mitigation strategies (the act of preventing or reducing the severity of impacts) were being discussed yet not implemented (Bassett et al. 2013). Without fast and genuine mitigation, adaptation is the only recourse in tackling anthropogenic climate change, and thus greater research has focused on this topic.

Both adaptation and mitigation were referenced as main categories of response to climate change by the Intergovernmental Negotiating Committee working on the draft of the United Nations Framework Convention on Climate Change (UNFCCC 1999); however, only mitigation was clearly defined. The subsequent uptick in the use of the term adaptation has coincided with considerable academic and policy debate on what it means. This issue was highlighted by Smithers and Smit who demonstrated that there is an 'inconsistent understanding of human adaptation to environmental variations' (1997: 129). Some have tried to clarify different types of adaptation, for example adaptation that is likely versus adaptation that is recommended (Smit et al. 1999); while others, such as Pittock and Jones (2000), have adopted a more practical approach to the issue and argue why adaptation is necessary, what should be prioritised and how research can contribute. The need for adaptation in the present has researchers looking to the past to understand what it is and how it can be generated.

Adaptation has long been associated with the Darwinian theory of evolution and the process of natural selection. It is understood to result from the combination of environmental forces and random genetic variations and those individuals whose characteristics were of greater survival value were 'selected'. Within the discipline of biology, adaptation is identified through the features shaped by natural selection, which in turn suggest the level of an organism's adaptedness (O'Brien et al. 1992). The philosophical definition of adaptation varies once again (O'Brien et al. 1992) while the term's use in the social sciences has not been universally accepted and it retains some negative connotations (Schipper et al. 2008b). Concerned with the long-term history of humankind, the theory of evolution is inherently attractive for archaeologists (Van Pool 2002); however, adaptation within archaeology, when directly addressed, varies from these other disciplines. This is likely due to the strong influence of anthropology on the development of archaeology and the increase in the study of adaptation in the context of archaeology and climate change.

Sarah Kerr

Adaptation in archaeological research can be considered the capacity for humans to adjust their cultural systems in response to external changes; while resilience is the capability to mitigate change (IPCC 2012a; IPCC 2012b; Cassar 2016; Løvschal 2022). These definitions are used in this chapter due to their applicability and comprehension in academic research, professional archaeology and government policy (European Union 2009; Daly 2019; Fluck 2016; Adaptation Committee 2016). They are understood as profoundly social concepts that are reflected in the material culture, architecture and landscapes preserved in the archaeological record. However, the discipline of archaeology lacks a universal definition of adaptation. Kirch (1980) suggests that there have been deficiencies in identifying the fundamental principles of the adaptation process while O'Brien and Holland (1992) suggest the term is used somewhat frivolously to explain the appearance and persistence of archaeological trends. Kirch (1980) suggests that the lack of agreement is in part due to the changing scale of archaeological research: the later decades of the twentieth century saw more local-scale studies of the past, rather than detecting large scale models of existence. McClain (2012) has highlighted a broader issue with assuming an understood 'common sense' in archaeological research (McClain 2012: 133). Part of this issue is the tendency not to establish a definition within each discussion on the topic. Through not transparently declaring the definition used, the subsequent interpretations are based on assumptions of which the reader is not aware. For these reasons, definitions as I am using them are being set out here.

Schipper and Burton (2008a) suggest that our current understanding of adaptation is now much complicated when compared to the earlier debates. They argue that, while the emergent discourse concentrates on what adaptation is, how it can be stimulated and what its limits are, the answers to these crucial questions remain unclear (Schipper et al. 2008a; Schipper et al. 2008b). Returning to today's imminent need for adaptation, this discourse has not accelerated a collective response to climate change and thus an 'adaptation deficit' endures (Burton 2008: 89). Greater discussion on what adaptation has looked like for past humans could therefore be useful in addressing these shortcomings.

The concept of resilience has faced some of the same challenges as that of adaptation. The concept emerged from ecology (Holling 1973) and Holling and Gunderson's (2002) definition has had considerable impact on resilience research. They established that resilience is the ability of an ecological system to absorb change while maintaining its structure and function (Holling et al. 2002). They represented this through two processes that describe change at multiple time and space scales: the adaptive cycle and panarchy. Others within ecology, such as Handmer and Dovers (1996), established that resilience could be separated into three categories: resilience to change, change at the margins and openness and adaptation, and suggest that all three may be required when faced with risk. Resilience tends to be used more sparingly than adaptation in the climate change context; however, this is changing, particularly within the discipline of archaeology, as climate change research becomes more widespread.

Redman and Kinzig extended Holling and Gunderson's (2002) work into archaeological discourse using the frameworks the latter had created (Redman 2005; Redman et al. 2003). More recently, archaeologists have approached the topic more directly and created definitions and approaches suited to the discipline and the material studied (Løvschal 2022). This has assisted in moving away from epistemological problems including the dichotomy between nature and culture; however, Løvschal (2022) argues that we remain in need of a new understanding of resilience, entirely separate from that which emerged from ecology. It may be understood in archaeological research, and so is used in this chapter, as the ability to change and reorganise while retaining the same characteristics such as function, structure, identity and feedbacks.

Greater exploration of the interactions between archaeology and climate changes, both past and present, has brought adaptation and resilience research to the fore, providing something of a balance to the former focus on climate-induced social collapse (Yoffee et al. 1991; Tainter 1988). Within collapse studies, vulnerability is an important concept yet is rarely referred to explicitly. Vulnerability can be defined as exposure to a hazard or the potential for negative outcomes (Nelson et al. 2012: 198). It could be viewed alongside adaptation as a counter-concept; that is, to overcome vulnerabilities, there must be adaptation however these concepts could be considered more closely linked. For example, awareness of vulnerability might actually improve adaptability and strengthen resilience (Heitz 2021). There is greater recognition of vulnerability, *and* the prevailing ability and willingness to adapt behaviour in the human past (Middleton 2017; Yasur-Landau et al. 2024). Therefore, constant threads in past human-climate dynamics were adaptation and resilience. It is a challenge to adapt in a suitable way to the rapidity of anthropogenic climate change currently faced (Schipper et al. 2008b) but one stepping-stone in that direction may be understanding how past societies adapted to changes within their immediate environment.

Case Study: Nørre Vosborg

Nørre Vosborg is a manor house located in Vemb, western Jutland, approximately twenty kilometres from Holstebro, Denmark ([Map 1](Map 1)). Jutland today is characterised by very flat land, as with much of Denmark. Predominantly at one metre above sea level, it is low-lying, very wet and marshy, with a history of pastoral farming. Prior to the industrial period's large-scale drainage, Jutland was characterised by wetlands and heathlands (Olwig 1996; Kristensen 2003).

Map 1. Map of Jutland identifying the location of the current Nørre Vosborg manor house, the Nissum Fjord and the North Sea. Map data © 2025 Google.

Nørre Vosborg is located in a low-lying watery landscape between a number of important watercourses. It lies less than three kilometres east of the Nissum Fjord whose western boundary connects to the North Sea. During the early modern period there was no physical separation between the fjord and the sea; as such, the sandy isthmus (narrow strip of land separating two bodies of water) is a relatively young formation (Kock 2015). The Nissum Fjord occupies seventy square kilometres and has an average depth of one metre, increasing to a maximum of three metres at its centre (Geopark Vestjylland 2024a). It is fed by a number of streams including the River Storå, Denmark's second longest river. Nørre Vosborg is located within the Vestjylland UNESCO Global Geopark, an area of international geological significance, comprising almost 5,000 square kilometres, half of which is within the North Sea (UNESCO 2024; Geopark Vestjylland 2024a).

Nørre Vosborg in Vemb is an early modern manor house, whose earliest extant building dates to the late sixteenth century. However, there are three other manor-house sites along the western coast of Jutland, which all share the name Vosborg: two sites are called Nørre (North) Vosborg and two are called Sønder (South) Vosborg. The four Vosborgs, all in close proximity to one another and all with the same name, derive from one manor, now lost. The first Vosborg was

The Moving Manors and Adaptation in Sixteenth-Century Denmark

divided into two manor houses, owned by two branches of the same family. Subsequently both these manor houses were moved as a result of environmental events in the sixteenth century. The four Vosborgs will be referred to as Vosborgs 1–4, according to the chronological order of their construction, with the extant Nørre Vosborg in Vemb referred to henceforth as Vosborg 3. This unusual story of the manor houses that were moved was researched and discussed as part of the CHICC project's workshops in Jutland (CORDIS 2023). The below narrative, which is in part the product of those workshops, will be conveyed, then followed by a discussion on the evidence for adaptation and resilience.

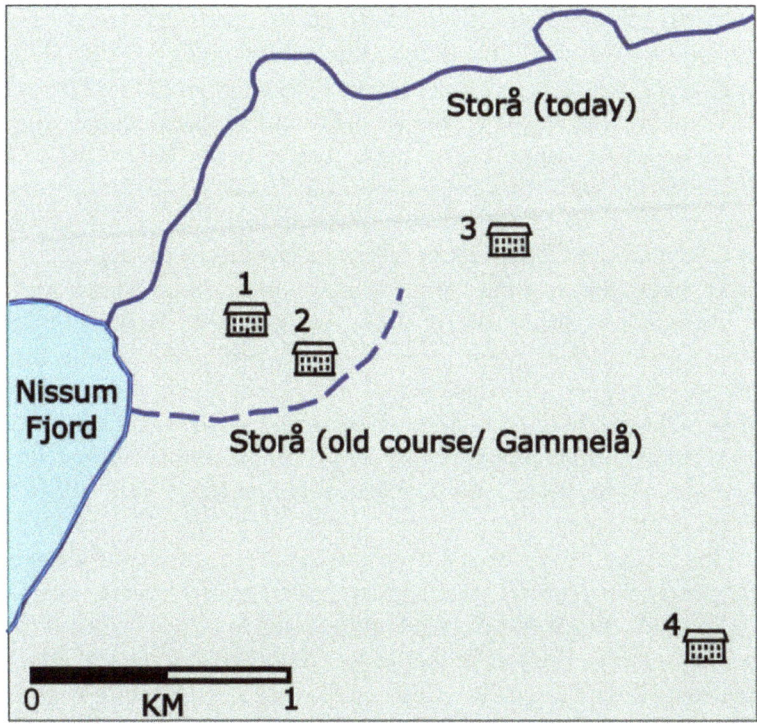

Map 2. Simplified drawing showing the position of the four (1–4) Vosborg manor houses. This also shows the original course of the Storå called the Gammelå (English: Old Course). Source: author's own adaptation from Kock 2015.

The first Vosborg manor house built along the edge of the River Storå was associated in 1299 with King Erik Menved (Regent 1286–1319) before being transferred to nobleman Niels Bugge in the fourteenth century (d. 1359) (Kock 2015) (Map 2). The course of the River Storå today is the result of partial draining and riverine management strategies – as such its original course, called the Gammelå, was further south than its current route (Map 2). The manor house (Vosborg 1) was occupied for more than two centuries until, in c. 1551, while under the ownership of Jytte Podesbusk and Knud Gyldenstjerne, it was divided into two parts in connection with an inheritance (Kock 2015). A second manor house, Vosborg 2, was built by Gregers Holgersen Ulftands (Podesbusk's nephew) a few hundred metres to the south-east of the original Vosborg 1 (Kock 2015). To differentiate between these, they were called Nørre Vosborg to refer to the original Vosborg 1 in the north and Sonder Vosborg to refer to Vosborg 2, slightly to the south (Map 2).

Neither Vosborg 1 or 2 is extant, yet both were explored from 2008 through archaeological excavation. The subsequent publications (Bøgh et al. 2014b; Bøgh et al. 2014a) included reconstruction drawings which depict relatively wealthy and high-status manor houses befitting the Danish nobility (Figures 1 and 2). Vosborg 1 comprised two raised platforms, one supporting the main stone-built house with the second supporting an outer courtyard. Access was granted from the east first to the outer courtyard which then led via a bridge to the inner platform. The eastern edge of the Nissum Fjord was at the rear of the main house, however the site today is at some distance from the body of water. This is due to the creation of the Felsted Kog in the nineteenth century: a drainage project which strove to create more agricultural land. The results of the excavation suggest that the late medieval manor house was constructed c. 1341, as indicated by the fell-dates of the timber poles which supported the bridge between the two platforms (Figure 1). The excavation also revealed that the overall water level was lower during this period (Bøgh et al. 2014a; Kock 2015; Kock 2019).

The remains of Vosborg 2 were also excavated and it was identified as four ranges around a courtyard upon a raised platform constructed of sand (Figure 2). The gateway was not located but presumed to be through the southern range, closest to the Gammelå. The ranges comprised at least one storey over a basement, whose foundations consisted of stone sills set into the earth. The ranges were constructed of various materials including timber and brick with internal clay floors, while the centre of the courtyard was a stone-packed mettled surface (Kock 2015). Although merely summarising the excavation reports, this demonstrates the greater extent of intact archaeological layers at Vosborg 2 when compared to Vosborg 1.

In the mid-sixteenth century, Vosborg 1 and 2 were extant, occupied, near to one another and close to the Gammelå; however, by the mid-1590s a catastrophic storm surge had impacted both manors (Bøgh et al. 2014a; Bøgh et al. 2014b; Geopark Vestjylland 2024b; Kock 2015). The end of the sixteenth

Figures 1 and 2. Reconstruction drawings of Vosborgs 1 (top) and 2 (bottom) from the publication of the excavations. Both manor houses were constructed on raised platforms. Drawing by Erik Sørensen, courtesy of his estate; as originally published in Bøgh et al. 2014a, pp. 137, 258.

century was a period of concentrated climate events, characterised by falling temperatures and significantly more unsettled weather patterns, which damaged both the natural and built environment (Kock 2019). This was a discernible climatic shift particularly when compared to the previous centuries, when Vosborg 1 was first occupied and the manor house constructed, which was comparatively milder and without the risk of storm surges (Kock 2019).

The extreme weather of the sixteenth century was not restricted to Denmark; rather England, Holland, France, Poland and Germany also experienced extreme and unusual climatic events particularly during the 1590s (Pfister et al. 1999; Clark 1985). Previous research suggested that the areas impacted were experiencing a cyclical crisis that afflicted premodern societies (Clark 1985); however, more recent explanations suggest that the last decade of the sixteenth century was an anomalous period that saw some of the worst weather ever recorded in the northern hemisphere (Parker 2018). This has been attributed to a severe episode of the Little Ice Age (LIA) which more broadly included volcanic eruptions, reduced solar activity and multiple El Niño events causing an increased frequency of extreme climatic events (Parker 2018). The onset of the LIA in Northern Europe, in the fourteenth century, brought heavy precipitation and falling temperatures. The subsequent crop failures led to inadequate food supply and loss of livestock, thus contributing to The Great Famine, between 1315 and 1322 (Jordan 1996).

In Denmark, the LIA lasted for approximately 500 years with societal impacts including land desertion, a decreased workforce which contributed to the Late Medieval Agrarian Crisis (Svensson 2019), a slowing of urbanisation and reduced standard of living for much of the population (Scott et al. 2019; Primeau et al. 2019). The second half of the LIA is evident in contemporary writings. The weather diaries compiled by astronomer Tycho Brahe (1546–1601) between 1585 and 1600 describe colder winters and wetter summers than previously experienced. The winter months were more frequently dominated by winds from the east while there was a higher frequency of snowfall suggesting longer periods of high pressure with cold air advection from the east or northeast (Pfister et al. 1999). Kock (2019) summarises that the west coast of Jutland was so heavily impacted by the LIA c. 1600 that this was a 'time of the storm surges'.

The 2008 excavations revealed that Vosborg 1 was destroyed by the 1593 storm surge while Vosborg 2 was badly damaged to a lesser extent. This is supported by contemporary accounts, which record the catastrophic event occurring on Christmas Eve and pleas to Christian IV, King of Denmark-Norway, for financial assistance to rebuild (Bøgh et al. 2014a; Bøgh et al. 2014b). The published excavation results include a reconstruction drawing of the event: it shows a destroyed Vosborg 1 in the foreground with Vosborg 2 in the background

The Moving Manors and Adaptation in Sixteenth-Century Denmark

Figure 3. Reconstruction drawing showing Vosborg 1 in the foreground destroyed after the 1593 storm surge and Vosborg 2 in the background badly damaged. Drawing by Erik Sørensen, courtesy of his estate; as originally published in Bøgh et al. 2014a. p. 312.

badly damaged (Figure 3). This usefully illustrates the excavators' impression of the differing impacts of the storm surge on each manor house.

The archaeological evidence at Vosborg 1 revealed large sand banks that appeared initially as part of the ramparts. Further exploration revealed they were actually sand deposited by the force of the 1593 storm surge. It contained building materials within, including bricks, timber and shredded peat, as well as materials from the nearby water courses, such as water-rolled stones (Kock 2015). The best-preserved section of either platform was that closest to the Nissum Fjord: at the western extent of the inner platform. This seems somewhat contradictory as this was the first part of the manor house impacted by the storm waters; however, the excavation revealed that this section was more reinforced than the other edges of the platforms (Bøgh et al. 2014a: 126–27). With distance from the western edge of the inner platform less extant material could be detected. The excavation therefore suggested that the storm water obliterated most of the platforms and consequently the buildings that sat upon them. The force of the storm surge was further identified through the recovery of the corner of the stone-built house which sat on the inner platform (Bøgh et al. 2014a: 114–16). It appears the corner of the house collapsed to the east, which allowed the building's height of six metres to be identified. Little occupation material was rediscovered in comparison to excavations of contemporary manor houses or castles (Bøgh et al. 2014a: 138). This is attributed to the force of the storm and the subsequent destruction of most of the manor house. The excavators suggested most occupation material was swept away or even crushed by the force of the storm (Bøgh et al. 2014a: 138). There was, however, some material culture recovered including partially glazed barrel tiles (sometimes referred to as Monk and Nun style tiles) probably from the roof of the stone-built house (Bøgh et al. 2014a: 171).

Vosborg 1 and 2 were close to one another, the Nissum Fjord and the Gammelå (Map 2). They were further similar in that they consisted of houses built upon constructed platforms. Where they differ is in how they were impacted by the storm surge of 1593. While Vosborg 1 was destroyed, Vosborg 2 was merely damaged and this allowed the occupants to salvage some material and reuse that material in the construction of a new manor (Vosborg 4, discussed below). The archaeological exploration of Vosborg 2 revealed that more of the manor house was preserved within the landscape – for example, the size and shape of the platform were identifiable in cropmarks (Figure 4). Vosborg 2 appears to have consisted of one platform with buildings arranged in a courtyard layout. The buildings were constructed of varying materials including timber and brick with tiled roofs. Small finds confirmed that Vosborg 2 was constructed in the mid-sixteenth century, after Gregers Holgersen Ulftands inherited part of the original manor. These include green-glazed tiles probably from a stove, window glass and a coin from 1552 (Bøgh et al. 2014a: 268). The material culture suggests it was high-status manor, as expected of the early modern nobility, but also that it was only briefly occupied before the destructive storm surge occurred.

Vosborg 1 and Vosborg 2 were both impacted by the storm surge in 1593 although the extent of the damage varied at each manor house. Within the archaeological remains, recovered during excavations, there is evidence of resilience and adaptation. Both examples reveal different ways in which the occupants' lives incorporated resilience and adaptation as they managed living in the watery landscape that was at times unpredictably violent. These specific examples will be discussed below.

The Moving Manors and Adaptation in Sixteenth-Century Denmark

Figure 4. Aerial image looking north showing the remains of Vosborg 2 in cropmarks. Source: Holstebro Museum.

Discussion: Detecting Adaptation in the Past

Adaptation, as accepted in archaeological research, is the capacity to adjust in response to external changes while resilience is the ability to retain some characteristics in the face of modification. These definitions will be a lens to explore the excavations of Vosborg 1 and 2 and used to consider the lives of the occupants and their responses to the 1593 storm surge.

At Vosborg 1, the stone-built house was atop a platform constructed of sand, close to the Nissum Fjord in the west and the Gammelå (original course of the River Storå) to the south (Map 2 and Figure 1). The edge of the platform was reinforced with timber poles and planks in a grooved layout, which suggests a deliberate construction method aiding greater resistance against the water (Kock 2015). The platform edge was double-walled and in-filled to make it robust. The timber bridge between the inner and outer platforms was dendrochronologically dated to c. 1341 and there was evidence of several stages of repairs carried out between the fourteenth and sixteenth centuries (Bøgh et al. 2014a: 127; Kock 2015). This could suggest other climatic events occurred prior to 1593 leading to the need for restoration, or may simply be indicative of general wear. Either way it demonstrates how people responded to the environment and how they acted to make themselves resilient. This is evidence of continuation, overcoming either wear-and-tear or damage and maintaining the site for occupation. By reinforcing the platform, several generations of occupants were exerting their capacity to mitigate. This evidence suggests ongoing resilience and adaptation over 200 years. The 1593 storm surge presented the occupants with a disastrous scenario to which they adapted. After Vosborg 1 was destroyed, construction began at Vosborg 3, the current and extant Nørre Vosborg, described below. Although the site of Vosborg 1 was abandoned after the storm surge, highlighting its vulnerability, the building of Vosborg 3 demonstrates the occupants' capacity to adjust. They did so while maintaining elements of their way of life, such as their occupation of and residence in Jutland.

At Vosborg 2 there is little evidence of resilience through maintenance as seen at Vosborg 1. This may be due to the short occupation of Vosborg 2 as it was built c. 1551, just over forty years before its abandonment in 1593. Resilience and adaptation are detected at this site in the evidence for material being removed and possibly reused for the construction of a new manor house, Vosborg 4 (Kock 2015: 20). The excavation revealed instances of stone-robbing in response to the 1593 storm surge. Stone-robbing is when stone is removed for a new purpose. This is usually attributed to economic or practical incentives; that is, rather than sourcing materials and incurring the cost of doing so, materials that can be reused, are removed and repurposed. Stone-robbing was a widespread practice throughout the medieval and early modern periods even in areas where stone was readily available and plentiful. Pre-cut stone was expensive, while importing non-local varieties, often used for articulation of apertures, was a costly endeavour as well as a status indicator. While finance was likely a motivator, there were also social and cultural meanings to a building's fabric which led to the reuse of its material components.

At Vosborg 2 there were two clear instances of stone robbing. The four ranges surrounded a courtyard constructed of a mettled surface; that is, a stone-packed surface. These are usually relatively small stones, often naturally rounded cobbles or water-rolled pebbles, rather than deliberately dressed (hand-cut). At Vosborg 2 the mettled surface had been removed after the manor was damaged (Kock 2015: 20). In addition, pits which had held large stone sills (supporting stones which provided a dry course to prevent rising damp) were also revealed (Kock 2015). These were the foundational stones of the buildings, providing not only a dry course but support for the materials laid above. The empty pits indicated that the sills therein had been removed. There were, therefore, two very different types of stone removed: the small, readily available stones which created the mettled surface of the courtyard plus the large, fundamental stones necessary for constructing a dry and robust building. Further building fabric, including large quantities of brick, was rediscovered in the silted-up remains of the Gammelå. This suggests that intact and useful material was transported by boat away from the damaged site (Kock 2019). This suggests that the occupants

The Moving Manors and Adaptation in Sixteenth-Century Denmark

extracted from the damaged manor house any material that could be removed or possibly any material that would assist in the quick construction of a new manor house. By transporting the material from Vosborg 2 along the Gammelå, the occupants employed what made their manor house vulnerable – the watery landscape – to aid their resilience.

Material culture found in-situ at Vosborg 2 suggests that the manor house was considerably damaged and not everything could be taken. This correlates with contemporary accounts from Aalborg and Copenhagen which recorded that the manor house was severely damaged and the owner was to receive royal compensation to reconstruct fallen buildings (Kock 2019). A tile depicting the moon god Luna, glazed stove tiles, window tiles and a coin minted in 1552 were rediscovered through excavation (Bøgh et al. 2014a: 268–79). Some of these items, such as the decorated tiles, appear to have been abandoned by the occupants of Vosborg 2 as they were broken, while those which survived the storm surge were removed. The majority of the material culture left at the manor house, and subsequently recovered by archaeologists, was within the basement level which could suggest it was left because the basement had flooded as a result of the storm surge. This demonstrates the coexistence of resilience and vulnerability when the manor house was impacted by the storm surge.

Further evidence of adaptation and resilience is indicated through how the occupants responded after the 1593 storm surge and the construction of Vosborg 3 and Vosborg 4. After Vosborg 1 was destroyed, it was rebuilt from scratch at the current and extant Vosborg 3. This is now called Nørre Vosborg and it is located one kilometre inland from the original 1 and 2. It is a multi-period complex, which is architecturally significant due to its courtyard ranges. The earliest known building dates to the turn of the seventeenth century suggesting that it was constructed immediately after the storm surge. The stepped gable of eastern range is indicative of the early modern period (Figure 5). The speed with which the new manor house was constructed indicates the family's wealth and access to resources; that is, the tools equipping them with the ability to adapt. It further implies a readiness to adapt, which is a key feature of resilience. The earliest range at Vosborg 3 was built upon a constructed platform within a series of banks and moats to create a manor house similar in form to Vosborg 1. This change of location and continuity of layout can be viewed as adaptation and resilience; that is, the manor house was built anew while following, or even replicating, the construction approaches utilised at Vosborg 1 and 2.

Figure 5. Photograph of the extant Vosborg 3 (Nørre Vosborg 3). The range with the stepped gable is the earliest range and dates to the end of the sixteenth century. Source: Sarah Kerr.

The adaptation detected did not appear to cease at the end of the sixteenth century, shortly after the storm surge; rather, it seems recollection of the earlier manor houses was maintained by the occupants of Vosborg 3. This is demonstrated through the additional ranges added in the seventeenth, eighteenth and nineteenth centuries to create the courtyard plan seen today (Bøgh et al. 2014a). While the creation of courtyard houses built upon raised platforms can be, in part, attributed to the fashions of the late medieval and early modern periods and the practicality of living in such a way, it connects the manor houses from the earliest iteration of Vosborg 1 built c. 1341 and the last addition to Nørre Vosborg 3 in the nineteenth century. The continuation in the form (courtyard layout) and function (elite residence) of the building, and the identity it displayed, is evidence of resilience as these were maintained after the change (the storm surge) occurred. These tangible and intangible connections across time and space, enacted as a response to an extreme environmental event, can be determined as evidence of adaptation.

The final manor house in the sequence is Vosborg 4, built further inland than any of the previous manor houses and two kilometres away from Vosborg

The Moving Manors and Adaptation in Sixteenth-Century Denmark

1 and Vosborg 2. Unlike Vosborg 3, it is not extant and nor has it been explored archaeologically; thus a courtyard plan cannot be determined. In an area called Sonder Vosborg there is a modern farm complex of the same name. While there are no extant historic building remains, there are remnants of a considerable moat, thought to have been constructed c. 1600 and no later than 1604, surrounding a modern garden. This suggests that in the vicinity of today's Sonder Vosborg there was an early modern manor house, Vosborg 4, constructed shortly after the 1593 storm with recycled materials from Vosborg 2, as indicated from the stone robbing discussed above (Bøgh et al. 2014a: 21; Kock 2015: 19; Kock 2019).

As with the adaptation of Nørre Vosborg 1 to Nørre Vosborg 3, the transition from Sonder Vosborg 2 to Sonder Vosborg 4 continues the use of the name 'Vosborg'. Sustaining the name was imperative to representing the family's identity. The manor house was a focal point of the estate representing the family's power over the tenants and as such the elite's relative higher status. The name of the manor, as well as the family's name, was indicative of their permanence in the area. This was vital for maintaining control over the estate and its tenants but also displaying legacy beyond the boundaries to the remainder of early modern Denmark. Once again vulnerability is evident here, coexisting with adaptation and resilience. If the family did not maintain the Vosborg name, with its longstanding identity, they might have appeared as *nouveau riche* who lacked the prestige of longevity. Adaptation and resilience are also detected in the reuse of the Vosborg name. Sustaining the name created a connection between Vosborg 2 and Vosborg 4; it reinforced a continuation of the family's status, power and control. Therefore, the family reorganised the manor's physical surroundings yet retained the identity embedded in the name of their manor. This intangible connection to the previous Vosborg manor house occurred as the physical manor changed through rebuilding; as such, it is evidence of mitigating the drastic changes that occurred as a result of the storm surge.

The archaeological evidence points to the resilience of the elite families who owned the Vosborg manors and reveals little about that of the estates' tenants. This is in part due to the focus of the archaeological excavations – the manor houses – rather than any other part of the estate, such as peasant houses, agricultural land use or material culture discovered elsewhere. However, it also highlights the differing abilities to adapt in the early modern period. The elites of the Vosborg manors had the social and financial resources to respond to the storm surge with change and continuity. They had far greater resources to do so and likely adapted at the expense of the lower-status occupants of the estate. This hints at some of the essential qualities required for effective adaptation.

The emphasis of this discussion has been on identifying adaptation and resilience in the archaeological and architectural remains of the Vosborg manors. However, there is also evidence for vulnerability and these concepts are not sharply defined or easily disentangled. The vulnerability of the manors and their occupants is manifested archaeologically through much of the same evidence that points to their resilience. The visible impact of the storm surge, such as the fallen façade, destroyed material culture, flooded basement and so on indicates a highly vulnerable site. The buildings were too close to the Nissum Fjord and North Sea, exposed to high water levels and not sufficiently robust to withstand storm conditions. These physical conditions made the occupants vulnerable to the surrounding environment. This does not negate the resilience also evident in this case study – rather it demonstrates the complexity of identifying these concepts within the archaeological record. This is demonstrated through the use of the Gammelå to transport material from the damaged Vosborg 2 to the new, more resilient site of Vosborg 4. The watery landscape surrounding the Vosborgs contributed to both vulnerability and resilience.

Adaptation and resilience, as adopted in this study, are typified by the ability to absorb change while maintaining some things the same. The story of the four Vosborgs is centred on change but also continuity. The change is more immediately obvious: the destruction and damage of Vosborg 1 and Vosborg 2 and the rebuilding of new manor houses further inland away from the Nissum Fjord. However, continuity is also evident most tangibly through the resurrection of various building materials at Vosborg 2 such as the mettled surface and the stone sills. This stone was probably reused to aid the construction of Vosborg 4, built c. 1600 after the storm surge impacted Vosborg 2, heightening the sense of tangible continuity between the manors. There is also continuity detected in the reuse of the name Vosborg, specifically Nørre Vosborg for 1 and 3 and Sonder Vosborg for 2 and 4. Vosborg derives from Old Norse, meaning 'mouth of a water outlet' (Nørre Vosborg 2024; Nordvestkysten 2024). The reuse of the name at Vosborg 3 and 4 is despite the meaning of the term being somewhat incompatible with the new locations; that is, Vosborg 3 and 4 were built at a considerable distance from both the Gammelå and the Nissum Fjord. It appears that, while the physical distance was extended with the new constructions, the intangible relationship between the watercourses and the occupants was sustained into the seventeenth century, suggesting both continuity and resilience.

Conclusion

This chapter suggests that there is evidence of adaptation and resilience in sixteenth-century Denmark, demonstrated through the example of Nørre Vosborg. It accepts that there are overlapping and sometimes conflicting definitions of adaptation and resilience, some of which are discussed here. It was established

The Moving Manors and Adaptation in Sixteenth-Century Denmark

that clearly stating a specific definition is important for clarity within research and therefore it was highlighted that definitions used in this chapter draw on the main consensus within archaeological research and policy. Adaptation – that is, the ways in which humans adjust their cultural systems in response to a changing external environment – and resilience, the ability to change and reorganise while retaining some characteristics, were detected in the Nørre Vosborg case study.

Nørre Vosborg 3 exists today as a manor house with extant remains dating to the end of the sixteenth century; however, it is part of a larger story involving four manor houses, Vosborgs 1–4. A destructive storm surge impacted the west coast of Jutland in 1593 and caused considerable damage to the earliest two manor houses, Nørre Vosborg 1 and Sonder Vosborg 2. The archaeological and architectural evidence suggests that Vosborg 1 was destroyed and rebuilt at Nørre Vosborg 3, further inland from the Nissum Fjord and North Sea. Vosborg 2 was also impacted although not to the same extent as Vosborg 1. The excavation of Vosborg 2 suggests some material was removed, transported along the Gammelå and reused to build Sonder Vosborg 4, the furthest inland of any of the manor houses. It was discussed that this is evidence of adaptation and resilience. This is particularly evident in the act of absorbing change while maintaining some characteristics, as demonstrated through the abandoning of the original manor house while retaining some of its materiality. Similarly, the building of Vosborg 3 utilised the same form and plan that had existed at Vosborg 1, which also hints at the balance of change and continuity. The use of the name Vosborg across all four manor houses was also discussed. Meaning 'mouth of the water outlet', Vosborg 3 and 4 were built much further inland than the original two manor houses, creating a mismatch between the name and the new locations. This suggests resilience and adaptation as the occupants adapted to their new settlements with the previous locations embedded within the names of the manors.

Bibliography

Adaptation Committee. 2016. *UK Climate Change Risk Assessment 2017 Evidence Report*. London: Adaptation Sub-Committee of the Committee on Climate Change.

Bassett, T.J. and C. Fogelman. 2013. 'Déjà vu or something new? The adaptation concept in the climate change literature'. *Geoforum* 48: 42–53. https://doi.org/10.1016/j.geoforum.2013.04.010.

Bøgh, A., H. Henningsen and K. Dalsgaard. 2014a. *Få Nørre Vosborg i tid og rum. Borg og herresæde*. Aarhus: Aarhus universitetsforlag.

Bøgh, A., H. Henningsen and K. Dalsgaard. 2014b. *Få Nørre Vosborg i tid og rum. herregård i lyst og nød*. Aarhus: Aarhus universitetsforlag.

Borrero, L.A. 2014. 'Adaptation in archaeology'. In C. Smith (ed.) *Encyclopedia of Global Archaeology*. New York: Springer. pp. 23–26. https://doi.org/10.1007/978-1-4419-0465-2_251.

Burton, I. 2008. 'Climate change and the adaptation deficit'. In E.L.F. Schipper and I. Burton I. (eds), *The Earthscan Reader on Adaptation to Climate Change*. Abingdon: Routledge. pp. 89–97. https://www.routledge.com/The-Earthscan-Reader-on-Adaptation-to-Climate-Change/Schipper-Burton/p/book/9781844075317.

Calvin, K. et al. 2023. *AR6 Synthesis Report: Climate Change 2023*. Accessed June 24, 2024. https://www.ipcc.ch/report/ar6/syr/.

Cassar, J. 2016. 'Climate change and archaeological sites: Adaptation strategies'. In R.-A. Lefèvre and C. Sabbioni (eds), *Cultural Heritage from Pollution to Climate Change*. Bari: Edipuglia. pp. 119–127.

Clark, P. (ed.). 1985. *The European Crisis of the 1590s: Essays in Comparative History*. London: Harper Collins Publishers Ltd.

CORDIS. 2023. 'Culture, heritage and identities: Impacts of climate change in North West Europe'. Accessed 27 June 2024. https://cordis.europa.eu/project/id/895147.

Corrigan, T. 2017. 'Defining adaptation'. In T. Leitch (ed.) *The Oxford Handbook of Adaptation Studies*. Oxford: Oxford University Press. pp. 23–35.

Costanza, R., L. Graumlich, W. Steffen et al. 2007. 'Sustainability or collapse: What can we learn from integrating the history of humans and the rest of nature?' *AMBIO: A Journal of the Human Environment* 36 (7): 522–527.

Daly, C. 2019. *Built & Archaeological Heritage Climate Change Sectoral Adaptation Plan: Prepared under the National Adaptation Framework*. Dublin: Department of Culture, Heritage and the Gaeltacht. https://assets.gov.ie/75639/a0ad0e1d-339c-4e11-bc48-07b4f082b58f.pdf.

Desjardins, S.P.A., T.M. Friesen and P.D. Jordan. 2020. 'Looking back while moving forward: How past responses to climate change can inform future adaptation and mitigation strategies in the Arctic'. *Quaternary International* 549: 239–248.

European Union. 2009. *White Paper. Adapting to Climate Change: Towards a European Framework for Action*. Brussels: Commission of the European Communities. https://www.eea.europa.eu/policy-documents/white-paper-adapting-to-climate.

Fankhauser, S. 2017. 'Adaptation to climate change'. *Annual Review of Resource Economics* 9 (2017): 209–230.

Fluck, H. 2016. *Climate Change Adaptation Report*. Swindon: Historic England. https://historicengland.org.uk/research/results/reports/28-2016 .

Folke, C. et al. 2010. 'Resilience thinking: Integrating resilience, adaptability and transformability'. *Ecology and Society* 15 (4). https://www.jstor.org/stable/26268226.

Geopark Vestjylland. 2024a. 'Nissum Fjord. Geopark Vestjylland'. Accessed 24 June 2024. https://www.geoparkvestjylland.com/geopark/geosites/nissum-fjord.

Geopark Vestjylland. 2024b. 'Nørre Vosborg. Geopark Vestjylland'. Accessed 25 June 2024. https://www.geoparkvestjylland.dk/geopark/geosites/noerre-vosborg.

Handmer, J.W. and S.R. Dovers. 1996. 'A typology of resilience: Rethinking institutions for sustainable development'. *Industrial and Environmental Crisis Quarterly* **9** (4): 482–511.

Heitz, C. 2021. 'On theorizing vulnerability for archaeology'. Accessed 24 January 2025. https://zenodo.org/records/5513463.

Holling, C.S. 1973. 'Resilience and stability of ecological systems'. *Annual Review of Ecology, Evolution, and Systematics* 4 (1973): 1–23.

Holling, C.S. and L.H. Gunderson. 2002. 'Resilience and adaptive cycles'. In L.H. Gunderson and C.S. Holling (eds), *Panarchy. Understanding Transformations in Human and Natural Systems*. Washington: Island Press. pp. 25–62. https://hdl.handle.net/10919/67621.

IPCC. 2012a. 'Glossary of terms'. In C.B. Field, V. Barros, T.F. Stocker et al. (eds), *Managing the Risks of Extreme Events and Disasters to Advance Climate Change Adaptation*. Cambridge: Cambridge University Press. pp. 555–564. https://archive.ipcc.ch/pdf/special-reports/srex/SREX-Annex_Glossary.pdf.

IPCC. 2012b. *Managing the Risks of Extreme Events and Disasters to Advance Climate Change Adaptation: Special Report of the Intergovernmental Panel on Climate Change*. Cambridge: Cambridge University Press. https://www.cambridge.org/core/product/identifier/9781139177245/type/book.

Jackson, R.C., A.J. Dugmore and F. Riede. 2018. 'Rediscovering lessons of adaptation from the past'. *Global Environmental Change* 52: 58–65. https://doi.org/10.1016/j.gloenvcha.2018.05.006.

Jordan, W.C. 1996. *The Great Famine: Northern Europe in the Early Fourteenth Century*. Princeton, NJ: Princeton University Press.

Kerr, S. 2020. 'The future of archaeology, interdisciplinarity and global challenges'. *Antiquity* 94 (377): 1337–1348. https://doi.org/10.15184/aqy.2020.138.

Kerr, S. 2023. 'Citizen science and deep mapping for climate communication: A report on CHICC'. *Journal of Community Archaeology and Heritage* **10** (2): 82–89. https://doi.org/10.1080/20518196.2022.2051139.

Kerr, S. and F. Riede. 2022. 'Upscaling local adaptive heritage practices to internationally designated heritage sites'. *Climate* **10** (7): 102. https://doi.org/10.3390/cli10070102.

Kirch, P.V. 1980. 'The archaeological study of adaptation: Theoretical and methodological issues'. *Advances in Archaeological Method and Theory* 3: 101–156.

Kock, J. 2015. 'Nørre Vosborg – en kamp mod elementerne'. *Skalk* 3: 18–28. https://pure.au.dk/portal/en/publications/n%C3%B8rre-vosborg-en-kamp-mod-elementerne.

Kock, J. 2019. 'Stormflodernes tid – klima, natur og flytningerne af Nørre Vosborg i 1500-tallet'. Accessed 26 January 2025. https://danmarkshistorien.dk/vis/materiale/stormflodernes-tid-klima-natur-og-flytningerne-af-noerre-vosborg-i-1500-tallet.

Kristensen, S.P. 2003. 'Multivariate analysis of landscape changes and farm characteristics in a study area in central Jutland, Denmark'. *Ecological Modelling* **168** (3): 303–318. https://doi.org/10.1016/S0304-3800(03)00143-1.

Løvschal, M. 2022. 'Retranslating resilience theory in archaeology'. *Annual Review of Anthropology* 51 (2022): 195–211. https://doi.org/10.1146/annurev-anthro-041320-011705.

McClain, A. 2012. 'Theory, disciplinary perspectives and the archaeology of later medieval England'. *Medieval Archaeology* **56** (1): 131–170. https://doi.org/10.1179/0076609712Z.0000000005.

Middleton, G. 2017. 'The show must go on: Collapse, resilience, and transformation in 21st-century archaeology'. *Reviews in Anthropology* 46: 1–28. https://doi.org/10.1080/00938157.2017.1343025.

Nelson, M.C., M. Hegmon, K.W. Kintigh et al. 2012. 'Long-term vulnerability and resilience: Three examples from archaeological study in the southwestern United States and northern Mexico'. In J. Cooper and P. Sheets (eds), *Surviving Sudden Environmental Change. Answers From Archaeology*. Denver: University Press of Colorado. pp. 197–222. https://www.jstor.org/stable/j.ctt1wn0rbs.13.

Nordvestkysten. 2024. 'Nørre Vosborg Manor. Visit Nordvestkysten'. Accessed 27 June 2024. https://www.visit-nordvestkysten.com/northwest-coast/whatson/norre-vosborg-manor-gdk606918.

Nørre Vosborg. 2024. 'The Story of Nørre Vosborg'. Accessed 27 June 2024. https://nrvosborg.dk/the-story/?lang=en.

O'Brien, M.J. and T.D. Holland. 1992. 'The role of adaptation in archaeological explanation'. *American Antiquity* **57** (1): 36–59.

Olwig, K.R. 1996. 'Environmental history and the construction of nature and landscape: The case of the "landscaping" of the Jutland heath'. *Environment and History* **2** (1): 15–38. https://doi.org/10.3197/096734096779522464.

Parker, G. 2018. 'History and climate: The crisis of the 1590s reconsidered'. In C. Leggewie and F. Mauelshagen (eds), *Climate Change and Cultural Transition in Europe. Climate and Culture*, Volume 4. Leiden: Brill. https://brill.com/display/book/edcoll/9789004356825/B9789004356825_006.xml.

Pfister, C., R. Brázdil, R. Glaser et al. 1999. 'Daily weather observations in sixteenth-century Europe'. *Climatic Change* **43** (1): 111–150. https://doi.org/10.1023/A:1005505113244.

Pittock, B. and R. Jones. 2000. 'Adaptation to what and why?' *Environmental Monitoring and Assessment* **61** (1): 9–35. https://doi.org/10.1023/A:1006393415542.

Primeau, C., P. Homøe and N. Lynnerup. 2019. 'Temporal changes in childhood health during the medieval Little Ice Age in Denmark'. *International Journal of Paleopathology* 27: 80–87.

Redman, C. and A. Kinzig. 2003. 'Resilience of past landscapes: Resilience theory, society, and the *longue durée*'. *Conservation Ecology* **7** (1): 14. https://www.ecologyandsociety.org/vol7/iss1/art14/.

Redman, C.L. 2005. 'Resilience Theory in archaeology'. *American Anthropologist* **107** (1): 70–77.

Riris, P. et al. 2024. 'Frequent disturbances enhanced the resilience of past human populations'. *Nature* **629** (8013): 837–842.

Schipper, E.L.F. and I. Burton (eds). 2008a. *The Earthscan Reader on Adaptation to Climate Change*. Abingdon: Routledge. https://www.routledge.com/The-Earthscan-Reader-on-Adaptation-to-Climate-Change/Schipper-Burton/p/book/9781844075317.

Schipper, E.L.F. and I. Burton. 2008b. 'Understanding adaptation: Origins, concepts, practice and policy'. In E.L.F. Schipper and I. Burton (eds), *The Earthscan Reader on Adaptation to Climate Change*. Abingdon: Routledge. pp. 1–8. https://www.routledge.com/The-Earthscan-Reader-on-Adaptation-to-Climate-Change/Schipper-Burton/p/book/9781844075317.

Scott, A.B. and R.D. Hoppa. 2019. 'The ice age with little effect? Exploring stress in the Danish Black Friars cemetery before and after the turn of the 14th century'. *International Journal of Paleopathology* 26: 157–163.

Smit, B., I. Burton, R. Klein and R. Street. 1999. 'The science of adaptation: A framework for assessment'. *Mitigation and Adaptation Strategies for Global Change* 4: 199–213.

Smithers, J. and B. Smit. 1997. 'Human adaptation to climatic variability and change'. *Global Environmental Change* 7 (2): 129–146.

Svensson, E. 2019. 'Crisis or transition? Risk and resilience during the Late Medieval agrarian crisis'. In Niall Brady and Claudia Theune (eds), *Settlement Change Across Medieval Europe: Old Paradigms and New Vistas*. Leiden: Sidestone Press. pp. 171–181. https://urn.kb.se/resolve?urn=urn:nbn:se:kau:diva-74835.

Tainter, J. 1988. *The Collapse of Complex Societies*. Cambridge: Cambridge University Press.

Tubi, A., L. Mordechai, E. Feitelson, P. Kay and D. Tamir. 2022. 'Can we learn from the past? Towards better analogies and historical inference in society-environmental change research'. *Global Environmental Change* 76: 102570.

UNESCO. 2024. UNESCO Global Geoparks. https://www.unesco.org/en/iggp/geoparks/about.

UNFCCC. 1999. Report of the Intergovernmental Negotiating Committee for a Framework Convention on Climate Change on the work of the second part of its fifth session. New York. https://unfccc.int/documents/942.

Van Pool, T.L. 2002. 'Adaptation'. In J.P. Hart and J.E. Terrell (eds), *Darwin and Archaeology*. Westport: Bergin and Garvey. pp. 15–28

Yasur-Landau, A., G. Gambash and T.E. Levy (eds). 2024. *Mediterranean Resilience: Collapse and Adaptation in Antique Maritime Societies*. Sheffield, UK Bristol, CT: Equinox Publishing Ltd.

Yoffee, N. and G.L. Cowgill (eds). 1991. *The Collapse of Ancient States and Civilizations*. Tucson: University of Arizona Press.

Sarah Kerr

The Author

Sarah Kerr is a lecturer in archaeology and a member of the Radical Humanities Laboratory, University College Cork, Ireland. She obtained her Ph.D. from Queen's University Belfast, UK, and held Postdoctoral Fellowships at Katholieke Universiteit Leuven, Belgium, and Trinity College Dublin, Ireland, before she started teaching archaeology and heritage at The University of Sheffield, UK. She held a Marie Skłodowska-Curie Actions Fellowship at Aarhus University, Denmark, before joining UCC in September 2023. She is a medieval archaeologist and heritage specialist, interested primarily in the built environment and how buildings were the products of social norms and expectations and how, in return, they were agents that shaped everyday life. Her second monograph, *Late Medieval Lodging Ranges: The Architecture of Identity, Power and Space*, was published by The Boydell Press in 2023.

Chapter 6

ARCHITECTURAL CLIMATE CHANGE ADAPTATIONS IN LITTLE ICE AGE NORWAY c. 1300–1550

Kristian Reinfjord

Abstract

Dwellings interact with climates to suit different temperatures, rainfalls and rainfall conditions. Different technologies were adapted to buildings to manage colder climates and more snow and therefore water during the Little Ice Age. Adaptations are identified in the archaeological material, particularly in high-status buildings from fifteenth-century Norway. Medieval architecture alterations were entangled with climate changes. New technologies accommodated lower temperatures and water increase. Building campaigns dated to the period could also have been involved in several societal developments (e.g., technology, ritual, social patterns or consumption) that are also revealed in built environments. The examples presented correspond chronologically with the Little Ice Age, are secondarily added to a dwelling structure, are directly associated with changing climates and serve as a solution to climate-related problems. Four features are particularly significant instances of climate change adaptation and are here discussed: drains, brick rebuilding, tile stoves and stone cellars.

Introduction

As far as possible, humans seek comfortable dwelling conditions. Mild temperatures and low humidities are preferred for indoor climates. As with people today, medieval and early modern dwellers were conscious of their surroundings. This chapter argues that changing climates during the Little Ice Age could have initiated architectural adaptations to secure comfort. As such, climate was an actant in architectural technology networks. For example, a fourteenth-century manuscript from Sogn in Western Norway (*Diplomatarium Norvegicum* 4, 198) bears witness to a seller who was to deliver his house to a buyer in perfect

doi: 10.63308/63881023874820.ch06

condition. Erling of Lomheim left his house in good order for its new owner, including an attic, a cellar and an operative drainage, dug out and closed. The deed of sale, witnessed by Thorkel Tang, Thorberg and Paal in Hvam, assured dry and comfortable living environments. This written source tells us that drains were considered important building features. But were there other architectural devices associated with living conditions and climates? I here investigate climate-initiated technologies in medieval architecture as an argument for climate change adaptations in Little Ice Age Norway.

Dwellings interact with their climates and must be designed to suit different temperatures, humidities and snow or rainfall conditions. Sudden climatic changes could therefore affect building technologies, designs and strategies. How did the Little Ice Age, with its colder climates (see e.g. Pfister and Wanner 2021; Wanner, Pfister and Neukom 2022), manifest itself in early modern built environments? Longer winters and colder springs caused more snow. Melting snow caused more accumulated water in and around buildings, which had to be managed. I here argue that climate change can be studied in architectural adaptations and developments of then-existing medieval buildings and monuments. Several architectural adaptations are identified in the archaeological material, particularly in high-status buildings from fifteenth-century Norway. Medieval stone architecture alterations, executed in the fourteenth to sixteenth centuries, were entangled with climatic changes which brought colder temperatures and water increase as a result of more melting snow.

Climate history and architecture adaptation is a little-explored field of research and climate is given a subordinate role in architectural history (for overviews see Huhtamaa and Ljunqvist 2021; Ljungqvist, Seim and Huhtamaa 2021). However, some examples exist, such as Gunhild Eriksdotter's studies on the indoor climate of Baroque castles and early modern architecture in Sweden (Eriksdotter 2013a, 2013b; Eriksdotter and Legnér 2015). Studies of medieval architecture in Norway often focus on the first phases of building and their medieval chronologies (e.g. Fischer 1951; Hommedal 1998). Late medieval and early modern alterations in medieval architecture are seldom studied. Supplied with archaeological field reports, finds and building remains, architecture proves to be a fruitful approach to investigate socio-natural entanglements of the Little Ice Age. No attempt is made to show every example of architectural adaptations in the period, but carefully chosen examples show that climate was in fact a driving force in building technology.

Architectural Climate Change Adaptations in Little Ice Age Norway

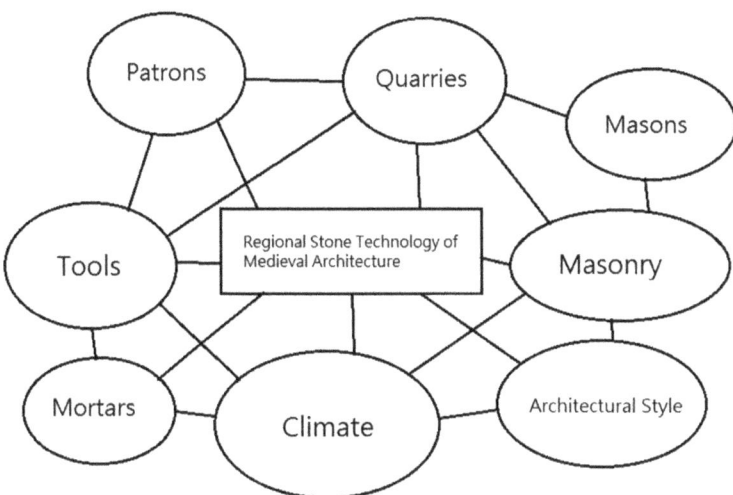

Figure 1. Climate as an actant in a network of architectural technology. Changing climates contributed to architectural adaptations and architecture, as did other 'building' materials and actants. Source: Kristian Reinfjord

Climate as Actant – Theorising Architectural Climate Adaptations

My research is grounded in an actor-network approach to the material culture of built environments, their technology and climatic implications. Entities such as climate, architectural style and technologies like heating devices and drains are seen as actants of a network. In architecture, configurations/actants are seen not just as the property of human actants, such as the masons or patrons of cathedrals, but rather as the effects of a networking activity between material and human actants that 'link in one continuous chain' (Latour 1993: 10–11). All actants together form a finished product, whether it is architecture or society, their relations, associations and alliances (Law 1991). This could apply to production or a technological process, as in the study of medieval stone technology. As suggested, climate is an important factor in architectural design. All entities link in a heterogeneous network where all actants are of importance in the finished product of architectural adaptations. The entanglement of humans, things and even external stressors, as well as their interactions and relationships, are all objects of interest and prerequisites for establishing a desired dwelling.

Buildings consist of different entities, of which climate could be one. Thus, it is not important what the different entities are, but what they do and, in turn, what is done to them. According to Bruno Latour (1996: 371), actants and entities can be distant or of no resemblance in the network and yet have actuality there. Distinctions of micro- versus macro-scale play no important part either, as the metaphor of connections replaces such distinctions. In early modern society, this could mean that an actant could come from abroad, such as a stylistic influence, or be local, such as a nearby stone quarry. As such, an entity could be a changing climate, rain increase or frost. At the same time, all entities would be of equal importance in the network of regional technologies. The actual physical size of actants should be of minor importance in the network. From a technological perspective, the large stone quarry could be of less importance to the result of a building than the sand grains in a mixed mortar for instance. Climate change affected architecture indirectly, by creating problems of water management and frost. Ideas of stylistic influence, existing in the heads of bishops or architects, were immaterial but played essential parts in erecting castles. Neither are actants inside or outside; rather 'the only question one may ask is whether or not a connection is established between two elements' (Latour 1996: 372). In such a manner, influences from abroad could be of greater importance than local influences, such as the tradition of stone building itself introduced to Norway from abroad in the medieval period.

By re-theorising how medieval society, and builders in particular, adjusted to climate change during the Little Ice Age, one can hypothesise that castles, halls and dwellings built in the medieval warm period were adapted to face colder weather, heavy rainfall and increased snow in the Northern Hemisphere during the late medieval and early modern periods. Building adaptations were actants in building networks and entangled with changing climates. It is my notion that these can be read in the archaeological material of secondary added architectural elements. As there are few or no written sources on climate change adaptations of medieval buildings in Norway, the material used here is found in the archaeology of *rebuilding* of Norwegian medieval stone architecture. Such buildings in the region were mostly erected in the period between 1130 and 1350, but were, however, changed to face changing climates and altered from 1350 to 1550. Most stone-building campaigns dated to the Little Ice Age period could also have been involved in several societal developments (e.g. technology, ritual, social patterns or consumption) also revealed in built environments. To be of value, the examples should correspond chronologically with the Little Ice Age, be secondarily added to a dwelling structure, be directly associated with changing climates and serve as a solution to climate-related problems. Four features are particularly significant instances of climate change adaptation: drains, brick rebuilding, tile stoves and stone cellars. These are found in adapted medieval castles of Norway.

Architectural Climate Change Adaptations in Little Ice Age Norway

Medieval Architecture Adapted: Climate-Initiated Technology Assets

As part of this project, I have selected some climate-adapted features in relevant buildings executed after around 1400 to show how the climate as an actant played a part in architectural adaptation. Material has been located both in standing building remains and ruins. Moreover, archaeological finds and excavation reports have been consulted. As wood was the main building material of medieval and early modern Norway, few preserved remains are found. Changing climate-related features could have been used in vernacular buildings as well, but are not preserved.

Stone buildings are found throughout Norway, and eight building complexes of kings, archbishops and bishops are of relevance. I here take advantage of relevant examples found in eastern Norway. These are castles and residences, all with dwelling compartments. In particular, the architecture changes radically in the early modern period in terms of room distribution and room size. It is uncertain if this change is climate-related. Nevertheless, large halls such as the thirteenth-century Royal Hall of Håkon Håkonsson in Bergen change during the fifteenth century into building complexes with many small rooms, as seen in Rosenkrantz Tower in Bergen, Akershus castle in Oslo or the Archbishop's residence Steinvikholm in Trøndelag. Early modern building complexes in Norway seem to have relied on many small rooms. These would have been easier to keep warm and dry in Little Ice Age Norway. These changes correspond with European trends but could have been chosen as adaptations to changing climates.

The late medieval and early modern periods also represent periods of architectural transition. Royal and ecclesiastical households moved from peripatetic institutions to ones based in fewer residences for longer periods. This important change of character involved more climate-driven building initiatives and adaptations. I identify four case studies as a response to architectural use and changed climates, read as selective environment designs (e.g. Hawkes, McDonald and Steemers 2002). Examples of four climate change-related adoptions are here presented, as found in the archaeological material: disposal of unwanted water, facilitating thermal comfort, efficient heating devices and facilitating frost-free storage.

Drains: Getting Rid of Unwanted Water

Different ways of leading unwanted water out of buildings and courtyards have been identified in the selected buildings. These can be dated to the late fifteenth century onwards and are found as secondary adaptations. In the simplest form, drains are laid out in courtyard cobblestone floors. Here long, open canals collect water and lead it out through openings in curtain walls. These canals are seen in the bishop's residence at Hamar, for instance. These canals are preserved in situ and are still working today, providing good examples of architectural climate adaptations. More complex drains are sometimes found within buildings. Closed canals of standing flagstones or brick were also installed after the first building phase to lead water out of castle dwellings. Examples are found in the archbishop's Steinvikholm in Trøndelag as well as the main tower of the bishop's residence in Hamar, eastern Norway. In both examples, canals were laid underneath wooden floors and led out of the buildings. In some cases, the canals led into water cisterns to be collected. In the corner tower at Hamar, secondary brick canals and closed drains led water away from dwelling spaces. Some of the examples can be studied in situ, but most remains are archaeologically documented during excavations. At the fourteenth-century Stonehouse at Granavollen, a drain was reinstalled in wood to lead water out of the building's cellar. At Steinvikholm, canals were even cut in soapstone and prefabricated to be installed after a fire in 1562 to enhance indoor comfort.

Examples reveal how the effects of changing climate materialised in buildings. At the Hamar bishop's residence, the building plot seems less favourable due to foundation and bedrock water challenges. Builders took action to tackle underground water issues and, despite the scarcity of material, the archaeological evidence from Hamar gives a glimpse into how early modern builders adapted their old buildings to lead unwanted water out of their dwellings. The medieval Hamar complex at Domkirkeodden is situated on a peninsula stretching into Lake Mjøsa. The buildings are situated on limestone bedrock with a variation in topsoil depth between ten centimetres and three metres (Pedersen 1999: 179). The limestone bedrock is often uncovered, with small exposed outcrops. The flaky limestone bedrock creates an uneven peninsula with natural stone diches collecting and channeling water. A small lake in the eastern end and small streams draining into Lake Mjøsa contributed to a challenging environment for erecting large stone buildings, and one especially vulnerable to climate change. The bishop's residence is located between the cathedral church and the small lake. It is cut into the bedrock in several places, and one would expect issues with unwanted water.

Distinct features of the castle courtyard include linear structures laid out on the cobblestone floor. These canals lead rainwater from the different building roofs within the castle to the courtyard and then further to the surrounding walls and out of the complex. By gravity flow, a hole in the northern side of the ring wall lets water out, and the cobblestone floor canal reaches it perfectly. Even today, these canals demonstrate great efficacy during heavy rains. Similar stone floors cover the tower cellar, a vaulted room cut into the bedrock. Several niches for storage of quality goods indicated the desire to keep the cellar dry. A

Architectural Climate Change Adaptations in Little Ice Age Norway

wooden cistern from a secondary building phase on the eastern side of the cellar room collects rainwater. Excavation photos from the 1950s show a wooden floor on top of the floor's stone cover, making a dry surface. No other drainage system seems to have been in use in the tower cellar.

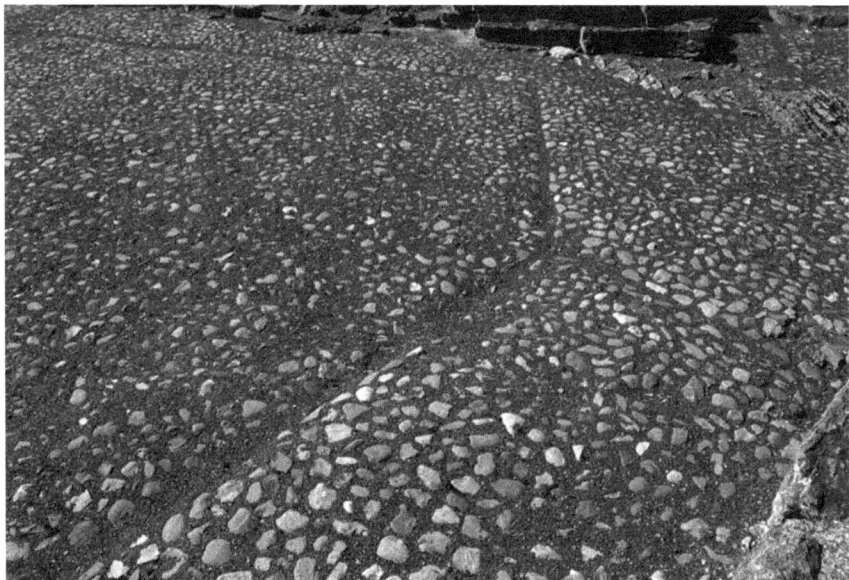

Figure 2. Inserting drains down into cobblestone floors to secure a drier courtyard was an architectural adaptation of the sixteenth century. Here as part of the Hamar bishop's residence. Source: Anno Domkirkeodden.

Another stone building in the Hamar bishop's residence, named the 'town hall', shows a more complex draining system incorporated into its cellars. Placed outside the fifteenth-century eastern curtain wall surrounding the complex, this large stone building was torn down when the castle was remodeled around 1450 (Sæther 1990: 21). Only two cellars survive, and the two rooms and a staircase show a vaulted ceiling and wall niches. Nicely cut details and portals suggest a high-status position in the castle. A document from 1359 (*Diplomatarium Norvegicum* 4, 399) mentions that the town hall lay within the bishop's residence structure. The Hamar chronicle also describes the building as a town hall (Arnesen 1937). The 'town hall' building has at least two phases and dates to the castle's first (c. 1190) and second phase (c. 1250), and the several wall niches suggest a storage function (Reinfjord 2023). When storing paper, wine or fine foods you do not want a damp cellar. The building, measuring 6 by 6.5 metres, cuts into the bedrock, and the floor is 2.4 metres beneath ground level.

On the cellar floor, we see a 25-centimetre-deep basin for collecting water. The 6 by 4 metre basin was covered by a wooden floor, which was laid on recesses around the cellar bottom and supported by foundation stones within the basin. On the southern wall, a drainage ditch leads water out of the building. However, the ditch only leads the top water from the basin, as indicated by its placement higher on the wall. One indication that the drainage was executed after the principal construction is the rough-cut stones in the wall. Between this large, vaulted cellar of the town hall and the eastern curtain wall, a gravity-flow canal forms a drainage structure to the southwest. This ditch was probably built to lead water away from the building's roof. This drain was constructed with flat stones at the bottom and on top and with standing flagstone walls. There is no contact between the drain and the building structure, making exact dating difficult. The drain breaks an older and smaller ditch, probably belonging to an older building phase, which suggests a secondary installment.

In the northeastern part of the bishop's residence lies the ruin of a big, square stone building. The structure was excavated in 1980–83 and interpreted as a 'servant's quarter' or a 'corner tower'. It dates to the period around 1250 and was integrated in the second phase complex (Reinfjord 2023). Three rooms make up the building: first, an entrance with two stairs, one leading to the courtyard and one to the above stories; second, a southern cellar-like vaulted storage room with niches and a window cut into the bedrock; and a third northern room with a large rectangular area. The vaulted room was probably a storage area, while the northern room was probably a service quarter, based on finds of horse equipment. The southern vaulted room, partly cut into the bedrock, ran into problems with moisture coming from the bedrock surfaces on the south side. To manage these water issues, a drainage canal was extended from the southern room all the way underneath the limestone floor and through the outer northern wall of the building. The structure cuts through the dividing wall between the rooms to lead water out. The canal heading northeast connects with a drain coming from the entrance room. This drain seems secondary, an interpretation based on how the canal is led through the wall and on the use of brick. The drainage systems are made of parallel lines of standing medieval seized bricks and at some points standing limestones set in clay. The canal width is approximately ten centimetres. Transverse bricks close the structure on top. A slight slope towards the northeast secured water flow. The drain breaks the northern wall by parallel standing bricks. The drain was adjusted and customised within the building, by adding a second drain from the entrance room.

Architectural Climate Change Adaptations in Little Ice Age Norway

Figure 3. Secondary built closed drains could have been a good way to a drier indoor climate. Here is an example from the northeastern corner tower of the bishop's residence in Hamar. Source: Anno Domkirkeodden.

Another example of drainage considered here is found in a dwelling stone house at Gran, located on the western side of Lake Mjøsa, about seven miles north of Oslo. Closely connected with the two medieval twin churches from the mid-1100s and a stone tower, the stone house was built around 1300 (Rosborn 2014). Earlier, the building was a corner tower as part of a larger structure with a high surrounding wall. The building uses gothic masonry, rises two stories above ground, and measures 12 by 8 metres. Placed on a small raised area with a bedrock foundation, the stone house has a cellar cut into the ground. The bedrock at Gran is relatively loose and lets water easily into building cellars. The cellar has a barrel-vaulted ceiling, reached by inside stone steps from the first-floor room. In 1972–1973, excavations underneath the cellar floor revealed drainage ditches within and outside the building (Rosborn 1976). Excavations took place around the building in 2018 and confirmed this (Friis 2020: 30). The drainage ditches cut into the bedrock with a north–south orientation. Other smaller ditches led water from other parts of the room to the main ditch. The published excavation report from 1976 does not show how the inside building drainage was built, but we can assume that it resembles the outside extension.

The 2018 investigations carried out by the Museum of Cultural History in Oslo show that the drainage system continues outside, leading water out of the building (Friis 2020). The drainage runs through the building wall, underground and through a masonry arch. The drainage ditch also suggests that the cellar floor of the Stonehouse had a wooden floor, as shown in the bishop's residence at Hamar. The system cuts deep into the bedrock and consists of a flat flagstone floor and standing stonewalls. The 1972 excavation showed a hollow log in the bottom of the ditch, the log was covered with roof tiles. On top of the structure, together with a clay pipe fragment, lay a medieval coin struck by Erik of Pommern in the 1430s. This suggests that the drain had been updated and elaborated to keep the cellars dry. However, in 2018, when the drainage was rediscovered, radiocarbon dating showed that the log was part of a 1700s restoration of the system.

Rebuilding in Brick: Facilitating Thermal Comfort

Medieval stone halls constituted a category of building used for both dwelling and political and formal purposes. These uses required good thermal comfort. As the material suggests, at least some of these were rebuilt in the fifteenth and sixteenth centuries with smaller openings (windows and doors) to help conserve warmth. Two examples serve to illustrate the point. The southern wing hall of the bishop's residence at Hamar was rebuilt with brick around 1500 (Reinfjord 2023). It is important to note that this was a dwelling quarter of the complex. Here, bricks were added to rework windows, as brick was used for openings that slanted inwards. The two portals of the preserved cellar rooms were also shaped with brick. The brick portals are probably secondary, corresponding with the

Architectural Climate Change Adaptations in Little Ice Age Norway

rebuilding of the windows. Gothic-style door jambs in Mjøsa limestone with toolmarks of the Hamar building lodge were reworked into the secondary door arrangements. Between the jambs, bricks were applied to cover the wall cores on both sides, which may be interpreted as an attempt to improve insulation. The portal stones seem to belong to the Gothic phase four (i.e. after 1430) of the residence, and they could have been altered by the use of brick. This is also seen in the reworked windows from the early sixteenth-century phase. Further, an internal wall staircase was closed by bricks in the eastern cellar opening.

Figure 4. In eastern Norway brick use in rebuilding contributed to warmer dwellings. At the southern wing hall at Hamar, all openings were narrowed down in order to make warmth stay inside rooms. Source: Kristian Reinfjord.

Kristian Reinfjord

Older Gothic openings were reduced, a possible strategy chosen to facilitate thermal comfort. Brick use was a rare technology in the region of eastern Norway and is, to my knowledge, often found in thermal comfort projects, as the rebuilding of openings suggests. Similar adaptation strategies can also be studied in the Akershus castle in Oslo, where the state business areas of King Christian IV had stoves and fireplaces installed during the seventeenth century (Sinding-Larsen 1924: 94). In the same castle, several windows were rebuilt smaller during the early modern period, as seen for instance in the Romerike hall or a closed door in the castle cellar (Sinding-Larsen 1924: 62, 69). Once cold air was kept out, indoor temperatures were raised to keep the Little Ice Age outside.

Tiled Stoves: Installing More Efficient Heating Devices

Tiled stoves, installed in dwellings during the fifteenth and sixteenth centuries, represented an important new heating technology and materialisation of climate actants. This technology was an improvement on medieval open fireplaces, since the tiled stove device consisted of a closed fireplace with a chimney. The oven collected heat in its tiles more effectively, since it had internal smoke canals. Remains of secondary installed fireplaces and tiled stoves can be identified in the archaeological material in Norway as glazed tile fragments. These are often decorated with green or orange glaze and iconographic surfaces. This fragmentary material is scattered throughout the studied buildings. It is contemporary with climate change during the Little Ice Age, often dating to the sixteenth century. Tile stoves cannot be studied in detail, considering the scarceness Norwegian material. Nevertheless, the frequency of tile in the period indicates that these tiled stoves were a climate adaptation made to the existing architecture. It may also have been a fashionable way to heat rooms, as well as part of architectural climate adaptation. I have found remains of tiled stoves in dwelling contexts dated to the early sixteenth century at Hamar and Oslo bishop's residences and in the sixteenth-century archbishop's Steinvikholm castle. This 'new' technology actant was a high-class choice of a new and more efficient way of heating rooms. Stove technology replaced open fireplaces in large open halls, as seen for instance in the archbishop's residence in Trondheim (Lunde 2023). Together with the building of smaller, energy-efficient rooms, tile stoves contributed to confronting changing climates. In addition to keeping residents warm, late medieval dwellers wanted their goods warm (or at least frost-free) during Nordic winters.

Architectural Climate Change Adaptations in Little Ice Age Norway

Figure 5. High-class home-owners installed tile ovens in their dwellings. These are found in archaeological contexts as glazed tile fragments. Here is an example from the Hamar bishop's residence (HKH-02017a-b), dated to the early sixteenth century. Source: Anno Domkirkeodden.

Eastern Norwegian Stone Cellars: Facilitating Frost-free Storage

In parallel with the Little Ice Age, the building of underground vaulted stone cellars accelerated. As these structures date to the period in question, I here argue that frost-free storage facilities were valued from the fourteenth century onwards as adaptations to changing climates. New archaeological evidence shows (e.g. Stige and Lindhart Bauer 2018) that stone cellars became popular in late medieval period towns. Yet stone cellars were also a rural phenomenon, and these sub-ground structures served both stone and wooden houses. As temperatures dropped, it proved a challenge to keep wine and other high-status goods from freezing. Thus, cellars could be interpreted as an architectural adaptations to changing climates.

Ten medieval stone cellars from Little Ice Age eastern Norway offer a distinct group of building remains. The rural stone cellar is a special eastern Norwegian phenomenon. Cellars, which were often associated with bishops and canons, were structures indicating high status. Underground stone structures were built beneath wooden houses at the bishop's residence at Hamar, where at least three cellars remain. At Hamar, preserved stone cellars have also been interpreted as remains of the cathedral school or the canons' residences. They comprise a large sub-soil stone complex of three cellar rooms, perhaps serving large timber buildings (Sæther 2005). Also associated with the Hamar bishop are three stone cellars found as part of the bishop's residence at Storøya in Tyrifjorden, south of Oslo. Canons' residence cellars are preserved at the rectories at Ringsaker and Hoff, and were parts of high-status building complexes and rich parishes, as the high-end eastern Norwegian basilica churches on these sites indicate. A prominent church site was also Granavollen at Hadeland where the Stonehouse at Hadeland, built in the vicinity of St. Nicolai's and St. Mary's Churches at Granavollen, has a cellar preserved, most likely a part of a medieval canon or priest residence. All eastern Norwegian stone cellars are connected to bishops or wealthy canons and are closely connected to the cathedral-building milieu.

All the rural medieval cellars seem to have been built in the Gothic period of the fourteenth century by one or two generations of masons. All the cellars are barrel-vaulted, and the vaults start their inward slanting between 0.5 metres and 1.6 metres up the wall, which is a distinct technological feature of these structures. The free-standing vaults are closed at each end with walls to make rooms. Most cellars have niches for the safekeeping or storage of precious goods, and could provide good storage conditions throughout long winters for delicate consumables, such as wine. The cellars are stylistically homogeneous. As particularly seen at Ringsaker, the two cellars at Storøya and in the Stonehouse at Granavollen fall into the same style. Here, the big cellar rooms are accompanied by a small stair room providing access to a main room. The common two-room plan contrasts with the simpler one-room cellars, as seen at Hoff, where only a single room is preserved. Although a second room might have been used at Hoff, it has not been preserved due to a secondary use of the cellars as a vegetable storage room from the eighteenth century onwards. Few of the cellars have their original openings preserved today, since they were often reused. The structures were desired throughout the early modern period, and into the nineteenth century (Reinfjord 2022) in order to preserve root vegetables, an activity that may also indicate climate adaptation. The most exclusive stone cellars are found as part of the canons' residences at Domkirkeodden, which had more than three rooms originally. These rooms are among the largest cellar rooms built in this period. Additional rooms might be connected to the three preserved rooms, but they remain unexcavated. The different rooms might originate from other periods, but their masonry and building technique seem to indicate that they were built in a single phase. These cellars were also reused as parts of eighteenth-century dwellings at the Storhamar manor at Domkirkeodden after 1716.

Architectural Climate Change Adaptations in Little Ice Age Norway

Figure 6. Building underground rooms enabled frost-free storage. Stone cellars could have been a solution to preserve high-end goods through the harsh winters of the Little Ice Age and beyond. The picture shows the cellar of the Hamar town hall. Source: Anno Domkirkeodden.

In addition to the rural examples discovered in the Hamar diocese, stone cellars were also used in medieval towns, as seen in Oslo (Stige and Lindhart Bauer 2018) and Bergen (Ekroll 1997: 139). A common feature is that the cellars often consisted of more than one room, and it has been suggested that they are remains of stone buildings used by town residents, challenging the notion that only the King and the Church built in stone in Norwegian medieval towns (Stige and Lindhart Bauer 2018). These are, however, preserved only in fragments. An intact cellar – originally a ground floor room in the Oslo bishop's residence – is the stone structure today found under the Ladegården. This large stone room, paved with cobblestones, displays a four-part groin vault, supported by a centrally placed column. The large room (10 by 11 metres internally) has two portals in the southern and western walls as well as a western window (Dahlin 1990: 112). This cellar does not resemble the barrel-vaulted multi-room cellars of Hamar diocese, and it can be argued that the Oslo cellar belongs to a different tradition, but was as much a climate-related initiative.

Kristian Reinfjord

Conclusions: Adapting to Changing Climates

At the dawn of the Little Ice Age, upper-class patrons modified their dwellings to accommodate colder climates. The present article has sought to investigate the impacts of changing climate on architecture in the early modern period and which actions were taken to adapt buildings. Climate has been presented as an actant in architectural technology networks. Technological development in the period was driven by the desire for drier and warmer indoor environments. It can be stated that climate was an actant in technology networks and contributed to architectural adaptations. By building smaller rooms, architecture changed as a whole. This could be a plausible explanation for the transformation of large open halls into multiple small lodgings. In the Nordic countries, in particular, winters were especially harsh, and snow increase caused more melting water in spring. Late medieval nobles also changed their lifestyles and adopted more permanent residences during the winters. This change encouraged a stronger architectural focus on indoor comfort. Technology was used to tackle climate change. Such climate actants were bricks used to reduce windows and doors, drains and more efficient heating devices such as tiled stoves. Moreover, the use of underground stone cellars became more widespread during the late medieval period. Just like today, historical homeowners adapted to changing climates to be more comfortable. Climate was an actant in architectural networks and an actant in technology that initiated and played a part in architectural development as adaptations were initiated by changing climates.

In this way, analysing the Little Ice Age as an actant illustrates past climate change adaptations and their consequences. I have here drawn attention to some examples of building strategies in the early modern period. Only some assets have been discussed, but a more in-depth study of the material will broaden our knowledge of climate change adaptations in the Little Ice Age. The fourteenth-century western Norwegian Erling Lomheim fixed his drain in order to enhance his house value, just as we add extra layers of paint, and perhaps add a bowl of fresh limes in the kitchen to help the estate agent boost sales profit. In the Little Ice Age, dry and comfortable indoor climates were crucial, both to sell at high prices and to keep high-class residents satisfied.

Bibliography

Arnesen, A. 1937. *Hamarkrøniken. Med andre kilder til kunnskap om det gamle bispesæte ved Mjøsen*. Oslo: Cammermeyers boghandel.

Dahlin, E. 1982. *Middelalderens bispegård i Oslo. En bygningsarkeologisk undersøkelse*. Unpublished MA thesis. Oslo: University of Oslo.

Ekroll, Ø. 1997. *Med kleber og kalk: norsk steinbygging i mellomalderen*. Oslo: Samlaget.

Eriksdotter, G. 2013a. 'Did the Little Ice Age affect indoor climate and comfort? Re-theorizing climate history and architecture from the Early Modern Period'. *Journal for Early Modern Cultural Studies* **13** (2): 24–42.

Eriksdotter, G. 2013b. 'När slottet blev beboeligt. Inneklimatets betydelse på Skoklosters slott under Wrangels tid'. *Bebyggelseshistorisk tidsskrift* **66**: 8–29.

Eriksdotter, G. and M. Legnér. 2015. 'Indoor climate and thermal comfort from a long-term perspective. Burmeister House in Visby, Sweden, c. 1650–1900'. *Home Cultures* **12** (1): 29–54. https://doi.org/10.2752/175174215X14171914084692.

Fischer, G. 1951. *Norske kongeborger*, vol. 1. Oslo: Cappelen.

Friis, E.K. 2020. *Steinbygning fra middelalder, Gran prestegård, 167/1, Gran, Oppland*. Unpublished Archaeological field report. Oslo: Museum of Cultural History.

Hawkes, D., J. McDonald and K. Steemers. 2002. *The Selective Environment: An Approach to Environmentally Responsive Architecture*. London: Spon.

Hommedal, A.T. 1998. *Erkebispegården i Trondheim. Rapport over de bygningsarkeologiske undersøkingane til Dorothea og Gerhard Fischer 1952–1972*. Trondheim: Riksantikvaren.

Huhtamaa, H. and F.C. Ljungqvist. 2021. 'Climate in Nordic historical research – a research overview and future perspectives'. *Scandinavian Journal of History* **46**: 665–695. https://doi.org/10.1080/03468755.2021.1929455.

Lange, Chr. C.A. and C.R. Unger. 1857. *Diplomatarium Norvegicum. Oldbreve til kundskab om Norges indre og ydre forhold, sprog, slægter, sæder, lovgivning og rettergang i middelalderen*. Fjærde samling, första halvdel. Christiania: P.T. Mallings forlagshandel.

Latour, B. 1988. *The Pasteurization of France*. Cambridge: Harvard University Press.

Latour, B. 1993. *We Have Never Been Modern*. London: Harvard University Press.

Latour, B. 1996. 'On actor-network theory: A few clarifications'. *Soziale Welt* **47**: 369–381.

Law, J. 1991. *The Sociology of Monsters: Essays on Power, Technology and Domination*. London: Routledge.

Ljungqvist, F.C., A. Seim and H. Huhtamaa. 2021. 'Climate and society in European History'. *Wiley Interdisciplinary Reviews: Climate Change* **12**: e691. https://doi.org/10.1002/wcc.691.

Lunde, Ø. 2023. *Erkebispegården ved Nidarosdomen*. Trondheim: Museumsforlaget.

Pedersen, E.A. 1999. 'Tusen års historie i kirkegårdens dyp – fra utgravningene ved Hamar domkirkeruin 1988–1992'. *Universitetets Oldsaksamling årbok* 1999: 177–204.

Pfister, C. and H. Wanner. 2010. *Climate and Society in Europe: The Last Thousand Years*. Bern: Haupt Verlag.

Reinfjord, K. 2022. 'Middelalderkjelleren på Østre Toten prestegård'. In O. Nøkleby (ed.) *Prestegarden på Toten frå middelalderen til vår tid*. Lena: Hoff menighetsråd. pp. 43–51.

Reinfjord, K. 2023. *Stone Building. Organization and Development of Construction Technology in Eastern Norway c. 1130–1537*. Unpublished Ph.D. thesis. Bergen: University of Bergen.

Rosborn, S. 1976. *Stenhuset och Tvillingkyrkorna i Gran, Hadeland. Ett medeltida kyrkokomplex i bygnadsarkeologisk belysning*. Unpublished report. Oslo: Riksantikvaren.

Rosborn, S. 2014. *Søsterkirkene og steinhuset på Granavollen: studier i eldre norsk steinbyggerkunst*. Toten: Stiftelsen Steinhuset på Gran.

Sinding-Larsen, H. 1924. *Bidrag til Akershus' slotts bygningshistorie i de første 350 aar 1300–1650 paa grunnlag av den bygningsarkeologiske undersøkelse 1905–1924*. Oslo: Oppis forlag.

Stige, M. and E. Lindhart Bauer. 2018. 'Middelalderbyen Oslo – av stokk og stein'. *Collegium medievale* 31: 71–101. https://niku.brage.unit.no/niku-xmlui/handle/11250/2590240.

Sæther, T. 1990. 'Et nyutgravd stykke av Hamars middelalder'. *Fra kaupang og bygd* 1990: 7–28.

Sæther, T. 2005. *Hamar i middelalderen*. Hamar: Domkirkeodden.

Wanner, H., C. Pfister and R. Neukom. 2022. 'The variable European Little Ice Age'. *Quaternary Science Reviews* 287: 107531. https://doi.org/10.1016/j.quascirev.2022.107531.

The Author

Kristian Reinfjord Ph.D. is Head of Cultural History and Senior Curator at Anno Domkirkeodden Museum. He is an archaeologist specialising in medieval buildings and material remains of the period. His interests include vernacular architecture, heritage studies and conservation of Norwegian built heritage.

LITTLE ICE AGE CLIMATE

Chapter 7

THE IMPACT OF WILDFIRE AND CLIMATE ON THE RESILIENCE AND VULNERABILITY OF PEASANT COMMUNITIES IN SEVENTEENTH-CENTURY FINLAND

Jakob Starlander

Abstract

This chapter explores the impact of wildfire and climate on the resilience and vulnerability of peasant communities in Finland. It examines the socio-economic consequences of forest and settlement fires by analysing several different source categories, including local district court protocols, tax records and seventeenth-century legislation. The occurrence of fire disasters is compared with reconstructions of climatic conditions during the century and the chapter estimates the relative impact of climate anomalies on the frequency of fire disasters, as well as establishing different factors of resilience and vulnerability of the Finnish rural population.

Introduction

The most central source of light throughout most of human existence was the daylight of the sun, which structured and limited the work and activities of people. Yet not all was dark when the sun had set beyond the horizon. Candlelight and sparkling fireplaces provided illumination during the dark hours of the day. For millennia, fire has been an essential cornerstone of human life, affecting the way we live, what we eat, our behaviour and interaction with each other, and it has fundamentally shaped and changed our societies (Gowlett 2016). Furthermore, fire was instrumental in the industrial revolution and in generating favourable conditions for population growth and economic prosperity. However, despite all the positive advancements, fire can also be a treacherous ally which, if mismanaged, can lay waste and destroy more than just houses and property, but also whole cities and landscapes. Within historical research, the occurrence of fire disasters has been strongly linked with urbanisation and industry, and during the early modern period (ca. 1500–1800), most of the largest and most disastrous urban fires took place during the seventeenth century, such as the Great Fire of London in 1666 or that in Aachen 1656. Earlier research has emphasised that many of the most severe urban fire disasters occurred during the Little Ice Age (LIA), a period when it was generally very cold. The explanation for this paradoxical relationship can be found in the climatic variability connected to the LIA, which contained several episodes and individual years of severely dry and hot summers (Garrioch 2018, 2024; Pfister and Wanner 2021; Zwierlein 2021).

Recent findings explaining the relationship between climatic conditions and the occurrence of urban fire disasters have revealed that the socio-economic disturbance and political instability often following such disasters resulted in a higher level of vulnerability and in disruptions of local governments' ability to implement precautionary measures for fire prevention (Garrioch 2024). Whereas these findings can tell us much about how early modern *urban* spaces were affected by fires and climatic variability, there are to date almost no studies that have considered these circumstances in *rural* environments during the LIA.

As such, this chapter will explore the impact of wildfire and climate on the resilience and vulnerability of peasant communities in Finland (then a part of the Swedish Realm) during the seventeenth century by examining the socio-economic consequences of forest and settlement fires. This will be done by analysing several different source categories, including local district court protocols, tax records and seventeenth-century legislation. Moreover, the occurrence of fire disasters will be compared with reconstructions of climatic conditions during the seventeenth century in order to estimate the relative impact of climate anomalies on the frequency of fire disasters, as well as to establish different factors of resilience and vulnerability of the Finnish rural population.

Resource Areas and Sources: North Ostrobothnia and Lower Satakunta

Two Finnish regions will be under investigation in this chapter, namely North Ostrobothnia and Lower Satakunta. Finland was an integrated part of the Swedish Kingdom from the thirteenth century until 1809 when it was lost to Russia. During this time, and especially during the early modern period, Finland and the Finnish population constituted a very important part of the Swedish Kingdom's economy in terms of taxes levied on the rural population and because of the growing tar industry that skyrocketed during the seventeenth century (Virrankoski 1973; Villstrand 1992; Starlander 2023). The production of tar was particularly intense in the region of Ostrobothnia, and it soon became the Swedish Kingdom's third most important export product after iron and copper.

The region consisted mainly of peasant farmer households whose primary food source was based on field cultivation combined with a longstanding tradition of slash-and-burn agriculture (Fi. *kaskiviljely*; Sw. *svedjebruk*). Nevertheless, the production of tar successively came to supplement the peasant household income to such a degree that it developed into a customary economic practice that diversified the means of subsistence (Starlander 2023: 245). The region of North Ostrobothnia saw a particularly intense increase in tar production towards the end of the century. The region provides an interesting case for study in that tar was produced in the peasants' forest commons, thus in an environment rich in combustion-sensitive materials. Since the production of tar included a process of pyrolysis, similar to that of charcoal production, it was a highly hazardous enterprise that could easily get out of hand if not carefully monitored. Similar to North Ostrobothnia, the rural population in Lower Satakunta consisted mainly of peasant farmers (Koskinen 2017: 90), but it was an economic-geographic region more focused on agriculture and cattle breeding than tar production (Huhtamies 2004: 25). However, slash-and-burn agriculture was common practice throughout the century, which increased the risk of fire disasters.

The two regions make for an insightful comparison due to the similarities and differences that can be recognised between them. In the case of the former, fire was a universal element essential for making everyday life in all rural localities functional, be it at the farm or in the forest. Not only was it the only source of light during the dark hours, but it was also used to prepare food, as a source of heat, to dry grain and to clear lands and meadows (Sw. *ängsrörjning*) to name a few. In terms of differences, the two regions contrast with each other in the ways in which the peasant communities engaged in agricultural and rural industrial practices, as the North Ostrobothnian peasants were deeply involved with tar production for the sustenance of their households and Lower Satakunta more occupied with slash-and-burn agriculture.

The sources analysed in this chapter are local district court protocols, tax records and seventeenth-century legislation. Together, they offer exceptional possibilities to study grassroots-level events in relation to governmental responses to increasing trends of fire disasters in the countryside. The local court, serving as a conflict-solving arena for the rural population (Larsson 2016; Starlander 2023), was the place to which people went when a fire had struck either a village or a forest in order to establish how the fire had started, what the damage was and what the level of economic compensation (Sw. *brandstod*) was to be given to the affected party. As such, they contain detailed information about the set of occurrences that led to the fire and what happened afterwards. The tax records contain information about whether an affected party was able to resume the same level of subsistence after a fire disaster since they indicate if the homestead became deserted (Sw. *öde*) after a fire or not (Huhtamaa et al. 2022). Lastly, Swedish legislation contains information about how this development was perceived by Swedish authorities during the seventeenth century, as well as instructions and regulations on how fires were supposed to be controlled and handled.

Resilience and Vulnerability

Whether at the peasant homestead or in the forest, the risk of fires starting and spreading out of control was a constant feature of early modern rural life. In this context, measures had to be taken in order to strengthen the resilience of peasant communities and individual households. Establishing the level of peasant resilience in relation to any natural or anthropogenic disaster is not altogether straightforward since the concept cannot be universally defined. Nevertheless, previous research has defined the concept in terms of a society's ability to withstand disruptions and adapt during times of transformation, whilst maintaining the same function, structure, identity and feedback mechanisms (Soens and De Keyzer 2022: 3; see also Walker et al. 2004; Soens 2018; van Bavel et al. 2020). Examples of such adaptation and resilience mechanisms can, for example, be found in medieval communities of the Flemish Coastal Plain that experienced recurrent episodes of flooding, which led to substantial investments being made in water control systems (Soens 2011). Closely related to resilience, although at the other end of the spectrum, is the level of vulnerability of a society. Whilst rural societies could demonstrate different capacities for resilience, this could easily be undermined by structural constraints rooted in economic inequality, social injustice and climatic unpredictability. This can be exemplified by a study done by Heli Huhtamaa et al. (2022), which shows that freeholding peasants in Ostrobothnia during the seventeenth century were better off than tenants of the Swedish Crown (Sw. *kronobönder*) in times of sustenance crises because of more advanced social networks and higher levels of wealth. As such, it is important to acknowledge the dynamic interplay between resilience and vulnerability and that, while a society as a whole could be considered resilient, there might still be individuals or groups that remained or were made more vulnerable and exposed to hazards (Soens 2018).

Climate Variability and Impact on Forest Fires

There are many ways in which a fire disaster can start, and there are many factors to consider when seeking to explain the circumstances around their origin, spread and intensity. One is the anthropogenic factor, meaning human agency. Nevertheless, any kind of environmental disaster can have both exogenous and

The Impact of Wildfire and Climate

endogenous explanations – that is, be a consequence of human actions and/or natural phenomena (Fara 2001; De Keyzer and Van Onacker 2022). Furthermore, it should be emphasised that one does not necessarily exclude the other, meaning, for example, that if a fire is started because of human mismanagement, the speed with which it spreads and the severity at which it burns is often influenced by climate conditions.[1] Nevertheless, in order to be able to say anything about the reasons for *why* fire disasters happened in seventeenth-century Finland, it is first necessary to establish *when* they occurred, which can be done through a quantitative examination of the court records.

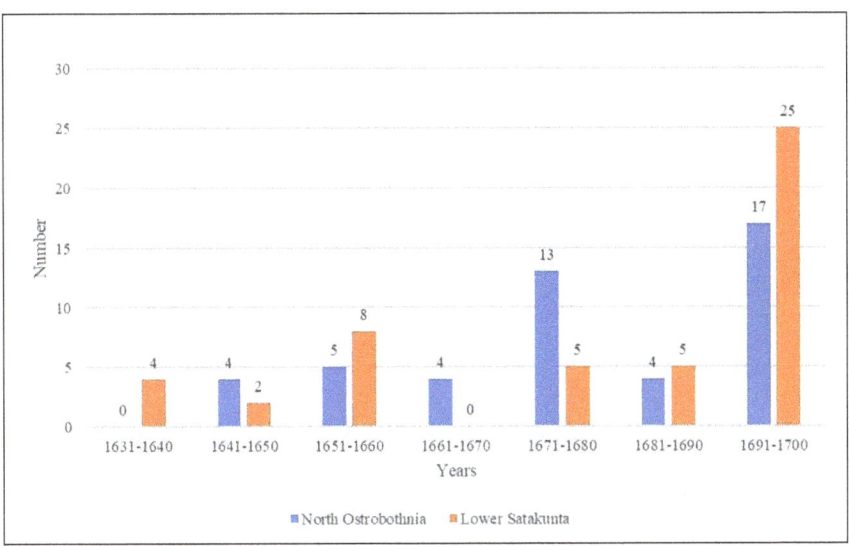

Figure 1. Number of reported forest fires in North Ostrobothnia and Lower Satakunta during the seventeenth century in ten-year increments. Data source: NAF, Court records.

Figure 1 shows the number of reported forest fires in North Ostrobothnia and Lower Satakunta between 1631 and 1700. Whereas the number of reported forest fires remained fairly stable for most decades of the century (on average one forest fire every other year), the 1670s and 1690s are clear breaks from that trend. During the 1670s, there is a sharp increase in reported forest fires in North Ostrobothnia, whereas the same development is not noticeable in Lower Satakunta, and during the last decade of the century, reports of forest fires reached an all-time high in both regions. How can these anomalies be explained in relation to climate variability?

---

1. For further reading on the issue of anthropogenic and natural disasters, within disaster theory, a long-debated issue has been how 'natural' natural disasters actually are, see Mauelshagen 2015: 177–179.

During the 1670s, nine of the total eighteen (fifty per cent) reported forest fires occurred in 1678. By examining the average summer temperatures (June, July, August) during the period (Figure 2), it is clear that the year 1678 was noticeably warmer than the preceding summers of the same period. Similarly, during the 1690s, 37 of the 42 reported forest fires (88 per cent) occurred

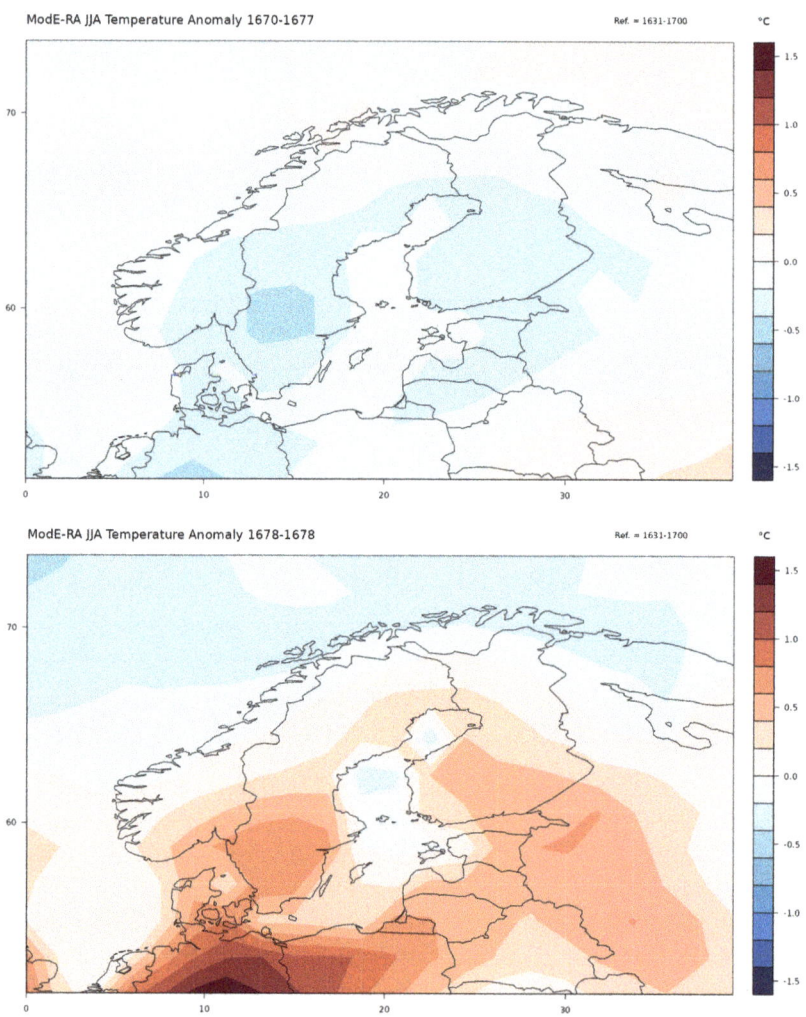

Figure 2. Mean temperature during JJA (June, July, August) of 1670–1677 (top) and mean temperature during JJA during 1678 (bottom). Data source/map adapted from Warren et al. 2024.

during the first four years of the decade and, when comparing the summer temperatures of these years to the following six years of the decade (Figure 3), the difference is even more striking. As such, there is a clear correlation between distinctly warmer summers and the occurrences of forest fires.

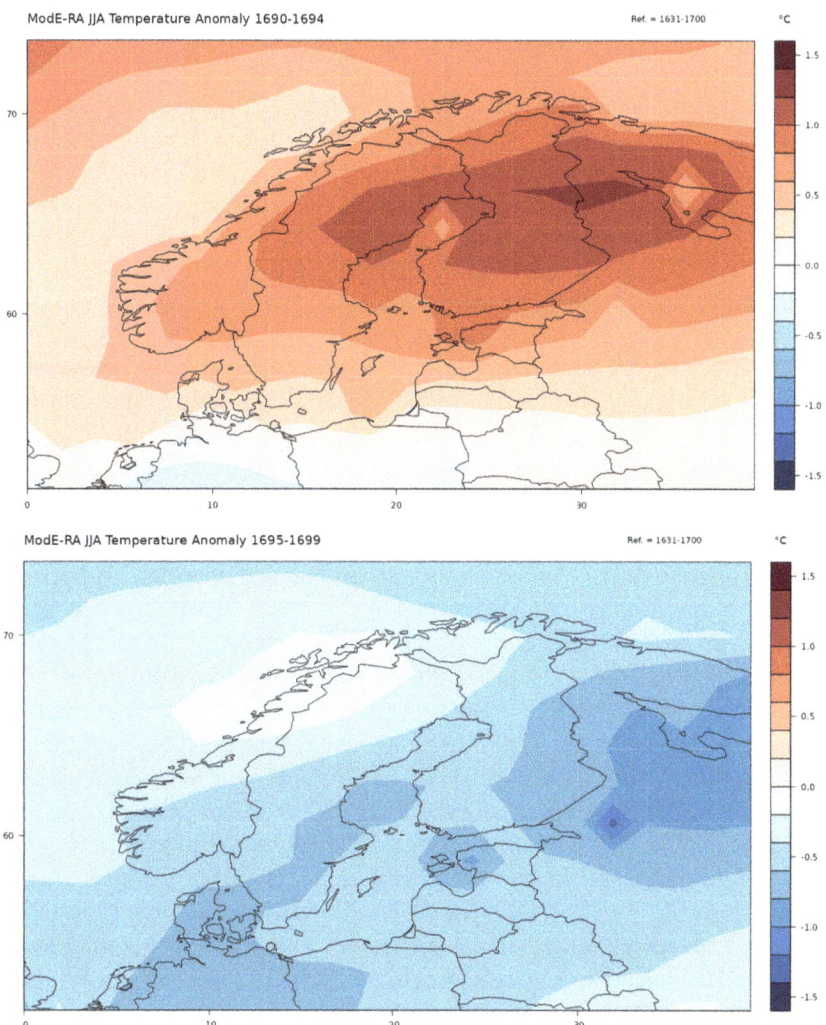

Figure 3. Mean temperature during JJA (June, July, August) of 1690–1694 (top) and mean temperature during JJA during 1695–1699 (bottom). Data source/map adapted from Warren et al. 2024..

Jakob Starlander

Regional reconstructions of temperature variability across the globe have shown how different regions during the last millennia experienced periods of cooling and warming both similarly and differently (Ljungqvist 2010). However, as mentioned above, the LIA (about AD 1250–1860) was not without episodes of warm and hot summers, such as the extreme hot spell and drought experienced in Switzerland in 1540 (Wetter and Pfister 2013). Fluctuations such as these have been attributed to different underlying mechanisms in the climate system, including reduced solar activity, ocean currents, atmospheric circulation patterns, orbital cycles and volcanic eruptions (Ljungqvist 2010; Wanner et al. 2022; Skoglund 2023: 39). Whilst it is difficult to explain the exact cause of the seventeenth-century Finnish hot spells, the anomalies identified during certain periods and years accentuate the non-uniformity of climate behaviour, since long-term cooling trends do not necessarily preclude short-term warm anomalies. Furthermore, considering that the number of reported forest fires in the two regions differed distinctly during the 1670s, it is also important to highlight that regional and societal differences, as well as local weather patterns or geographical features, might play a role in the occurrence of forest fires.

As mentioned above, whilst a quantitative examination of the local court records can provide the *when*, it is also important to address the *why*, which warrants a qualitative examination of the court records. There are several accounts in the court records that explain the development and anomalies presented above. For example, in the parish Kalajoki in North Ostrobothnia in 1678, it was explained to the court how a watermill belonging to Henrich Henrichsson had been consumed by a rapidly spreading forest fire. Although it was uncertain how the fire had started, the peasant Sigfred Henrichsson stood accused and admitted that he had caused the fire, although 'claimed his innocence, that he had not been able to extinguish the fire in the great drought that was then present'.[2] Whilst the LIA consisted of long periods of cold conditions, it has been established that exceptionally hot summers and prolonged episodes of drought still occurred (Pfister 2018; Pribyl 2020; Garrioch 2024), and even though the cold of the winter months could extend well into spring, a severely dry summer could nevertheless occur in the same year. This is demonstrated by another account from July 1678 when the two peasants Anders Jönsson and Henrich Sigfredsson described the speed with which 'a hasty and unforeseen forest fire' had consumed ninety barrels of tar that they had produced. As if that was not enough, it was further explained that 'the cold had done great damage

---

2. Kalajoki parish, 12 and 13 Aug. 1678, Nation Archives of Finland (NAF), Court Records, KO a:20, 1678–1678, act 86. Original text: 'sin del uti warande oskyldigheet, at han Eldhen icke mechtat ell:r förmådt i den stora Tårka då war, uthsleckia'.

The Impact of Wildfire and Climate

to their fields and thus they were unable to pay taxes this year, and it was said that Henrich was completely impoverished, and Anders was little better'.[3]

The court records from the 1690s provide further grass-roots information explaining *why* most forest fires occurred during the first half of the decade. In Lower Satakunta, a particularly severe forest fire was discussed in Kokemäki parish in October 1691, where it was explained how all the outlying lands (Sw. *utmark*) of Harjavalta village were completely burnt due to a mismanaged swidden (Sw. *svedja*). It was further explained that 'timber and bark forest as well as some meadows were destroyed by fire',[4] and the laymen of the court added that:

> ... smoke had risen from that part all summer since the fire had been simmering in the earth, but finally in one day extended over half a mile, which had been so fierce that it had broken away from the trees over the great Kumo river and started to burn, but had been put out by the people present, which nevertheless had done irreparable damage to the property of Harjavalta and the neighbouring villages...[5]

The consequences of the Kokemäki fire of 1691 were so severe and extensive that the court concluded that the damages 'cannot be estimated'. The explanation as to *why* this had been possible was 'because it was a very dry summer' and that 'even though it had been raining, the fire in some places smouldered in the earth'.[6] Apart from the Kokemäki forest fire, there are several other accounts that provide information of the particularly hot and dry summers of the 1690–1694 period from both Lower Satakunta and North Ostrobothnia. For example, in the neighbouring parish Eura in Lower Satakunta in 1692, Tomas Michelsson stood accused of letting the fire from his swidden spread,

---

3. Liminka parish, 18, 19 and 20 Aug. 1678, NAF, Court Records, KO a:19, 1678–1678, act 187. Original text: 'kölden giordt stoor skadha, på deras åker, och dem således odugeligit falla, skatten i åhr kunna ährläggia, sades och denna Henrich wara ahldehles uthfattig och Anders föga bettre.'
4. Kokemäki parish, 8 and 9 Oct. 1691, NAF, Court Records, II KO a:6, 1691–1691, act 250–251. Original text: 'timber och näfwerskogene sampt några ängiar wara af eldh fördärfwade'.
5. Kokemäki parish, 8 and 9 Oct. 1691, NAF, Court Records, II KO a:6, 1691–1691, act 250–251. Original text: 'röök slåtts på den kantten upgå heela Sommaren hwarest elden i jorden legat och Pyttiat, men omsider på een dag fahrit öfwer half mijhl wägh, som warit så häfftigh att den ifrån trään slagit öfwer stoora Cumo Elfwen och begyntt brinna, men blifwit af tilstädes warande fålck släckt. Som lijkwähl på Hariawalda och näst om liggiande byars ägor giort een obotelig skada'.
6. Kokemäki parish, 8 and 9 Oct. 1691, NAF, Court Records, II KO a:6, 1691–1691, act 250–251. Original text: 'som intet skall kunna verderas'; 'efter som det war muckit tårr sommar'; 'oansedt dett sedan regnar:/ elden på någre ställen Kyttiar/Pyttiar i Jorden'.

which had caused great damage to the forests in the area where he lived. This was considered as particularly reckless considering the 'great ... drought' at that time.[7] In North Ostrobothnia in 1693, the peasant Michel Mattsson in Kalajoki parish stood accused of having unleashed a fire as he was producing tar and was asked how come he singlehandedly dared to ignite a tar pit 'especially in such dry and hot weather'[8].

Following the quantitative and qualitative examination of the local court protocols, the answers to *when* and *why* forest fires occurred can be explained by the interplay between anthropogenic activities and climate variability. The slash-and-burn agriculture practised by peasant communities in Lower Satakunta and the tar industry of North Ostrobothnia were both activities that increased the risk of forest fires occurring. Similarly, episodes of dry and hot summers distinctly added to this risk as they produced environmental conditions where fires could more easily take hold and spread, with devastating economic and environmental consequences. As such, it can be established that the different forest-related industries practised by rural populations in Finland increased the level of vulnerability in their communities. However, peasant communities did not remain idle in face of this development but took active measures to limit the spread of forest fires in any way they could.

The practice of slash-and-burn agriculture and tar production played a significant role in the rural population's household economy in Finland throughout the early modern period, as well as throughout the Baltic region (Myllyntaus et al. 2002; Tomson et al. 2015). This gave them a unique relationship with and understanding of their surrounding environment in terms of the ways in which fire could be used. Since the late nineteenth century, prescribed burnings have become a common way of reducing the risk of wildfires occurring by clearing areas of materials with high combustion content (Cogos et al. 2020). However, using fire to fight forest fires was not a new strategy in the nineteenth century. This is well illustrated by the court records as a fire was quickly spreading in the parish of Huittinen in Lower Satakunta in 1693. As the peasant Jacob Matsson saw the fire approaching his homestead, he quickly started a new fire 'in order thereby to avert a greater danger, which would run against the ... fire'. In this way, he was able to save his homestead and his family from economic ruin, which

---

7. Eura parish, 3 and 4 June 1692, NAF, Court Records, II KO a:7, 1692–1692, act 93. Original text: 'stoor ... torka'.
8. Kalajoki parish, 26 and 28 Aug. 1693, NAF, Court Records, KO a:14, 1693–1693, act 157–158. Original text: 'i synnerheet uthi sådan tårr och heet wäderlek'.

well demonstrates the peasants' level of resourcefulness.[9] It is evident, therefore, that whilst mismanagement of fire and accidents was in some sense inevitable, the rural peasant population possessed knowledge of how to effectively quench fire with fire. Nevertheless, not all were as lucky and, in order to increase the general level of resilience for peasant communities, laws made available to them by the central Swedish government were important, as we now shall see.

Swedish Forest Legislation

The first exclusively forest-oriented legislation passed in the Swedish Kingdom was the royal forest ordinance of 1647. It was conceived from the state's desire to secure enough resources for the growing iron and copper industries on the Swedish mainland and was soon followed by a second forest ordinance in 1664 (Karlsson 1990; Starlander 2023: 76–77). Parallel to this legislative development, a growing fear of wood shortage was felt among legislators, not only in Sweden, but across Central Europe (Wing 2015; Warde 2018). Since the development of detected forest fires increased during the century, and because the intention of the first forest ordinances was not to deal with this growing threat, this would eventually lead to the creation of Sweden's first forest fire ordinance in 1690 (Starlander 2024). Whereas the focus of the previous ones (in terms of forest fires) had been to provide legal instructions as to how to accurately identify the perpetrator and issue fines accordingly, which contained serious penalties comparable to, for example, Spanish medieval forest regulations where death penalties were sometimes stipulated (Wing 2015: 60), the ordinance of 1690 instead contained clear instructions on how to thoughtfully and structurally subdue and extinguish forest fires.

The forest fire ordinance contained detailed instructions on how to act when a fire was raging, including how to call upon neighbours to assist in putting it out, that word of the fire should pass through all villages of the parish, that at least one man from each village should assist, that the chief constable (Sw. *länsman*) must be informed, that the church bell must be rung, how to clear areas so that the fire could not spread, and that axes, spades, buckets and other implements should always be at hand.[10] It is difficult to know exactly how the new law was implemented, although the court records are remarkably enough able to give us a glimpse of how it could unfold. In North Ostrobothnia in August 1691, a large forest fire had been raging in the parish of Saloinen for four weeks, which

---

9. Huittinen parish, 21 and 22 July 1693, NAF, Court Records, KO a:13, 1693–1693, act 357–359. Original text: 'i mening att der medh afwäria en större fara, som skulle löpa emot ... eldh'.
10. *Kongl. May.tz Förordning, Angående Skogz-eldar* (1690).

had caused great damage to the forests, meadows and tar production sites in the general area. As the discussion about who had started the fire was conducted, word of its continued 'fierce spreading' reached the courtroom, which incited the court to order the chief constable and all laymen of the court that were present to hastily go 'to the fire with every man in the villages, to seek to subdue and extinguish it, so far as they next court meeting want to escape the penalty and fine contained in the Royal Ordinance concerning forest fires'.[11]

The success of the new legislation in lessening the general occurrence of forest fires in Finland is difficult to assess. However, by returning to the comparison between climate reconstructions and the number of reported forest fires during the 1690s, it is noteworthy that most reported fires occurred during the warmest years of the decade following the declaration of the forest fire ordinance in 1690. Thus, it can be interpreted that the new instructions had no impact. However, considering the 'in-action' court case from 1691, it is evident that the new legislation was indeed adhered to, which could also suggest that even more fires and worse socio-economic and environmental consequences might have occurred had the new ordinance not been decreed when it was. Furthermore, it can be determined that the growing trend of fire disasters, which have been shown to have a correlation with the climatic variabilities identified during the century, motivated legislators to act and implement new legislation aimed at diminishing the effects of this development.

Fire Disaster Relief

Western Europe saw a general rise in relief systems being implemented during the seventeenth century, such as the Elizabethan Poor Laws of 1597 and 1601 in England or community-based poor relief systems in France and the Netherlands (Birtles 1999; van Leeuwen 1994; Hindle 2004; Healey 2024). The legal framework provided by the local courts in Finland was a decisive instrument that the local population could use for their own benefit in times when fire disasters occurred. It has been shown how the court records contain information about *why* and *when* fire disasters occurred in seventeenth-century Finland, but the records also contain information about the extent of support that was given to those who suffered from a fire disaster. Laws that were instituted in order to provide support and relief to those affected by fire in rural areas existed from the fourteenth century with Magnus Eriksson's Law of the Realm, which declared

---

11. Saloinen parish, 11 and 12 Aug. 1691, NAF, Court Records, KO a:11, 1691–1691, act 207–208. Original text: 'häfftigt grasserar'; 'till eelden med hwar man i byiarna, att sökia dämpa och släckia den, så framt dhe nästa ting wela undfly dhes straff och böter som Kongl. Förordning angående Skogs Eelden innehåller'.

The Impact of Wildfire and Climate

that fire support (Sw. *brandstod*) was a mandatory duty that all had to fulfill and that '[n]o one should be free from this'.[12] Even though it is difficult to establish the extent to which this law was practised up until the seventeenth century, new legislative action was taken in 1642 with a renewed version of the Beggar Regulation, and later in 1681 with the Swedish king Karl XI's House Inspection Ordinance. These created revived pathways for rural populations to apply for support after a fire disaster. This can be seen in Figure 4, which shows the number of fire support applications at local courts in North Ostrobothnia (in total 434) and in Lower Satakunta (in total 711) during the seventeenth century.

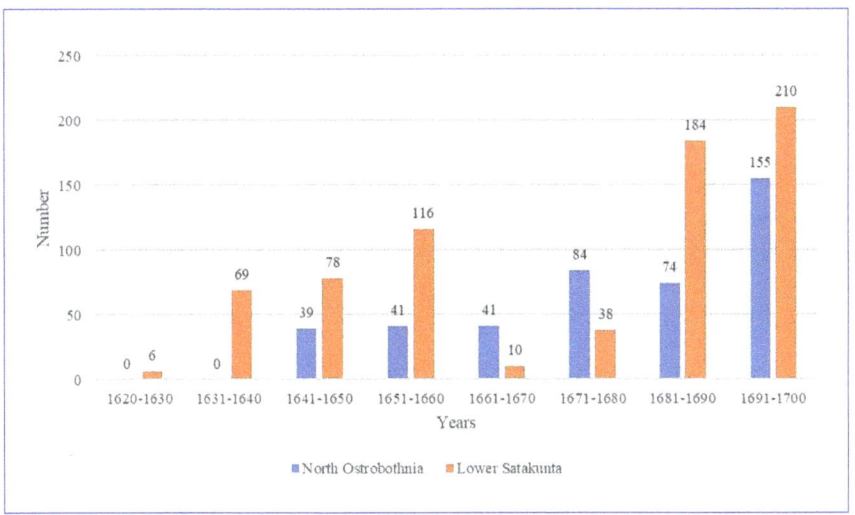

Figure 4. Number of fire support applications in North Ostrobothnia and Lower Satakunta during the seventeenth century in ten-year increments. Data source: NAF, Court records.

The new legal framework was quickly utilised by the peasant population as the number of applications steadily grew in both regions throughout the century. It has been explained in earlier research that the level of economic compensation given to a single household could reach as much as two-thirds of that which was lost in the fire, although these numbers are misleading since the socio-political status of certain individuals, such as clergymen, could lead to a much higher compensation than the general average (Virrankoski 1973: 550–552). Nevertheless, through a detailed investigation of one fifth (in total 229) of the fire support applications in Lower Satakunta (161) and North Os-

---

12. *Magnus Erikssons landslag*, Åke Holmbäck and Elias Wessén (1962), p. 120. Original text: 'Det skall ingen vara fri från.' (The cited text is not the original old Swedish text, but a modern Swedish translation.)

trobothnia (68), it can be established that approximately a third of the value of what was lost in a fire was given to the applicant by the parish community in which he or she lived.[13] However, in particularly devastating cases, it was impossible to assess the damage that followed from a fire. This can be exemplified by a case put forth by the peasant Olof Korfoinen from Kalajoki parish in North Ostrobothnia who in 1680 explained how 'for him last midsummer night, while the people were asleep, his whole farm and barnyard, with the little property he had there, and his daughter, aged eight years, who had lain down in the evening in a fodder barn near the farm, not knowing what and how the fire had been started, were burnt up by the fire'.[14]

In order to establish the level of impact rural fire disasters had on peasant households in terms of being able to endure and resume the cultivation of their homesteads, contemporary tax records are useful. In the Swedish Kingdom, if a peasant household was not able to pay its taxes for more than three consecutive years, the family was evicted and marked as '*öde*' (En. deserted) in the tax records. However, an '*öde*' marking did not always mean that its occupants had been evicted, since they sometimes remained on the farm (Huhtamaa et al. 2022). Nevertheless, since the court records contain information of the date of the fire disaster, the name of the fire support applicant, the village he or she came from and descriptions of the devastation left behind, an investigation of the tax records makes it possible to determine whether the fire support was sufficient to avoid eviction. However, the individual who filed the fire support application was not always the same person registered in the tax records, which means that the result of the comparison is limited. Of the 229 applications studied in detail, 56 applicants in Lower Satakunta and 32 applicants in North Ostrobothnia can be traced.[15] The results show that only four households (seven per cent) in Lower Satakunta and one household (3.2 per cent) in North Ostrobothnia were marked as '*öde*' up to five years after the fire event occurred. Moreover, the tax records also indicate what kind of land the peasant owned or rented, and by comparing the tax records and court records, it is evident that different kinds of landownership were of no consequence in terms of receiving fire support,

---

13. NAF, Court records.
14. Kalajoki parish, 25, 26 and 27 Aug. 1680, NAF, Court Records, KO a:25, 1680–1680, act 64. Original text: 'för honom förledhne Midsommars Tijdh om Natten, medhan fållket Såfwo igenom wådha opbrunnit Hehla hans man och fäägårdh, medh hans ringa der ägandhe lösöhror, såsom och Hans dåtter om sine åtta åhr gammall, hwilken sigh om aftonen lagdt uthi een dher wedh gården hafwande foodherlada, Icke wetandhes huru och på hwadh sett Elden löös kommit'.
15. NAF, Turun ja Porin läänin tilejä, Asiakirjat (1633–1809); NAF, Pohjanmaan läänin tilejä, Asiakirjat (1635–1776).

The Impact of Wildfire and Climate

and that it provided equal opportunities for all peasant families to get back on track. This can be contrasted with the findings of Huhtamaa et al. (2022) who have shown that subsistence crises in Ostrobothnia generally hit tenants of the Swedish Crown worse than freeholders during the same period. This does not necessarily mean that peasant communities regarded the consequences of different disasters differently, but rather that the accessibility of legislation and the local courts provided a legal structure and platform where these issues could be effectively discussed and help could be given, which ultimately lowered the level of vulnerability and strengthened the resilience of peasant communities.

Conclusions

The purpose of this chapter has been to explore the impact of wildfire and climate on the resilience and vulnerability of peasant communities in Finland during the seventeenth century by examining the socio-economic consequences of forest and settlement fires in two regions of the Finnish countryside. This has been done by quantitatively and qualitatively examining local district court protocols, tax records and Swedish legislation in relation to climate variability during the period. The analysis has shown that reports of fire disasters were recurrent throughout the century and thus something the rural population in both regions were accustomed to. Two periods of anomalies can be noted in the 1670s and 1690s. The explanation for this can be found through the comparison of climate reconstructions and the content of the court records – namely, that certain years of both periods contained unusually warm and dry summer periods. The main causal factor of the fire disasters was the peasant populations' agricultural and forestry-related activities, slash-and-burn agriculture in Lower Satakunta and tar production in North Ostrobothnia. Both activities increased the level of vulnerability of their communities since they included the use of fire in a combustion-rich environment.

As in other regions on the European continent throughout the Early Modern Period, it has been shown how the ability to adapt to changing environmental and climatic conditions was crucial, as in seventeenth-century Spain where sharing and cooperation, rationing and food market exchanges were essential coping strategies used by communities in order to cope with repeated episodes of drought (Grau-Satorras et al. 2021). It has also been shown, for example, how European farmers on a generalised level adjusted the usage of crops towards those that were more resistant to cold (Pei et al. 2016). In the Finnish case, measures and strategies to reduce the frequency of and extinguish fires were taken at different levels, by individual peasant households, peasant communities and Swedish state legislators. The examination of fire support applications and tax records has shown that the relief and assistance given by peasant community members was in almost all cases enough for peasant families to resume the cultivation of their homesteads. Cooperation between peasants and communities, as well as between peasants and local authorities, made their societies more resilient against fire disasters.

Bibliography

Birtles, S. 1999. 'Common land, poor relief and enclosure: The use of manorial resources in fulfilling parish obligations 1601–1834'. *Past & Present* **165** (1): 74–106. https://doi.org/10.1093/past/165.1.74.

Cogos, S., S. Roturier and L. Östlund. 2020. 'The origins of prescribed burning in Scandinavian forestry: the seminal role of Joel Wretlind in the management of fire-dependent forests'. *European Journal of Forest Research* 139: 393–406. https://doi.org/10.1007/s10342-019-01247-6.

De Keyzer, M. and E. Van Onacker. 2022. 'Vulnerabilities avoided and resilience built. Collective action, poor relief and diversification as weapons of the weak (The Campine, Belgium, 1350–1845)'. *Continuity and Change* 37 (1): 13–42. https:.

Downey, S.S., M. Walker, J. Moschler et al. 2023. 'An intermediate level of disturbance with customary agricultural practices increases species diversity in Maya community forests in Belize'. *Communications Earth & Environment* 4 (428): 1–13. https://doi.org/10.1038/s43247-023-01089-6.

Fara, K. 2011. '"How natural are 'natural disasters"? Vulnerability to drought of communal farmers in Southern Namibia'. *Risk Management* 3: 47–63. https://doi.org/10.1057/palgrave.rm.8240093.

Gowlett, J.A.J. 2016. 'The discovery of fire by humans: a long and convoluted process'. *Philosophical Transactions of the Royal Society B: Biological Sciences* 371 (1696). http://dx.doi.org/10.1098/rstb.2015.0164.

Garrioch, D. 2019. 'Towards a fire history of European cities (late Middle Ages to late nineteenth century)'. *Urban History* 46 (2): 202–24. https://doi.org/10.1017/S0963926818000275.

Garrioch, D. 2024. 'Large fires and climatic variability in urban Europe, 1500–1800'. *Climates and Cultures in History* 1: 3–27. https://doi.org/10.3197/whp-cch.63842135436332.

Grau-Satorras, M., I. Otero, E. Gómez-Baggethun and V. Reyes-García. 2021. 'Prudent peasantries: Multilevel adaptation to drought in early modern Spain (1600–1715)'. *Environment and History* 27 (1): 3–36. https://doi.org/10.3197/096734019X15463432086964.

Healey, J. 2024. 'Social discipline and the refusal of poor relief under the English old poor law, c. 1650–1730'. *The Historical Journal* 67 (5): 920–942. https://doi:10.1017/S0018246X23000651.

Hindle, S. 2004. *On the Parish?: The Micro-politics of Poor Relief in Rural England c. 1550–1750*. Oxford: Clarendon.

Huhtamaa, H., M. Stoffel and C. Corona. 2022. 'Recession or resilience. Long-range socioeconomic consequences of the 17th century volcanic eruptions in the far north'. *Climate of the Past* 18: 2077–92. https://doi.org/10.5194/cp-18-2077-2022.

Huhtamies, M. 2004. *Knektar och bönder: knektersättare vid utskrivningarna i Nedre Satakunda under trettioåriga kriget, Svenska litteratursällskapet i Finland*. Svenska litteratursällskapet i Finland, Helsinki.

Karlsson, P-A. 1990. *Järnbruken och ståndssamhället: Institutionell och attitydmässig konflikt under Sveriges tidiga industrialisering 1700–1770*. Stockholms Universitet.

Koskinen, U. 2017. *Satakunnan historia V Maakunnan synty: (1550–1750)*. Pori, Satakunnan Museo/Porin kaupunki.

Larsson, J. 2016. 'Conflict-resolution mechanisms maintaining an agricultural system. Early Modern local courts as an arena for solving collective-action problems within Scandinavian civil law'. *International Journal of the Commons* 10 (2). https://doi.org/10.18352/ijc.666.

Ljungqvist, F.C. 2010. 'A regional approach to the Medieval Warm Period and the Little Ice Age'. In S. Simard and M. Austin (eds), *Climate Change and Variability*. Rijek, Croatia: Sciyo. pp. 1–26.

Magnus Erikssons landslag, i nusvensk tolkning av Åke Holmbäck och Elias Wessén. 1962. Stockholm: Nord. bokh. (distr.).

Mauelshagen, F. 2015. '10: Defining catastrophes'. In K. Gerstenberger and T. Nusser (eds), *Catastrophe and Catharsis: Perspectives on Disaster and Redemption in German Culture and Beyond*. Camden House: Boydell and Brewer. pp. 172–190. https://doi.org/10.1515/9781782046783-012.

Myllyntaus, T. and T. Mattila. 2002. 'Decline or increase? The standing timber stock in Finland, 1800–1997'. *Ecological Economics* 41 (2): 271–288.

Pei, Q., D.D. Zhang, H.F. Lee and G. Li. 2016. 'Crop management as an agricultural adaptation to climate change in early modern era: A comparative study of Eastern and Western Europe'. *Agriculture* 6 (3). https://doi.org/10.3390/agriculture6030029.

Pfister, C. 2018. 'The "Black Swan" of 1540: aspects of a European megadrought'. In C. Leggewie and F. Mauelshagen (eds), *Climate and Culture, Climate Change and Cultural Transition in Europe*. Leiden: Brill. pp. 156–93. https://doi.org/10.1163/9789004356825_007.

Pfister, C. and H. Wanner. 2021. *Climate and Society in Europe: The Last Thousand Years*. Bern: Haupt Verlag, 2021.

Pribyl, K. 2020. 'A survey of the impact of summer droughts in southern and eastern England, 1200–1700'. *Climate of the Past* 16: 1027–41. https://doi.org/10.5194/cp-16-1027-2020.

Skoglund, M. 2023. *Climate and Agriculture in the Little Ice Age: the case of Sweden in a wider European perspective*. Diss. Uppsala: Sveriges lantbruksuniversitet.

Soens, T. 2011. 'Threatened by the sea, condemned by man? Flood risk, environmental justice and environmental inequalities along the North Sea Coast 1200–1800'. In G. Massard-Guilbaud, G. and R. Rodger (eds), *Environmental and Social Justice in the City. Historical Perspectives*. Knapwell, Cambridge: The White Horse Press

Soens, T. 2018. 'Resilient societies, vulnerable people: Coping with North Sea floods before 1800'. *Past & Present* **241** (1): 143–177. https://doi.org/10.1093/pastj/gty018.

Soens, T. and M. De Keyzer. 2022. 'From the resilience of commons to resilience through commons. The peasant way of buffering shocks and crises' *Continuity and Change* **37** (1). https://doi:10.1017/S026841602200008X.

Starlander, J. 2023. *Tar and Timber: Governing Forest Commons in Seventeenth Century Northern Finland*. Dissertation Uppsala, Sveriges lantbruksuniversitet. https://urn.kb.se/resolve?urn=urn:nbn:se:slu:epsilon-p-120857.

Starlander, J. 2024. 'Rural inferno: Environmental and socio-economic consequences of wildfires in seventeenth-century Western Finland'. *Climates and Cultures in History* **1**: 65–69.

Tomson, P., R.G.H. Bunce and K. Sepp. 2015. 'The role of slash and burn cultivation in the formation of southern Estonian landscapes and implications for nature conservation'. *Landscape and Urban Planning* **137**: 54–63. https://doi.org/10.1016/j.landurbplan.2014.12.015.

van Bavel, B., D.R. Curtis, J. Dijkman et al. (eds), *Disasters and History: The Vulnerability and Resilience of Past Societies*. Cambridge, UK: Cambridge University Press.

van Leeuwen, M.H.D. 1994. 'Logic of charity: Poor relief in preindustrial Europe'. *The Journal of Interdisciplinary History* **24** (4): 589–613. https://doi.org/10.2307/205627.

Villstrand, N.E. 1992. 'Med stor möda i en hop gropar i marken: tjärbränning kring Bottniska viken under svensk stormaktstid'. *Historisk tidskrift för Finland* **77**.

Virrankoski, P. (ed.) 1973. *Pohjois-Pohjanmaan ja Lapin historia. 3, Pohjois-Pohjanmaa ja Lappi: 1600-luvulla*. Oulu.

Wanner, H., C. Pfister and R. Neukom. 2022. 'The variable European Little Ice Age'. *Quaternary Science Reviews* **287**.

Warde, P. 2018. *The Invention of Sustainability: Nature and Destiny, c. 1500–1870*. Cambridge: Cambridge University Press. https://doi.org/10.1017/9781316584767.

Warren, R., N.E. Bartlome, N. Wellinger, J. Franke, R. Hand, S. Brönnimann and H. Huhtamaa. 2024. 'ClimeApp: data processing tool for monthly, global climate data from the ModE-RA palaeo-reanalysis, 1422 to 2008 CE'. *Climate of the Past* **20**: 2645–62. https://doi.org/10.5194/cp-20-2645-2024.

Wetter, O. and C. Pfister. 2013. 'An underestimated record breaking event – why summer 1540 was likely warmer than 2003'. *Climate of the Past* **9**: 41–56. https://doi.org/10.5194/cp-9-41-2013.

Wing, J.T. 2015. *Roots of Empire: Forests and State Power in Early Modern Spain, c.1500–1750*. Brill, Leiden.

The Impact of Wildfire and Climate

Zwierlein, C. 2021. *Prometheus Tamed: Fire, Security, and Modernities, 1400 to 1900*. Leiden and Boston: Brill.

Unpublished sources

National Archives of Finland (NAF), Court Records, I KO a:6, 1644–1649
NAF, Court Records, KO a:19, 1678–1678
NAF, Court Records, KO a:20, 1678–1678
NAF, Court Records, II KO a:6, 1691–1691
NAF, Court Records, II KO a:7, 1692–1692
NAF, Court Records, KO a:11, 1691–1691
NAF, Court Records, KO a:13, 1693–1693
NAF, Court Records, KO a:14, 1693–1693
NAF, Court Records, KO a:25, 1680–1680
NAF, Pohjanmaan läänin tilejä, Asiakirjat (1635–1776)
NAF, Turun ja Porin läänin tilejä, Asiakirjat (1633–1809)
Kongl. May.tz Förordning, Angående Skogz-Eldar, sampt deras förekommande och släckiande. Gifwen Stockholm den 10 Novembr. Åhr 1690 (1690). Stockholm, Printed Kongl. Booktryckerijet, Sal. Wankijfs Effterlefwerska.

The Author

Jakob Starlander has a Ph.D. in Agrarian History from the Swedish University of Agricultural Sciences (SLU), Uppsala, Sweden. He has been a postdoctoral researcher at the Department of Economic, Social and Environmental History, Institute of History, Bern University, Switzerland. He is now a postdoctoral researcher at the Division of Agrarian History, SLU, Uppsala, Sweden.

Chapter 8

NORTHERN ICELAND TEMPERATURE VARIATIONS AND SEA-ICE INCIDENCE c. AD 1600-1850

A.E.J. Ogilvie and M.W. Miles

Abstract

This paper considers variations in the occurrence of sea ice off the coasts of Iceland, and compares these with air temperatures on land, particularly for the north of Iceland, for the period c. AD 1600-1850. Data are drawn from Iceland's rich treasury of historical records on climate and weather. For the most part, cold air temperatures on land and the incidence of sea ice correlate well, but this is not always the case. Periods with low temperatures and high sea-ice incidence include the early 1600s, the 1690s, the 1750s, the 1780s and the mid-1800s. A distinct mild period with little sea ice occurred during c. 1640 to c. 1680. Subsequent to our main study period, the most severe years of the nineteenth century were likely to have been 1858-1892. High sea-ice incidence is also evident in, e.g., the 1880s and 1910s, in contrast to the climate amelioration recorded in Europe. The most notable feature of Iceland's climate is its variability, thus making it problematic to ascribe a single distinct period reflecting a 'Little Ice Age'.

If it so happens that someone drifts north past the island (Iceland) he will see before him the blue-green ice of immense breadth and horrible extent which reaches out over all the oceans surrounding the northern coasts of Greenland. And some people think that it was the proximity of the ice that caused those who first discovered Iceland to give it the name they did, either because they first encountered it near Iceland off the coast or because great quantities of it reached the shores of the land itself. The Icelanders who have settled on the northern coasts are never safe from this most terrible visitor

(translated from the description of Iceland written c. 1590 by Oddur Einarsson. See Ogilvie 2022).

doi: 10.63308/63881023874820.ch08

Northern Iceland Temperature Variations and Sea-Ice Incidence

Introduction

This paper will compare the presence of sea ice (or lack of it) off the north coasts of Iceland with variations in temperature. The data used are drawn from historical documentary records with some reference to early instrumental records. Iceland is unusually rich in historical records, including many that provide detailed evidence of weather and climate. Observations on the connection between the incidence of sea ice off the coasts of Iceland and a fall in temperature on land, in particular in northern Iceland, the region most affected by sea ice, have a long history and were documented in some detail by Icelandic writers as early as in the late-sixteenth century (as seen in the epigraph to this paper). In this paper, the relationship between surface air temperatures and sea ice will be explored over a 250-year period, from c. 1600 to 1850. Analysis of the period 1850 onwards is ongoing; however, for context, some information for this period will be noted in the conclusions.

Sea Ice and Climate

The climate of Iceland is influenced by the country's geographical location at the intersection of cold Polar air and warmer Atlantic air, and the relatively warm Irminger and North Atlantic currents (IC, NAC) and the colder East Iceland current (EIC) as seen in Map 1. Iceland experiences frequent fluctuations in these different air masses and ocean currents. This is one of the main causes of the variability of the climate of Iceland. Its vulnerability to climate impacts in the past, and potentially in the future, is a consequence of this variability (Ogilvie and Jónsson 2001). Sea ice is also an integral part of Iceland's climate. Most of the ice that reached Iceland's shores in the past drifted south on the East Greenland Current as icebergs from glaciers on Greenland's east coast and as ice floes of frozen seawater. In the past, sea ice occurred most frequently off the coast of Iceland from late winter to early spring, but during severe seasons it could remain far into the summer and even the autumn. It could also occur in the early winter or earlier. The months of April and May were likely to have the greatest extent of ice, and September to December the least. The ice most frequently affected the north, northwest and east coasts. It rarely reached the south coast except in extreme years and almost never penetrated to the southwest. Map 1 shows the difference between the extent of ice in a mild or severe sea-ice year. In the former the ice stayed close to Greenland, and in the latter the ice could hold the island of Iceland in an icy grip.

Sea ice can take two main forms. One is the icebergs that break off from glaciers on land and then continue to drift on ocean currents. The other is the ice that forms on the surface when the temperature of the ocean becomes so

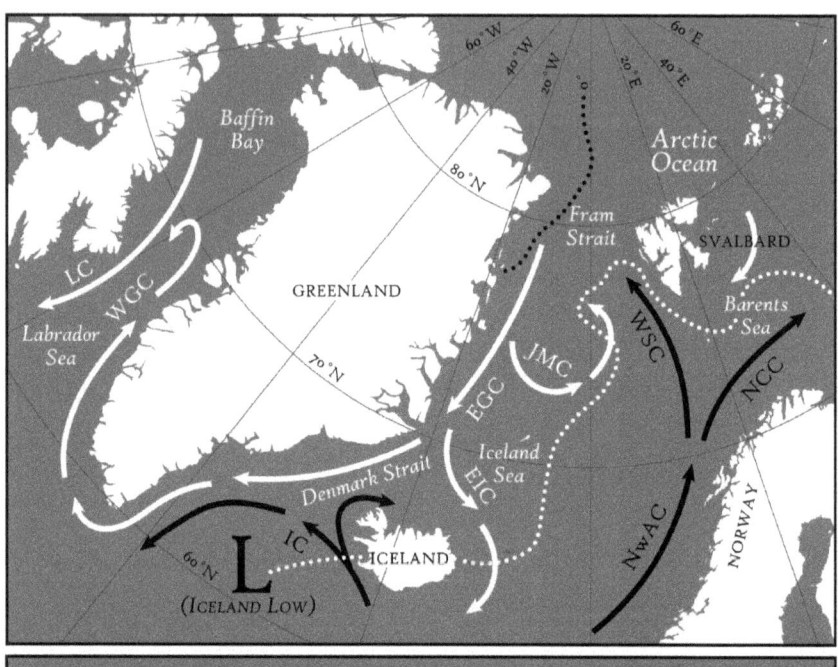

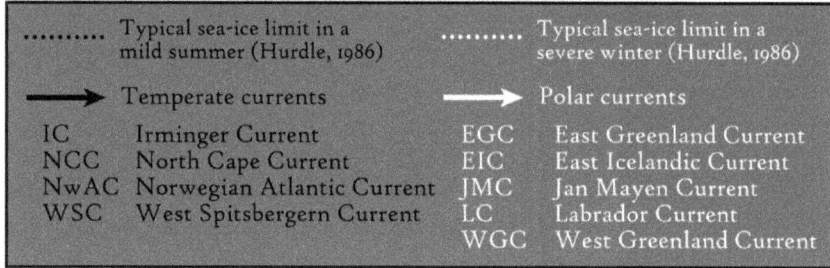

Map 1. Map of Iceland's location with surrounding North Atlantic Ocean currents and sea-ice limits. Map by Geoffrey Wallace (G. Wallace Cartography & GIS).

cold that it freezes. This usually occurs when the water temperature reaches -1.8 degrees Celsius. However, for the water to reach this freezing point, it is not just the immediate surface layer that has to be at this temperature – the top 20 to 350 metres of the surface mixed layer must also reach it (Ogilvie 2017). Sea ice has an aesthetic as well as a climatic aspect. Visually, it is one of nature's wonders, from the towering icebergs, 'mast high', to the pancake ice that stretches as far as the eye can see, to the striking aspect of its many colours,

Northern Iceland Temperature Variations and Sea-Ice Incidence

from deathly white to 'as green as emerald'.[1] Its beauty and strange elemental quality have inspired many writers and poets, from Samuel Taylor Coleridge to Mary Shelley to Charles Dickens (see, e.g., Potter 2007). Many Icelanders have described the ice in literary form, notably Mattías Jochumsson (1835-1920), poet, playwright and novelist, whose description of the ice as Iceland's 'ancient enemy' has remained a personification of ice still in use by Icelanders today (Ogilvie et al. 2021). The deadly elements of the ice are well-known, however, and it is proving to be a fickle friend, even to those cultures that have been able to use it to their advantage in the past (Ogilvie et al. 2021). The current lack of ice in Icelandic waters reflects the overall changing conditions in Arctic sea-ice distribution.[2] In past centuries, the presence of sea ice off the coasts of Iceland had numerous negative effects – e.g., lowering temperatures, preventing fishing and the arrival of trading vessels – and some positive ones, such as bringing a supplementary food source in the form of whales and seals, as well as driftwood, a vital resource in a country with few trees. On the whole, therefore, it would seem that the current situation is of benefit to Iceland. However, sea ice is a complex phenomenon and diminishing sea ice may have consequences as yet unknown. As one of the key drivers of the climate system of the Earth, it is one of the most important and variable components of the planetary surface and is the key to understanding many basic questions concerning regional and global climate change (Wadhams 2017; Miles et al. 2020).

Sources: Historical Records and Early Meteorological Observations

The information on sea ice and temperature presented in this paper is drawn primarily from historical documentary records. As regards these records, care has been taken to ensure that only reliable data are included in the analysis (see e.g., Bernard 2000). Thus, all sources used have been subject to rigorous historical analysis and are described in detail in previous work (Ogilvie 1982 et seq.) For the seventeenth century, the most useful sources are what are termed the 'Later Icelandic Annals' in reference to the medieval annals that were compiled in monasteries in Iceland and in many other European countries. These annals were laconic in the extreme, generally giving just a sentence or two to describe a year. The later annals are very different in that they can take many pages to describe events in a single year. They tend to give particular attention to weather events. In total, there are some forty of these annals and they were

---

1. 'And ice, mast high, came floating by, / As green as emerald'. From *The Rime of the Ancient Mariner* by Samuel Taylor Coleridge (1772–1834).
2. See NSIDC 2024 (https://nsidc.org/news-analyses/news-stories/arctic-sea-ice-has-reached-minimum-extent-2024).

written in many parts of the country (*Annálar 1400–1800*; Ogilvie 2005). For the current analysis, whenever possible, emphasis is placed on accounts from the north of Iceland. Coverage by these annals continues through to the end of the nineteenth century. From c. 1700 onwards, the primary sources for the climate and sea-ice reconstructions used here are official government reports. The government in question was the Danish government, as Denmark's rule dominated Iceland in effect from c. 1536 to 1944 when Iceland became a free and independent republic.

With the introduction of the Absolute Monarchy in Denmark in 1660, control over Iceland tightened. Part of the reason was undoubtedly the hope of exploiting Iceland's potential resources. As this was the time of a burgeoning interest in nature and science, it is possible that there was also a genuine desire to learn more about the colony. Whatever the cause, the result was that, from around 1700, the Governors of Iceland were required to send reports on the state of the country. To facilitate this task, officials, here termed 'Sheriffs' in each of the islands 22 counties were also required to write reports. These were written annually, sometimes more frequently, until 1894. They give detailed information on topics such as: weather and climate; sea ice; general environmental conditions; fisheries; trade; and agriculture, as well as socio-economic conditions (Ogilvie 2008). They are in manuscript form, in Danish,[3] in Gothic script and are held in the National Archives in Reykjavik. While there is a large variety of different historical records available for Iceland for the eighteenth and nineteenth centuries, these particular sources are most valuable for the climate reconstructions described here. The Icelandic word for 'county' is *sýsla* (pl. *sýslur)* and Map 2 shows the individual counties of Iceland. Emphasis here is on the reports from the northern counties of Húnavatnssýsla, Skagafjarðarsýsla, Eyjafjarðarsýsla and Thingeyjarsýsla. As local observers, Icelanders also wrote diaries and treatises that focus entirely on the weather. These are further supplemented by the accounts of environmental conditions written by visitors to Iceland in the eighteenth and nineteenth centuries, early newspapers and correspondence with foreign scholars (Ogilvie 2005). A list of all sources consulted may be found in Ogilvie 1982: 476–486.

---

3. Most of the Icelandic historical sources mentioned here are written in Icelandic. These have been translated by the first author, as have the Sheriffs' reports, written in Danish.

Northern Iceland Temperature Variations and Sea-Ice Incidence

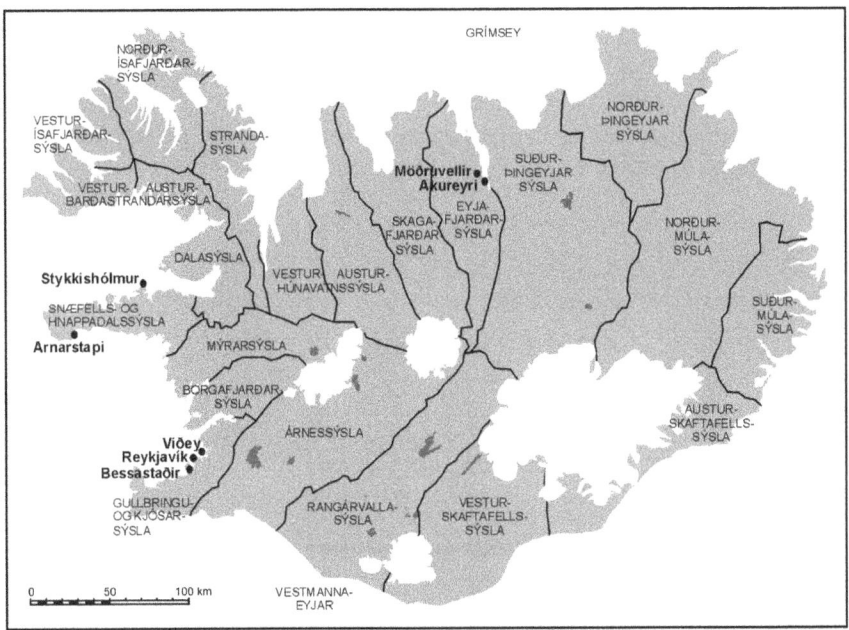

Map 2. Map of Iceland and its 22 traditional counties (sýsla/plural sýslur). It may be noted that the boundaries varied. Also shown are the residences of the District Governor and the Governor (Viðey, Bessastaðir (now the President's residence) and Möðruvellir. Also the capital, Reykjavík, and the northern capital, Akureyri. Map by Geoffrey Wallace (G. Wallace Cartography & GIS) after original by Astrid Ogilvie.

In addition to the historical records, climate information is available from early instrumental records. Comparison of the instrumental records with the documentary data is part of ongoing research by the authors and they are therefore noted here. The availability of these records is in large measure due to the work of the Icelandic meteorologist Trausti Jónsson. Of particular value is his extension of the well-known Stykkishólmur record, originally dating to 1846, back to 1820 with the inclusion of previously unused data (Sigfúsdóttir 1969; Jónsson T. 1989; Jónsson and Garðarsson 2001; Hanna et al. 2004, 2006). The principal series from the north of Iceland that is of relevance here is the series compiled by Hans Jacob von Scheel (1779–1851) a member of a Danish coastal surveying team with his co-worker Captain Frisak (1773–1834) at Akureyri from 1 September 1807 to 18 August 1814. There are both pressure and temperature measurements. The digitisation of this series is described by Jónsson and Garðarsson (2001).

Previous and Ongoing Research

Perhaps because of the importance of the weather for the Icelandic population in the past coupled with the high rate of literacy, many discourses were written in Iceland on variations in climate and climate impacts from the late-sixteenth century onwards (Ogilvie 2005). However, the seminal works in recent times are by Thorvaldur Thoroddsen (1855–1921), Lauge Koch (1892–1964) and Páll Bergþórsson (1923–2024). Thoroddsen was a dedicated geologist and geographer who devoted his life to explaining the natural world of Iceland. Among his many works is a compilation of climate and sea-ice events from AD 865 to 1900 (Thoroddsen, 1916–17). That this included unreliable information was first noted by Vilmundarson (1972) and subsequently explained in some detail by Ogilvie (1982 et seq.) In defence of Thoroddsen, it can be assumed that he intended simply to present the information that he found in the historical and literary sources available to him. He was not a historian, so his purposes did not involve source analysis.[4] While revisiting Thoroddsen's work for this paper, it was noticeable that he also favoured information that came primarily from the south of Iceland. It is also noteworthy that he did not consult the principal sources used here, the Sheriffs' reports from c. 1700–1894. It is remarkable that, for all his work on the past climate of Iceland, Thoroddsen concluded that, since the settlement of Iceland, the climate had never changed significantly: 'Then, as now, the sea ice came to the coasts, the glaciers were the same, the desert areas the same and the vegetation the same' (translated from Thoroddsen 1908–1922, vol. II: 371).

Lauge Koch was a Danish geologist and Arctic explorer. His principal work *The East Greenland Ice* (1945) concerns primarily sea ice, as the title suggests, and has a significant focus on the ice that reached the shores of Iceland on the East Greenland Current. From his research, Koch also concluded that cold periods need not necessarily be due to the presence of sea ice (Koch 1945: 247). In more recent times, the Icelandic meteorologist, Páll Bergthórsson, considered the connection between sea ice and temperature and, although Icelandic scholars and writers had noted this earlier, it was not until Bergthórsson´s research that statistical proof of the correlation was established (Bergþórsson 1969a). By comparing the temperature data from the sites of Stykkishólmur in the west of Iceland and Teigarhorn in the east (judging them to be representative of Iceland as a whole) with the incidence of ice off the coasts during the period 1846 to 1919, Bergthórsson found a significant correlation between temperature and sea ice. A similar comparison of ice and temperature during 1920 to 1969 did not

---

4. For an early critique of his work, see the article from 1915 by the Norwegian historian Edvard Bull, also discussed in Ogilvie and Jónsson 2001: 22-23.

produce quite such a good result. However, it was still statistically significant. Bergthórsson went on to infer temperatures during earlier periods based on assumed presence of ice. This aspect has been criticised by Vilmundarson (1972) and Ogilvie (1982 et seq.). However, Bergþórsson's basic conclusions regarding the connections between sea ice and air temperature cannot be disputed and they have paved the way for continued research into the topic. As regards Iceland's climate in the past, Bergþórsson was of a similar opinion to Thoroddsen, believing that the climate had not changed significantly: 'It is thus most likely that a nearly continuous cold period stretched from the late twelfth century until the present century' (translated from Bergþórsson, 1969b: 343). As the climate and sea-ice information presented by Koch and Bergþórsson is drawn primarily from Thoroddsen's compilation, his errors and omissions are replicated in their works. Nevertheless, the pioneering nature of all these studies must be emphasised and acknowledged.

Recent research on the past climate of Iceland is voluminous and therefore a discussion on this is beyond the scope of this paper. However, two projects may be mentioned. One concerns long-term sea-level pressure datasets from Iceland for the period from 1820 to 1999 produced for the project *Atmospheric Circulation Classification and Regional Downscaling* (ACCORD 2001) led by the Climatic Research Unit at the University of East Anglia in collaboration with the Icelandic Meteorological Office and which has relevance for temperature and sea-ice correlations. The other relevant project (acronym ICEHIST) involves current collaborative research by the present authors (see Acknowledgements). This considers aspects of the variability of Iceland's climate by advancing understanding of ice-ocean-atmosphere interactions, with a focus on abrupt anomalies in sea-ice export from the Arctic Ocean through Fram Strait and along the East Greenland Current. Sea-ice and freshwater anomalies originating in the Arctic can have far-reaching and long-lasting effects in the subpolar North Atlantic. A notable example occurred with the 'Ice Years' of 1965–71 associated with what has been termed the 'Great Salinity Anomaly' (Dickson et al. 1988). There are fragmentary indications of earlier such events; however, this remains essentially unexplored. An important scientific result from this project will be to ascertain the occurrence and extent of such outstanding anomalies, through identifying and understanding GSA-like events in previous centuries. This research is providing a historical perspective and contributing to a better understanding of changes in sea ice linked to ocean circulation and temperatures, marine ecosystems, and climate across the northern North Atlantic and adjacent areas. (See also Andrews et al. 2009 and Miles et al. 2014).

Methods

As noted above, the connection between the presence of sea ice off the coasts of Iceland, and lowered temperatures on land, in particular in the north of the country, has been observed by scholars and researchers both in the past and the present. This paper presents a further initial attempt to explore this relationship. At this stage of the investigation, it is deemed appropriate to merely present the descriptive evidence for past temperature variations as well as the presence of sea ice off the coasts. It is clear that the evidence is qualitative and relies on the subjective perceptions of observers in the past, as well as on the interpretative abilities of the current researchers. However, the use of such historical data and methods now has a long and distinguished history (see, e.g., Jones 2008; Pfister et al. 2018; White et al. 2025 and references therein) and the value of historical evidence for climate reconstruction has been much lauded (see, e.g., Nicholls 2010). As with the ICEHIST project noted above, such research may lead to advances in understanding ice-ocean-atmosphere linkages, in particular with sea ice as a key component and agent of change in ocean salinity and sea-surface temperatures in future climate scenarios (Muilwijk et al. 2024).

As regards weather terminology in the Icelandic language, this is far richer than English; however, a discussion of this topic is beyond the scope of this paper. Of course, it is clear that judging a season to be 'severe' or 'good' can mask a multitude of climatic nuances and has the effect of extreme simplification. Nevertheless, as a first step in evaluating the sea-ice/temperature relationship, it is judged to be valid. As noted earlier, sea ice was frequently more prevalent off Iceland´s coasts during the spring rather than the winter; however, for the sake of simplicity in the descriptions below main emphasis is placed on winter temperatures. This also reflects the emphasis placed on describing the winter season by past observers. The ice index presented in Figure 1 represents the number of seasons (winter, spring and summer) with sea ice present off the coast of Iceland per year. Sea ice only rarely appears in autumn so this season was not included in the index. Each number is weighted by the number of regions (north, south, east, west) that report ice. This is similar to Koch's weighted ice index. A comparison from 1601 to 1780 is shown in Ogilvie (1984: 145). The graphs are in broad agreement; discrepancies may be accounted for by Koch's use of erroneous data and the inclusion of further data for the current index.

Northern Iceland Temperature Variations and Sea-Ice Incidence

An Overview of Northern Iceland Temperature and Sea-Ice Variations c. 1600 to 1850

In the ACCORD project noted above, the period 1823–1999 was divided into five sub-periods each representing a 'temperature regime' (ACCORD, 2001: 390). Following this approach, we make suggestions for divisions for the period 1600 to 1850, as well as comparing temperature and sea-ice events in these years. The suggested thirteen divisions are: i) 1601–1605; ii) 1606–1640; iii) 1641–1680; iv) 1681–1700 v) 1701–1740; vi) 1741–1760; vii) 1761–1780; viii) 1780s; ix) 1790s; x) 1801–1820; xi) 1821–1850 (Figure 1). As may be seen, some of these years fit relatively easily into 20-year or longer periods, others do not and are categorised accordingly. Certain very interesting periods are quite brief, as is the case with the first one to be considered. This part of the analysis ends with the year 1850 as analysis of the instrumental temperature data is in preparation from 1851 onwards.

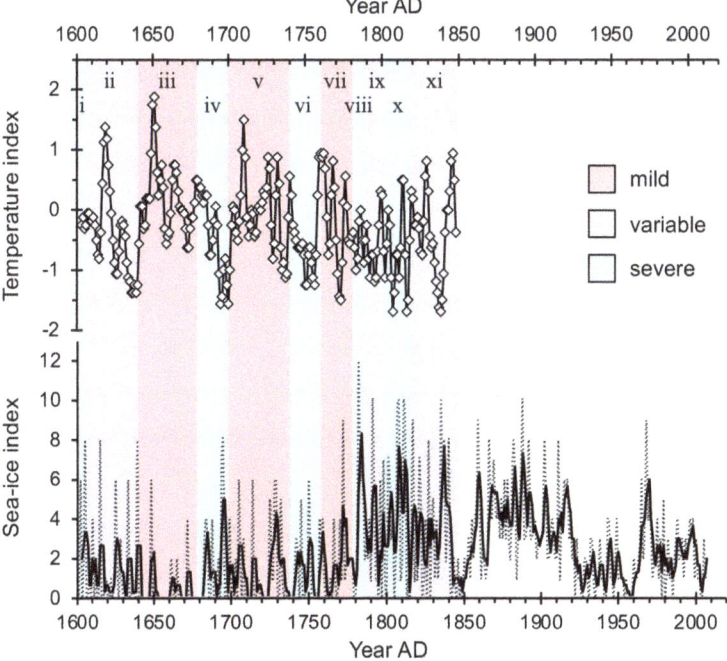

Figure 1. Upper series: A temperature index for Iceland based on documentary data showing variations during AD 1600–1850, smoothed with a 5-year binomial filter. Lower series: A sea-ice index for Iceland showing variations in the incidence of ice off the coasts during AD 1600–2005. The data have also been smoothed to highlight the lower frequency. Source: Astrid Ogilvie and Martin Miles.

i) **1601–1605: Severe with much sea ice.** The seventeenth century began with unusually severe years that were referred to in accounts for a long time afterwards. Years described as 'severe' were 1601, 1602 and 1605. Sea ice was described in 1602, 1604 and 1605. Examples of descriptions are as follows: For 1602 it is stated in the annal *Skarðsárannáll* (*Annálar 1400-1800*, I: 187): 'No one could remember such a harsh winter. It lasted from Christmas to St John's Eve (4 July N.S). There was sea ice until far into the summer.' For 1605 the same source states: 'A severe winter...Much ice came. It came to the east, came all around the coast round the eastern part and south around by Grindavík[5] by the end of the spring fishing season (12 May).' However, the year 1603 was described as 'good'. No mention is made of sea ice in 1601 but sea ice and severe temperatures are both described in 1602, 1604 and 1605. No sea ice is mentioned in 1603, the year that is described as 'good'. For this brief period, the presence of sea ice and severe years are in agreement.

ii) **1606–1640: Variable.** Following the severe years of the early part of the century, the next 35 years appear to have been extremely variable. Winters that were designated as mainly 'severe' were: 1605, 1610, 1612, 1614, 1615, 1616, 1625, 1627, 1628, 1629, 1630, 1633, 1634, 1636, 1637, 1639 and 1640. Sea ice was described in 1608, 1612, 1615, 1618, 1621 (only to the north of Iceland), 1624, 1625, 1626, 1628, 1629, 1630, 1633, 1634 ('some ice'), 1636 ('some ice'), 1638 ('not land fast') and 1639. Of these, 1615, 1625, 1628, 1633 and 1639 were described as 'very severe' ice years. 1625 was also an unusually harsh winter and Björn Jónsson, the author of *Skarðsárannáll*, wrote a poem about it called *Svellavetur* ('Ice winter' Jónsson, B. 1913). It is interesting to note that Icelandic scholars appeared to correspond frequently with their counterparts in other countries. In 1628, the scholar Árngrímur Jónsson wrote to the Danish physician and natural scientist Ole Worm (1588-1654): 'The Greenland (or Iceland) ice has lain around our coast until far into August and the merchant scarcely got into the harbour before he had to return home. In the meantime, we are suffering from the lack of hay and are sorely pressed by the highly unfavourable climate' (Benedikstsson 1948: 151). Years when both sea ice and severe winters were described were: 1608, 1612, 1615, 1618, 1622, 1625, 1628, 1629 (but only 'some ice'), 1633, 1634 (but only 'some ice'), 1636 (but only 'some ice') and 1639. Winters that were described as 'good'

---

5. Grindavík, with its natural harbour, has traditionally had an economy based on fisheries. In November 2023 it was affected by the eruption of the Sundhnúkur volcano. The town was evacuated and its future remains uncertain due to continuing volcanic and seismic activity.

Northern Iceland Temperature Variations and Sea-Ice Incidence

were: 1608, 1611, 1617 (in part), 1618, 1619, 1620, 1621, 1622, 1623, 1624, 1626 and 1635. During these 'good' years, sea ice was noted in 1608 (explicable by the severe spring that year), 1618, 1621 (but ice was only present off the north coast), 1622 (again, there was a severe spring), 1624 and 1626. While many severe years with much sea ice occurred during this period, the many years described as 'good' are notable. For the most part, the presence of sea ice and severe years (either winters or springs) are in agreement. Years where sea ice/temperature are not entirely in accord were 1634 and 1636 when the winter and spring were judged to be 'severe' but there was only 'some ice'.

iii) **1641–1680: Mild.** This forty-year period is notable for its many mild years and lack of sea ice. From 1641 to 1680 only four years were described as having severe ice conditions: 1648, 1665, 1672 and 1674. For 1651 and 1661 there was said to be 'some' sea ice. Years that were described as severe regarding to temperature were: 1641 (in part), 1644, 1648, 1654 ('quite hard'), 1655 ('severe but calm'), 1659, 1662, 1673, 1674 and 1680. So, only ten years were described as severe in a 39-year period, with most of these towards the end of the period. Years that were specifically described as 'good' were: 1642, 1643, 1646, 1647, 1649, 1650, 1651, 1652, 1653, 1656, 1657, 1658 ('good but stormy'), 1660, 1663, 1665 and 1679. Thus, a total of sixteen years. An example of a description is for the year 1647. 'A very good year in all things ... Frost hardly ever occurred and the weather was calm all winter with little precipitation ...' (the annal *Sjávarborgarannáll* in *Annálar 1400–1800*, vol. IV: 283–284). Another correspondent with Ole Worm, Gísli Magnússon, wrote to him on 26 September and stated: 'The previous winter (1647) appeared more like summer than winter' (Benediktsson 1948: 275). Clearly, temperature and sea ice were well in agreement during this period. An exception was the year 1651 when the winter and spring were described as 'good' but there was some sea ice present.

iv) **1681–1700: Variable then Severe.** During the first decade of this period, there was more sea ice off the coasts of Iceland than in any of the previous three decades. Severe ice years were described in 1683, 1684, 1685, 1688 and in 1691 there was sea ice 'present'. Severe temperatures were described in 1683 ('good then severe'), 1685 ('good then severe'), 1688 and 1691. Years with descriptions of good temperatures were the variable years of 1683 and 1685 then 1686, 1687, 1689, 1692 ('good then severe'), 1693 ('very good') and 1698. However, the 1690s brought extreme sea-ice years, perhaps similar to the early years of the century but this is likely to have been the coldest decade of the seventeenth century. Severe winters occurred in 1690, 1691, 1692 ('good then severe'), 1694, 1695, 1696, 1697, 1699 and in 1700. There were only two 'good' winters, in 1693 and 1698. Sea ice occurred in 1692, 1694, 1695 and 1699. The ice years of 1694 and 1695 were extreme. In 1694, the ice seems to have penetrated as far as the Westman Islands, off the south coast of Iceland. In 1695 the ice reached even further round the south coast, past Reykjavik.[6] There can be little doubt that temperature and sea ice correlated well during this period, with the exception perhaps of the year 1698 which was mainly good but sea ice was present.

v) **1701–1740: Relatively Mild.** This forty-year period was relatively mild, although not as mild as the years 1641–1680. During the first decade, sea ice was present in 1701, 1703 ('but not for long'), 1705 and 1706 ('but not for long'). During the second decade, 1714 was a severe ice year. Sea ice occurred in 1718 ('but not for long'). There were severe winters in 1701, 1703, 1706 ('mainly severe'), 1708, 1712 ('average to severe'), 1717, 1718 ('variable to severe'). 1720 was described as severe. Years described as 'good' are: 1702, 1704, 1705 ('tolerable'), 1707, 1709, 1710, 1711, 1713, 1714, 1716, 1719. During 1721 to 1740 sea ice occurred in 4 years: in 1726 ('some sea ice'), 1728 ('sea ice April to July'), 1729 ('April to August'), 1732 ('spring to end August') and 1733 ('all summer'). No sea ice was recorded during the most severe years of this latter decade.

vi) **1741–1760: Mainly very severe.** The years 1741 to 1760 included many very cold years, with much sea ice. During 1741 to 1750 there were seven years when sea ice was present (1741, 1742, 1743, 1745, 1748, 1749 and 1750). During the next decade, sea-ice years were: 1751, 1756, 1757, 1759 and with 'some ice' during 1760. The most severe ice-years were 1745 and 1756. In 1745 ice came early to the north and west and also reached the eastern and southern coasts. As regards temperature, this appears to

---

6. This information is from the annal *Vallanannáll*. The writer lived mainly in the north, but this year happened to be in the south of Iceland and wrote this description for 1695: 'From ... Christmas ... the weather was severe in the whole country with snow and northerly storms, layers of ice and severe frost all the time. Sea ice came to the north and lay to the time of the Althing (c. 9 July New Style). On 24 April a great deal of sea ice came into Nesjaflói (modern Faxaflói). Northerly weather had brought it to to the eastern and southern coasts of the country and it filled up all the bays in the south. Some of the ice drifted into the fishing grounds belonging to the people of Seltjarnarnes (now part of Reykjavik) and ice floes drifted into bays. It stayed to about the end of the fishing season (11 May). Sea ice had not come to Suðurnes for eighty years' (*Annálar* I: 429-430).

Northern Iceland Temperature Variations and Sea-Ice Incidence

have been the most severe year of the decade in all regions. Other severe winters in this decade in the north were 1742, 1747 and 1749. Good winters, or parts of winters were as follows: 1741 was 'good'; 1743 was 'reasonable' and 1744 was 'good on the whole' but the spring was very severe; 1746 was 'good' and 1748 was 'very good'. 1750 was 'good then severe'. On average, the 1750s were likely to have been one of the most severe decades of the period under consideration in terms of both temperature and sea ice, and is comparable to the severe 1690s. Cold years occurred one after the other from 1751 to 1757. It is likely that the year 1756 was one of the coldest years of this decade, if not the century. The minister and weather diarist Jón Jónsson the elder (1719–1795) noted 'severe weather with snow and lack of pasture all winter'. Social conditions were very difficult during the 1750s and a severe famine occurred during 1752-58 (Jónsson, G. 2023; Ogilvie and Sigurðardóttir, forthcoming). No sea ice appears to have been present during the severe year of 1754. Some sea ice was present during the 'mainly good' year of 1760.

vii) **1761–1780: Variable to Mild.** In comparison with the previous decades, the years 1761 to 1780 saw a return to milder conditions, albeit with several winters characterised as variable and some as severe. The characterisation 'variable' is given for this time period because descriptions of temperature seem to see-saw between 'good' or 'mild' and 'severe'. Sea ice was noted in thirteen years, but this was also variable: severe sea-ice years occurred in 1770, 1772 and 1775; there was 'much sea ice' in 1764 and 1766; 'some sea ice' in 1767, 1769 and 1777 and there was also said to be sea ice present in 1771, 1773, 1774, 1776 and 1778. As regards temperature, 1764, 1765, 1766, 1771 and 1774 were said to be variable. 1773 was said to be 'variable to severe' and 1780 was 'average then severe'. Winters characterised primarily as 'severe' or 'harsh' were 1761, 1770, 1772 and 1778. During 1761–1780 eight winters were characterised as 'good'. For this twenty-year period it may be seen that the 1760s was the milder of the two decades. During several of the winters that were described as 'variable' rather than severe (1764, 1766, 1771, 1773, 1774) there was sea ice present. However, the springs during these years were mainly characterised as severe. The years 1767, 1769, 1776, 1777 and 1778 had 'mainly good' winters but there was sea ice present in varying degrees during these years.

viii) **1780s: The Laki Fissure Eruption and the Famine of the Mist.** The 1780s were dominated by the Laki fissure eruption, known in Iceland as *Skaftáreldar* (from the fires seen on the Skaftá River adjacent to the eruption site). The eruption began on 8 June 1783 in a series of craters by the mountain Laki, southwest of the great glacier Vatnajökull in the district of Vestur-Skaftafellssýsla. The eruption lasted for eighty months and caused *Móðuharðindin* or 'the Famine of the Mist' (Bjarnar 1965). Some twenty per cent of the Icelandic population died in this famine (Gunnlaugsson et al 1984;[7] Demarée and Ogilvie 2001, 2016; Thordarson and Self 2003; Kleeman 2023; Ogilvie and Sigurðardóttir, forthcoming). However, the difficulties were compounded by severe weather and sea ice (Ogilvie 1986). There was ice off the coasts in varying degrees every year during 1781 through 1790, and these were mainly severe sea-ice years. In 1790 the ice reached the southwest coast, a true mark of severity. The year 1781 was characterised as mild, 1786 was 'tolerable' and 1787 was 'variable to good' but 1782, 1783, 1784, 1789 and 1790 were severe. 1788 was extremely snowy and 1785 seems to have been variable to severe.

ix) **1790s: Variable.** There was less sea ice in the 1790s than in the previous decade but there was long-lasting ice in 1792, 1794 and 1796. There was some ice in the spring in 1795 and 1799. For 1797 the sources specifically state that there was no ice. Severe and mild years were in balance with severe years in 1791, 1792, 1798. The years 1793, 1799 and 1800 were classed as good. 1794 was 'average', 1795 was variable as was 1797.

x) **1801–1820: Variable to Severe.** The winters (and springs) seem almost to alternate between severe and mild. 1801 was good to March then severe, with a severe spring. 1802 was extremely severe, 1804 was mainly good with a mainly severe spring. 1805 was good. 1806, 1807 and 1808 were mainly severe while 1809 was variable. 1810, 1811 and 1812 were mainly severe. Other severe winters were 1815, 1816, 1817 and 1819. 1813, 1814 and 1820 were classified as mild and 1818 was described as mild in the northwest, and harsher in the northeast. Severe sea-ice years were 1801, 1802, 1804, 1807, 1808, 1810, 1811, 1812, 1816, 1817, 1819. Some sea ice (not extensive) was seen in 1806, 1818 and 1820. It was stated that there was no sea ice in 1813, 1814 and 1815. It was in 1816 that the notorious 'Year Without a Summer' occurred in the wake of the eruption of Mount Tambora in Indonesia in 1815 (Stothers 1984). However, the year was not unusually severe in Iceland in comparison with other severe years around this time (Ogilvie 1992).

---

7. Some 32 contemporary reports regarding the eruption written by officials in Iceland (the Sheriffs noted earlier) are published in Gunnlaugsson et al. (1984: 299–417).

Northern Iceland Temperature Variations and Sea-Ice Incidence

xi) **1821–1850: Mild then Severe then Mild.** During the early part of this period several winters during the years 1821 through 1833 were described as mild (1821, 1823, 1828, 1829, 1830, 1831, 1832, 1833) or mainly mild. 1824 was variable and stormy. For 1825 floods and avalanches were noted. 1826 was 'stormy'. Winters described as 'severe' were 1822, 1827 ('partly severe'), then 1834, 1836, 1837 (in most northern districts), 1838 (but 'reasonably good' in Eyjafjarðarsýsla), 1839 ('mainly severe') and 1840. For the most part, the presence of sea ice and severe years are in agreement. The three years where this was not the case were 1822, 1823 and 1831. In 1822 there was a severe winter but, although the spring was mild, sea ice was present from May to July. 1823 was a good winter but although there was a severe spring, there was no sea ice in the north, and in 1831 the winter was mainly mild, and the spring was mild, but there was much ice from mid-January to mid-September. During the decade 1841-1850, several seasons were described as mild or 'not severe'. These included the winters of 1841, 1842, 1843 ('mild then severe'), 1844, 1846 ('not severe'), 1848 ('not particularly severe') and 1850 ('mainly mild'). Springs that were described as 'mild' were 1842, 1844, 1846 and 1849. There was no sea ice to speak of during this decade. No sea ice was reported during the 'frequent severe cold' during 1849.

Conclusions

As noted, previous researchers have concluded that the past climate of Iceland was more or less uniformly unfavourable, with heavy sea-ice incidence, in particular during the period 1600 to 1900 (e.g., Thoroddsen 1916/17; Koch 1945). However, careful scrutiny of the sources used by them, plus the addition of historical sources not included in their analyses, indicates that there was, in fact, a great deal of climatic variability from early settlement times to the present day and it is suggested that different climatic regimes or epochs may be discerned. Figure 1 shows sea-ice variations from 1600 to the present century. From this diagram, it is clear that the early and latter decades of the seventeenth century were years with much ice present. From c. 1640 to c. 1680 there appears to have been little sea ice off Iceland's coasts. During the period 1600 to 1850, the decades with most ice present were probably the 1780s, early 1800s and the 1830s. From 1840 to 1850 there was virtually no ice off the Icelandic coasts. As regards temperature variations, a cooling trend may be seen around the beginning and end of the seventeenth century. However, these periods are separated by a mild period from c. 1640 to 1680. It is interesting to note that this coincides with the cold period in central Europe that is often regarded as being at the height of the so-called 'Little Ice Age'. This may be regarded as a classic example of Iceland diverging from continental European weather systems (Meehl and van Loon 1979; Pfister and Wanner 2021). The early decades of the 1700s were relatively mild in comparison with the very cold 1690s, 1730s, 1740s and 1750s. The 1760s and 1770s include some mild years but also much variability. The 1780s are likely to have been the coldest decade of the century, but this was compounded by volcanic activity (Ogilvie 1986). The 1801s and 1830s were comparatively cold.

Historical sources for the period from 1850 onwards have been consulted by the authors but have not yet been analysed in depth. Nevertheless, some comments can be made regarding temperature and sea-ice variations up to the early twentieth century. From 1850 to 1855 there was little ice off the coasts. From 1855 to 1860 there was frequent ice again, although the incidence does not seem to have been as heavy as in the earlier part of the century. Further clusters of sea-ice years occurred again from c. 1864 to 1872. Several very heavy sea-ice years occurred during the 1880s, including 1881 and 1882. Some sea-ice years occurred in the 1890s and 1910s, but far fewer than in the 1880s. As regards air temperature, the 1880s were very cold on the whole. It was in the 1880s that Iceland experienced its last great subsistence famine (Harris 1978; Ponzi 1995). In the year 1888, poet and writer Matthías Jochumsson composed his famous poem *Hafísinn*, 'The Sea Ice' where the sea ice is personified as Iceland's ancient enemy. The high sea-ice incidence evident in the 1880s and 1910s - representing GSA-like events - is in contrast to the climate amelioration recorded in Europe in the mid-nineteenth century. Sea-ice incidence falls off dramatically from 1920 onwards, at the abrupt onset of the early Twentieth Century Warming (Wood and Overland 2009), with the exception of the 'Ice Years' from 1965–71.

The focus here has not been on climate impacts or socioeconomic aspects such as famines. (Regarding that topic, see, e.g, Ogilvie 2020; Júlíusson 2021; Jónsson, G. 2023; Ogilvie and Sigurðardóttir, forthcoming.) However, it is of interest that notable periods of famine did coincide with periods of severe years as well as with epidemics. Thus famines occurred in: the early 1600s; 1696–1702; during the smallpox epidemic of 1707–1709; during 1752–58; in association with the Lakagígar eruption and the severe years of the 1780s, in particular during 1784–1786. It is also of interest to note the conclusions of the ACCORD project described above. For the period 1823–1999 it was suggested that this could be divided into five sub-periods each representing a 'temperature regime' (ACCORD 2001: 390). These are: 1823–1857 (warmer then later cooling), 1858–1892 (the coldest period), 1893–1919 (slightly warmer), 1920–1964 (the warmest in this 1823–1999 period) and 1965–1999 (colder, primarily in the earliest part of this period, associated with the 'Ice Years'). The ultimate conclusion here is that there is clearly a strong relationship between air temperature and the presence of sea ice; however this is complex. Continuing research will help to solve the puzzle. Icelanders are, for the most part, grateful that the country's ancient enemy, the sea ice in the form of a 'fleet of silver'[8] that came to torment them, appears, for the present time, to have been vanquished. However, lack of ice brings its own problems, not least the effects that diminishing ice in the Arctic has on global climate.

Acknowledgements

We acknowledge support from the National Science Foundation of the USA (Award 212786) *Synthesizing Historical Sea-Ice Records to Constrain and Understand Great Sea-Ice Anomalies (ICEHIST)* PI Martin Miles, Co-PI Astrid Ogilvie. Support from the Centre for Advanced Study (CAS) in Oslo, Norway, which funds and hosts the research project 'The Nordic Little Ice Age' during the 2024/25 academic year is acknowledged by Astrid Ogilvie. We thank our Icelandic sea-ice colleagues Trausti Jónsson and Ingibjörg Jónsdóttir for collaboration over many years. Maps 1 and 2 were provided by Geoffrey Wallace (G. Wallace Cartography & GIS).

Bibliography

Manuscript Sources

Þjóðskjalasafn (National Archives), Reykjavík
    Bréf úr Húnavatnssýslu til Stiftamtmanns
    Bréf úr Skagafjarðarsýslu til Stiftamtmanns
    Bréf úr Eyjafjarðarsýslu til Stiftamtmanns
    Bréf úr Þingeyjarsýslu til Stiftamtmanns

Landsbókasafn (National Library), Reykjavík
    The Jón Jónssons´Weather Diary ÍBR 81-87 8vo, Lbs 332 8vo

Published Works

ACCORD. *E.U. Project Number ENV-4-CT97-0530. 2001.* University of East Anglia. https://crudata.uea.ac.uk/cru/projects/accord/.

---

8. The reference is to the poem *Hafísinn*, 'The Sea Ice' noted above, and part of the first verse (translated by Astrid Ogilvie and Níels Einarsson, see Ogilvie et al. 2022) reads 'Have you come, our country's ancient enemy? / You arrived upon the sandy shore / Before sailing ship, sun and urgent help / A fleet of silver, come to torment us!'

Andrews, J.T., D. Darby, D. Eberle, A.E. Jennings, M. Moros and A.E.J. Ogilvie. 2009. 'A robust, multisite Holocene history of drift ice off northern Iceland: implications for North Atlantic climate'. *The Holocene* 19 (1): 71–77.

Annálar 1400–1800. Annales Islandici Posteriorum Sæculorum I–VI. 1922–87. Reykjavík: Hinu Íslenzka Bókmentafélagi.

Benediktsson, J. (ed.). 1948. 'Ole Worm's Correspondence with Icelanders'. In Jón Helgason (ed.) *Bibliotheca Arnamagnæana I–VII*. Copenhagen: Munksgaard.

Bergthórsson, P. 1969a. 'An estimate of drift ice and temperature in Iceland in 1000 years'. *Jökull* 19: 94–101.

Bergthórsson, P. 1969b. 'Hafís og hítastig á liðnum öldum'. In Markús Á. Einarsson (ed.) *Hafísinn*. Reykjavík: Almenna Bókafélagið. pp. 94–101.

Bernard, H.R. 2000. *Social Research Methods: Qualitative and Quantitative Approaches*. Thousand Oaks: Sage Publications.

Bjarnar, V. 1965. 'The Laki eruption and the Famine of the Mist'. In Carl F. Bayerschmidt and Erik J. Friis (eds), *Scandinavian Studies*. The American-Scandinavian Foundation, University of Washington Press. pp. 410–21.

Bull, E. 1915. 'Islands Klima i Oldtiden'. *Geografisk Tidskrift* 23: 1-5.

Demarée, G.R. and A.E.J. Ogilvie. 2001. '*Bon baisers d'Islande*: Climatological, environmental and human dimensions impacts in Europe of the *Lakagígar* eruption (1783–1784) in Iceland'. In P.D. Jones, A.E.J. Ogilvie, T.D. Davies and K.R. Briffa (eds), *History and Climate: Memories of the Future*. New York: Kluwer Academic/Plenum Publishers. pp. 219–246.

Demarée, G.R. and A.E.J. Ogilvie. 2016. 'L'éruption du Lakagígar en Islande ou Annus mirabilis 1783'. In I. Parmentier (ed.) *Études et bibliographies d'histoire environnementale. Belgique – Nord de la France – Afrique centrale*. Sous la direction Namur: Presses Universitaires de Namur. pp. 117–157.

Dickson, R.R, J. Meincke, S.A. Malmberg and A.J. Lee. 1988. 'The "Great Salinity Anomaly" in the northern North Atlantic 1968–1982'. *Progress in Oceanography* 20: 103–15.

Gunnlaugssson, G.Á., G.M. Guðbergsson, S. Þórarinsson, S. Rafnsson and Þ. Einarsson. (eds). 1984. *Skaftáreldar 1783–84 Ritgerðir og Heimildir*, Reykjavík: Mál og Menning.

Hanna, E., T. Jónsson and J.E. Box. 2004. 'An analysis of Icelandic climate since the nineteenth century'. *International Journal of Climatology* 24 (10): 1193–1210. https://doi.org/10.1002/joc.1051.

Hanna, E.T., T. Jónsson, J. Ólafsson and H. Valdimarson. 2006. 'Icelandic coastal sea-surface temperature records constructed: Putting the pulse on air-sea-climate interactions in the northern North Atlantic. Part 1: Comparison with HadISST1 open ocean surface temperatures and preliminary analysis of long-term patterns and anomalies of SSTs around Iceland'. *Journal of Climatology* 19: 5652–5666.

Harris, R.L. 1978. 'William Morris, Eiríkur Magnússon, and the Icelandic famine relief efforts of 1882'. *Saga-Book of the Viking Society* 20: 31–41.

Jochumsson, Matthías. 1888. 'Hafísinn'. *Norðurljósið*. **6**, 6 April.

Jones, P.D. 2008. 'Historical climatology – a state of the art review'. *Weather* **63** (7): 181–186.

Jónsson, B. 1913. 'Svellavetur'. *Andvari* **38**: 104-110. https://timarit.is/page/4329603#page/n165/mode/2up

Jónsson, G. 2023. 'Hve Margir Dóu? Hungursneyð Átjánda Aldar og Nítjándu Aldar og Mannfall í Þeim' [How many died? Famines in the eighteenth and nineteenth centuries and loss of life in them]. *Saga Tímarit Sögufélags* **61** (2): 35-88.

Jónsson T. 1989. *Afturábak frá Stykkishólmi, Veðurathuganir Jóns Þorsteinssonar landlæknis í Nesi og í Reykjavík* [The observations of Jón Thorsteinsson in Nes and Reykjavík 1820–1854, and their relation to the Stykkishólmur series]. Icelandic Meteorological Office.

Jónsson, T. and H. Garðarsson. 2001. 'Early instrumental meteorological observations in Iceland'. In A.E.J. Ogilvie and T. Jónsson (eds), *The Iceberg in the Mist: Northern Research in Pursuit of a 'Little Ice Age'*. Dordrecht: Kluwer Academic Publishers. pp. 169–187.

Júlíusson, Á.D. 2021. 'Agricultural growth in a cold climate: The case of Iceland in 1800–1850'. *Scandinavian Economic History Review* **69**: 217–232. https://doi.org/10.1080/03585522.2020.1788985.

Kleeman, K. 2023. *A Mist Connection. An Environmental History of the Laki Eruption of 1783 and its Legacy*. Berlin: De Gruyter.

Koch, L. 1945. *The East Greenland Ice. Vol. 130, Utgave 3 af Meddelelser om Grønland, udgivne af Kommissionen for videnskabelige undersøgelser i Grønland*. København: C.A. Reitzels Forlag.

Meehl, G.A. and H. Van Loon. 1979. 'The seesaw in winter temperatures between Greenland and Northern Europe. Part III: Teleconnections with Lower Latitudes'. *Monthly Weather Review* **107**: 1095–1106.

Miles, M.W., D.V. Divine, T. Furevik, E. Jansen, M. Moros and A.E.J. Ogilvie. 2014. 'A signal of persistent Atlantic multidecadal variability in Arctic sea ice'. *Geophysical Research Letters* **41**. https://doi.org/10.1002/2013GL058084

Miles, M.W., C.S. Andresen and C.V. Dylmer. 2020. 'Evidence for extreme export of Arctic sea ice leading the abrupt onset of the Little Ice Age'. *Science Advances* **6**: 38. https://doi.org/10.1126/sciadv.aba432

Muilwijk, M., T. Hattermann, T. Martin and M.A. Granskog. 2024. 'Future sea ice weakening amplifies wind-driven trends in surface stress and Arctic Ocean spin-up'. *Nature Communications* **15**. https://rdcu.be/d8mFb.

National Snow and Ice Data Centre (NSIDC). 2024. Accessed 20 December 2024. https://nsidc.org/news-analyses/news-stories/arctic-sea-ice-has-reached-minimum-extent-2024

Nicholls, N. 2010. 'Why do we care about past climates, an editorial essay'. *Climate Change* **1**: 155–157. https://doi.org/10.1002/wcc.4.

Ogilvie, A.E.J. 1982. *Climate and Society in Iceland From the Medieval Period to the Late Eighteenth Century*. Unpublished PhD thesis. Norwich: School of Environmental Sciences, University of East Anglia.

Ogilvie, A.E.J. 1984. 'The past climate and sea-ice record from Iceland. Part 1: Data to A.D. 1780'. *Climatic Change* **6**: 131–152.

Ogilvie, A.E.J.. 1986. 'The climate of Iceland, 1701–1784'. *Jökull* **36**: 57–73.

Ogilvie, A.E.J. 1992. 'Documentary evidence for changes in the climate of Iceland, A.D. 1500 to 1800'. In R.S. Bradley and P.D. Jones (eds), *Climate Since A.D. 1500*. London: Routledge. pp. 92–117.

Ogilvie, A.E.J. 1992. '1816 – A year without a summer in Iceland?' In C.R. Harrington (ed.) *The Year Without a Summer? World Climate in 1816*. Ottawa: Canadian Museum of Nature. pp. 331–354.

Ogilvie, A.E.J. 2005. 'Local knowledge and travellers' tales: A selection of climatic observations in Iceland'. In C. Caseldine, A. Russell, J. Harðardóttir and O. Knudsen (eds), *Iceland – Modern Processes and Past Environments, Developments in Quaternary Science 5*. Amsterdam: Elsevier. pp. 257–287.

Ogilvie, A.E.J. 2008. '*Bréf sýslumanna til stiftamtmanns og amtmanns*: Environmental images of nineteenth-century Iceland from official letters written by district sheriffs'. In M. Wells (ed.) *The Discovery of Nineteenth-Century Scandinavia*. Norwich: Norvik Press. pp. 43–56.

Ogilvie, A.E.J. 2010. 'Historical climatology, *Climatic Change*, and implications for climate science in the 21st century'. *Climatic Change* **100**: 33–47.

Ogilvie, A.E.J. 2017. 'A brief description of sea ice'. In E. Ogilvie (ed.) *Out of Ice*. London: Black Dog Publishing. pp. 88–90. https://www.elizabethogilvie.org/outreach

Ogilvie, A.E.J. 2020. 'Famines, mortality, livestock deaths and scholarship: Environmental stress in Iceland c. 1500–1700'. In A. Kiss and K. Prybil (eds), *The Dance of Death. Environmental Stress, Mortality and Social Response in Late Medieval and Renaissance Europe*. London: Routledge. pp. 9–24. https://doi.org/10.4324/9780429491085

Ogilvie, A.E.J. 2022. 'Writing on sea ice: Early modern Icelandic scholars'. In K. Dodds and S. Sörlin (eds), *Ice Humanities: Living, Working, and Thinking in a Melting World*. Manchester: Manchester University Press. pp. 37–56. https://doi.org/10.7765/9781526157782

Ogilvie, A.E.J., B.T. Hill and G.R. Demarée. 2021. 'A fleet of silver: Local knowledge perceptions of sea ice from Iceland and Labrador/Nunatsiavut'. In E. Panagiotakopulu and J.P. Sadler (eds), *Biogeography in the Sub-Arctic: The Past and Future of North Atlantic Biota*. Chichester: Wiley. pp. 273–291. https://doi.org/10.1002/9781118561461.

Ogilvie, A.E.J. and R. Sigurðardóttir. Forthcoming. 'Living with ice and fire: Responses to natural hazards in early modern Iceland'. In D. Degroot, J.R. McNeill and A. Hessl (eds), *The Oxford Handbook of Climate Resilience*. Oxford: Oxford University Press.

Pfister, C. and H. Wanner. 2021. *Climate and Society in Europe*: Berne: Haupt Verlag

Northern Iceland Temperature Variations and Sea-Ice Incidence

Pfister, C., R. Brázdil, J. Luterbacher, A.E.J. Ogilvie and S. White. 2018. 'Early modern Europe'. In S. White, C. Pfister and F. Mauelshagen (eds), *The Palgrave Handbook of Climate History*. London: Palgrave Macmillan. pp. 265–295.

Ponzi, F. 1995. *Ísland Fyrir Aldamót – Iceland – The Dire Years*. Mosfellsbær: Brennholt.

Potter, R.A. 2007. *Arctic Spectacles: The Frozen North in Visual Culture, 1818–1875*. Seattle: University of Washington Press.

Sigfúsdóttir, A.B. 1969. 'Hitabreytiingar á Íslandi 1846–1968'. In M.Á. Einarsson (ed.) *Hafísinn*. Reykjavík: Almenna Bókafélagið. pp. 70–79

Stothers, R.B. 1984. 'The Great Tambora Eruption in 1815 and its aftermath'. *Science* **224** (4654): 1191–1198.

Thordarson, Th. and S. Self. 2003. 'Atmospheric and environmental effects of the 1783–1784 Laki eruption: A review and reassessment'. *Journal of Geophysical Research: Atmospheres*, **108** (D1): 4011.

Thoroddsen, Þ. 1908–1922. *Lýsing Íslands I–IV* [A Description of Iceland I–IV]. Copenhagen: Hinu Íslenzka Bókmenntafélagi.

Thoroddsen, Þ. 1916–17. *Árferði á Íslandi í púsund ár* [The Seasons in Iceland in 1000 Years]. Copenhagen: Hinu Íslenzka Bókmenntafélagi.

Vilmundarson, Th. 1972. 'Evaluation of historical sources on sea ice near Iceland'. In T. Karlsson (ed.) *Sea Ice, Proceedings of an International Conference*. Reykjavík: National Research Council. pp. 159–69.

Wadhams, P. 2017. *A Farewell to Ice: A Report from the Arctic*. Oxford: Oxford University Press.

White, S., D. Collet, A. Alcoberro et al. 2025. 'Climate, peace, and conflict – past and present: Bridging insights from historical sciences and contemporary research'. *Ambio*. https://doi.org/10.1007/s13280-024-02109-1.

Wood, K.R. and J.E. Overland. 2009. 'Early 20th century Arctic warming in retrospect'. *International Journal of Climat*ology **29**. https://doi.org/10.1002/joc.1973

The Authors

Astrid E.J. Ogilvie is a Research Professor at the Institute of Arctic and Alpine Research at the University of Colorado and a Senior Associate Scientist at the Stefansson Arctic Institute in Akureyri, Iceland. Her research focuses on the broader issues of climatic change and contemporary Arctic issues, as well as the environmental humanities. Her interdisciplinary, international projects have included leadership of the NordForsk Nordic Centre of Excellence project: *Arctic Climate Predictions: Pathways to Resilient, Sustainable Societies (ARCPATH)*; and *The Natural World in Literary and Historical Sources from Iceland ca. AD 800 to 1800 (ICECHANGE)*. She is currently a Fellow of the project *The Nordic Little Ice Age (1300–1900) Lessons from Past Climate Change (NORLIA)* at the Centre for Advanced Study at the Norwegian

Academy of Science and Letters. She is the author of some 100 scientific papers and has three edited books to her credit.

Martin W. Miles is a Senior Research Scientist at NORCE Norwegian Research Centre in Bergen, Norway, and the Institute of Arctic and Alpine Research, University of Colorado, Boulder, USA. He has nearly two decades of experience in university teaching and curriculum development in geography, climate and environmental science, and quantitative methods. His research specialities include climate-system variability and regime shifts, historical climate, paleoclimate, and sea ice. His regional areas of interest are the European-Atlantic Arctic and Subarctic, including Greenland and Svalbard. Methodological approaches include empirical analysis of multivariate data records and time-series analysis, using independent but complementary data sources such as historical observations, long instrumental time series and high-resolution paleo proxy records from biological (e.g., marine sclerochronology) and geological archives (e.g., marine sediments).

Chapter 9

INTEGRATING AGRICULTURAL VULNERABILITY AND CLIMATE EXTREMES. EIGHTEENTH-CENTURY NORWAY THROUGH THE WORKS OF JACOB NICOLAJ WILSE (1735–1801)

Ingar Mørkestøl Gundersen

Abstract

In this study, I have made use of agricultural and meteorological data from late eighteenth-century pastor Wilse in Spydeberg, southeastern Norway, to analyse the impact of climate extremes on a premodern farming society. His farm records from the 1770s are used to improve an existing GDD model and then tested, by using his measured weather data, on the warm and cold summers of 1783 and 1784 respectively. The improved GDD model demonstrates that the 1784 climate anomaly had the potential to severely affect the crops. Contemporary accounts from other parts of southeastern Norway support the model result by reporting widespread harvest failures. Even though Norway is particularly susceptible to climate variations, the importance of climate extremes for these events has been little discussed among Norwegian historians. However, an integrated approach can be used to move beyond mere correlation between climate and human proxies towards some level of causation and contribute with new insights on the role of climatic stress for sociopolitical changes in the past.

Introduction

The 'Little Ice Age' (LIA, ~1300–1850) describes one of the coldest periods of the Holocene in the Northern Hemisphere (Pfister et al. 2018; Wanner et al. 2022). Likely caused by a combination of reduced solar irradiance and increased volcanic activity, the LIA brought a colder climate, shorter growing seasons and several episodes of rapid multiannual cooling – so-called 'climate extremes'. Climatic variability and change put societies under pressure, even contributing to supra-regional famines and demographical crises (Ljungqvist et al. 2023). Norway went through several profound crises during the eighteenth and early nineteenth centuries, when widespread famine and epidemics resulted in population decline (Drake 1969; Sogner 1979). Most notable were the periods 1742–1743, 1773–1774 and 1809–1810. Although the causes were related to an intricate mix of social factors, including trade regulations, grain prices, harvest failures, poor social infrastructures and a wartime blockade (Drake 1969; Glenthøj and Ottosen 2014; Lunden 2002; Sogner 1976, 1979), they often coincided with severe cold-spells too (Dybdahl 2016). Even though Norway is situated at the margins of grain cultivation, and thus particularly susceptible to climate variations, the importance of climate extremes for these events has been little discussed among Norwegian historians.

This paper explores the role of climate extremes in harvest failures in eighteenth-century Norway by examining the pioneering work in agronomy and meteorology of Jacob Nicolaj Wilse (1735–1801), a prominent figure among the so-called Nordic Enlightenment intellectuals (Eriksen 2007; Ryan 2023; Stubberud 2016). Wilse's publications, as well as his scientific contributions to the Societas Meteorologica Palatina (SMP), provide a rare window on the interplay between climate variability and agricultural outputs during the LIA. This study utilises Wilse's detailed documentation and applies an improved GIS-based Growing Degree Days (GDD) model to analyse agricultural vulnerability to climate cooling in a premodern farming society in southeastern Norway during the cold summer of 1784. This study does not merely correlate climate variability and societal impacts, but integrates both climatic and historical evidence within a common socionatural framework. Furthermore, the study contextualises these findings within historical accounts of food shortages and demographic changes, thereby offering new insights in how climate extremes affected pre-modern Norway.

Wilse's Life and Legacy

Jacob Nicolaj Wilse was born in 1735 in Lemvig in Northern Jutland, Denmark. Both his parents came from educated families, and early on he became interested in emerging scientific disciplines such as botany, zoology, geography and medicine (Stubberud 2016: 33–35). His interests were of both practical and intellectual character, something that also defined his later career. Wilse combined his studies with experimentation, data collecting and practical know-how. He received his education at the University of Copenhagen, where he studied theology and science. He earned his degree in 1768 and, later that same year, he received his first priestly office at Spydeberg, in the present-day municipality

Integrating Agricultural Vulnerability and Climate Extremes

of Indre Østfold in Østfold county, southeastern Norway (Map 1 and Map 2). However, his legacy is primarily associated with his works in topography, agronomy and meteorology (Stubberud 2016). Particularly important were his efforts in developing standardised meteorological symbols that could replace wordy descriptions and thus make possible the quantification and comparison of weather data across borders and languages (Wilse 1778). He built on the works of Pieter van Musschenbroek (1692–1761) while improving the statistical qualities of van Musschenbroek's methods (Federhofer 2002).

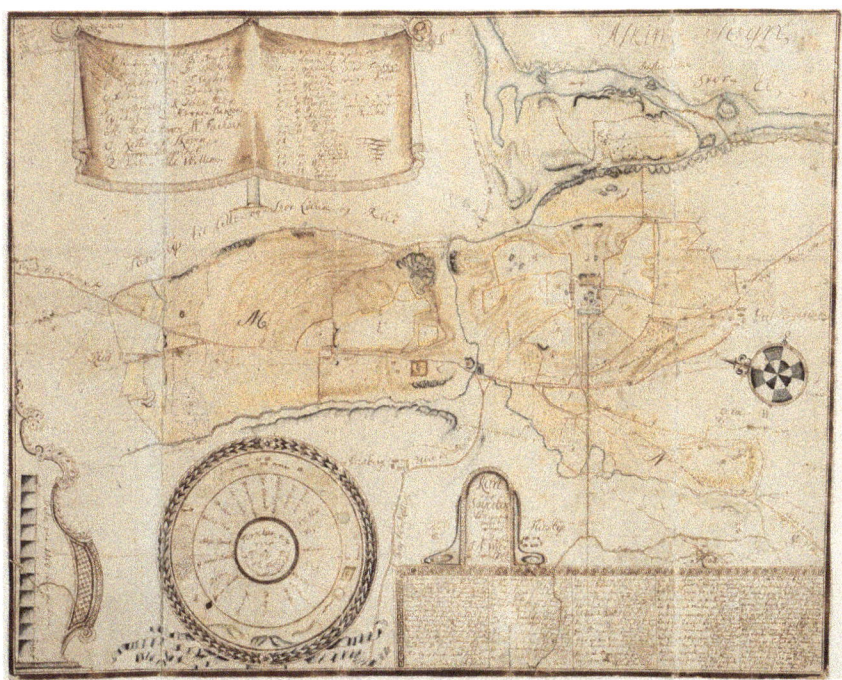

Map 1. Map by Jacob Nicolaj Wilse, 1768, covering Spydeberg rectory and associated fields. Source: Østfoldmuseene, BRM.06029 (CC BY-NC 4.0).

Along with his meteorological studies, Wilse took a keen interest in agriculture. He meticulously documented farming practices and harvests at his rectory, engaged in experimentation, and noted seasonal changes in weather and wildlife. His observations for the decade 1769–1779 were developed into a calendar that organised the dates for farm and household activities, such as sowing and harvesting, and informed the reader when the snow melt was likely to start, when floods usually occurred and when different migratory birds were expected to arrive in spring (Wilse 1991 [1779–1780]: 472–488). Thus, the calendar serves as a good example of his profoundly holistic approach to nature and society, which the biographer Tore Stubberud (2016) has described as almost ecological in its character. His efforts in meteorology earned him international recognition, and he was accepted to the SMP in 1783, thus becoming part of the most influential international meteorological network of his time (Winkler 2023). The same year he started reporting weather measurements from his rectory at Spydeberg to the society's headquarters in Mannheim, Bavaria. In 1786 he moved to the neighbouring parish of Eidsberg, also in Indre Østfold, and continued his observations throughout 1787.

Wilse was a leading representative of what Siobhan Moira Ryan (2023) has termed 'Nordic Enlightenment intellectuals'. These intellectuals often possessed an ecclesiastical background, but were firmly integrated into international science and brought a particularly strong interest in the emerging field of meteorology. Wilse was also an accomplished topographer, who published studies of Spydeberg and Eidsberg, as well as several shorter pieces based on his travels (Wilse 1790–1798, 1991 [1779–1780], 1992 [1793–1796]), adding to a rapidly growing body of topographical literature in Norway during the eighteenth century. Wilse seems to have been motivated by a personal ambition for international academic recognition (cf. Stubberud 2016: 70, 103, 108–117). However, his interest in science and agriculture also reflected a wider Enlightenment context, which fostered new ideas about social and scientific improvements. For instance, his topographic description of Spydeberg is concluded with a prophetic vision of how society may have progressed at Spydeberg 200 years into the future, where science and knowledge are put into systematic use to benefit humankind (Wilse 1991 [1779–1780]: 443–470). He portrayed an idealised future without misery, where modesty and rationality had transformed society and people lived in harmony with themselves and nature.

Many writers of the Nordic Enlightenment, including Wilse, were engaged in both agricultural questions and studies of climate (Eriksen 2007, 2020; Maliks and Johansen). This combination of interests may best be understood in light of the many subsistence crises during the seventeenth and eighteenth centuries, as well as the national interest in raising agricultural output to prevent similar crises in the future (Ryan 2023: 323; see also Lundstad et al., this volume). Wilse was aware of the intricate relationship between climate and farming, and he made several interesting remarks on how local climates affected farming conditions and working routines differently, even within small distances (Wilse 1991 [1779–1780]: 33). He personally experienced the ways that weather phenomena and climate anomalies associated with the LIA – including late springs, summer cold, continuous rain and night frost – could ruin crops. The years 1771–1775 were especially difficult, bringing repeated bad harvests both

Integrating Agricultural Vulnerability and Climate Extremes

in his parish and across the country as whole (Wilse 1991 [1779–1780]: 195). Wilse recorded his impression that the climate had indeed changed since 1759, and he wrote that this was common knowledge (Wilse 1991 [1779–1780]: 38). Thus, climate change seems to have been perceived by local populations at the time, and this perception must have raised awareness of how climate affected crops, particularly in a farming community such as Spydeberg. Similar discussions can be found elsewhere in Norway at the time, particularly the idea that the world was ageing (*mundus senescens*) and therefore provided less bountiful harvests than before (Eriksen 2007: 205–210). *Mundus senescens* was rooted in religious doctrine. According to some topographers, the theory was much discussed among commoners in late eighteenth-century Norway. However, it was questioned by leading intellectuals of the time, because it was lacking in evidence – evidence that the emerging field of meteorology might yet provide (Eriksen 2007: 207). This context of rising popular and scholarly interest in agricultural improvements and weather measurements highlights the historical significance of Wilse's investigations.

Wilse bore close witness to the causes and consequences of the 1770s crisis. He described how epidemics flourished due to malnutrition, causing high mortality rates and even killing a number of his colleagues in neighbouring parishes (Wilse 1991 [1779–1780]: 356). The disease was described as a violent fever that '… emerged almost like a plague after profoundly bad years, for instance 1741 and 1742, and 1772 and 1773 …' (Wilse 1991 [1779–1780]: 356, own translation). According to Wilse, long-term suffering gave rise to another epidemic in 1774, probably dysentery, which is described as a 'bleeding'. Wilse was also affected on a personal level. He married for the first time in 1770 but lost his wife to tuberculosis in 1783. In the meantime, the couple had lost four of their eight children, including two that died only a few days apart in July 1775. The cause of their deaths is unknown, but it was likely a consequence of poor living conditions and the many diseases that plagued the parish during and after the bad years of 1771–1775.

Wilse's Data and Agricultural Conditions at Spydeberg Rectory

Wilse has provided us with two datasets that are particularly useful for analysing climate impacts on agriculture during the LIA: farm statistics and the SMP weather measurements from Spydeberg and Eidsberg ([Map 2](#)). Unfortunately, they are not from the same decade. The farm statistics are from 1769–1779, while the SMP record covers 1783–1787. His daily weather observations from 1769–1780 are lost, and only a few general statistics for the ten-year period

Ingar Mørkestøl Gundersen

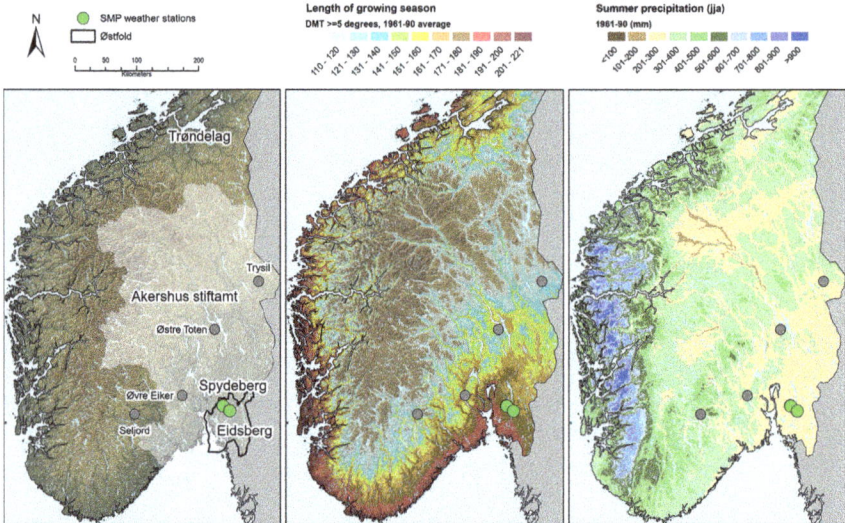

Map 2. Location of Wilse's SMP weather stations in Østfold county together with other placenames mentioned in the text (grey points), combined with climate data for the 1961–1990 reference period (weather data from the Norwegian Meteorological Institute [MET], published by Hanssen-Bauer et al. 2017). Southern Norway exhibits varying climate conditions ranging from a wet and long growing season in the west, to a dry and short growing season in the major river valleys in the east.

1771–1780 have survived in tables in his Spydeberg description (Wilse 1991 [1779–1780]: 489–497).

The farm statistics are listed in the same publication (Wilse 1991 [1779–1780]: 471–502) and include annual records for each crop (barley, rye, wheat, oats and peas), including ratios between sown and harvested (yield ratio) and numbers of hay loads. In addition, farm accounts (grains and livestock bought and sold) and dates for sowing and harvesting are provided.

The SMP records from 1783–1787 cover surface pressure, wind direction, wind force and air temperature. These were measured three times a day at the so-called Mannheim hours (07:00, 14:00 and 21:00 local time) (Winkler 2023). Precipitation was measured at other stations, but unfortunately not at Spydeberg/Eidsberg. The SMP datasets were published in twelve volumes between 1783 and 1795 in the *Ephemerides Societatis Meteorologicae Palatinae* (Ephemerides 1783–1794). The data from the Ephemerides have been compiled, corrected for instrumental errors, and computed into daily means by Pappert et al. (2021). Temperatures were originally measured in Réamur but converted

Integrating Agricultural Vulnerability and Climate Extremes

to Celsius in the latter study. The year 1783 is the warmest in Wilse's dataset, and 1784 is the coldest. Compared to data from the modern weather station at Spydeberg (MET id 3750), May to August (MJJA) temperatures in 1783 were 0.5 degrees Celsius warmer, while 1784 MJJA temperatures were −0.9 degrees Celsius colder than the 1961–1990 average.

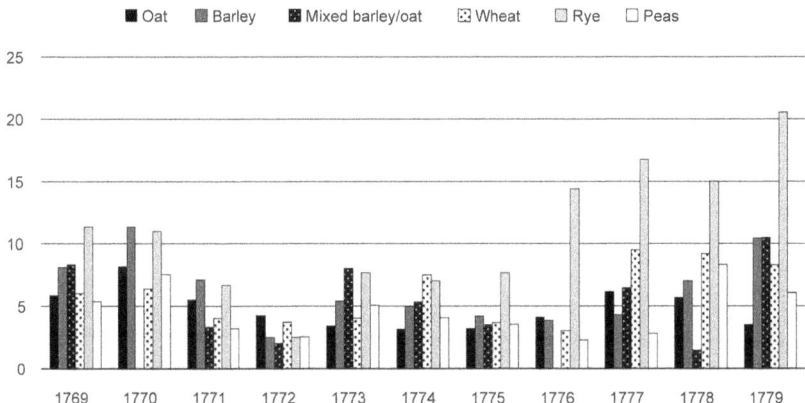

Figure 1. Yield ratio at Spydeberg rectory for main crops 1769–1779. Source: data from Wilse (1991 [1779–1780]).

Wilse's records reveal that the 1770s were a difficult decade at Spydeberg rectory, with consecutive poor harvests from 1771 to 1775.[1] Grains produced almost 85 per cent of the rectory's profits, making the institution heavily dependent on the outcome of harvests. The harvest almost completely failed in 1772, when production of all crop types fell below the 1769–1779 average. The year 1772 is also the only one in which grain sales resulted in a net loss for Wilse. This loss was compensated by selling livestock, but his action raised his total profit to only 3 *riksdaler* that year. This meagre profit stands in stark contrast to the 118.5 riksdaler he earned in 1770 entirely from grain sales. Forestry was an important source of income for many farmers but not for the rectory. The economic situation improved slightly over the next few years, but it was only towards the end of the decade that the farm became really profitable again.

A major challenge was not only the farm's dependency on grain, but also its narrow cultivation strategy. Oats made up 85 per cent of sown crops (91 per cent in 1772), meaning that, in years when oats failed, the whole enterprise was profoundly affected. According to Wilse (1991 [1779–1780]: 182), other

---

1. The early 1770s were difficult years in most of Europe (see Collet 2019).

farmers were even more dependent on oats. Other contemporary sources gave similar testimonies (Fleischer 2003 [1743]; Sigholt 2003 [1743]).

Judging from the farm records, oats were in general less productive than other species. However, their yields were more stable, with ratios between 1:3.2 and 1:8.1. Yields of other grains varied much more, making them less reliable, although more profitable in most years. The official protocols from 1723 and 1835 support this impression with higher yield ratios for other species than oats (Aschehoug 1890: 128–136). The year 1772 brought a poor harvest of all crops (Figure 1): none yielded more than four times the amount sown. Oats fared slightly better than the others, but nonetheless well below average.

The emphasis on oats also reflected local soil conditions. According to Wilse and other topographers, soils in Østfold were poor, leaving farmers dependent on limited supplies of manure (Fleischer 2003 [1743]; Wilse 1991 [1779–1780]: 165–167, 245–257). Livestock were not kept in great numbers and mainly for self-subsistence. Since oats are less soil-demanding than other crops, the emphasis on this one species was probably an attempt to adapt to the poor soil. Restricted access to manure, however, also locked farmers in a narrow subsistence strategy that left them vulnerable to weather conditions particularly harmful to oats. Moreover, experiments with organic farming show that focusing on a single grain lowers farming output over time, while a diversified strategy increases agricultural robustness. A mixed farming strategy, combined with regular fallowing, is important for soil recovery and preventing soil deterioration and plant diseases (Behre 1981; Bjørnstad 2012: 101–102; Frøseth 2004). According to Randi B. Frøseth (2004: 168), oats should only be cultivated every fifth year in organic farming, and barley and wheat every third year, in a cycle involving pasturing and grass production. Wilse (1991 [1779–1780]: 170) practised crop rotations that adjusted for soil conditions at each individual field; however, these rotations often included oats two years in a row. In short, the oat is a thrifty species, has few diseases and is able to grow in nutrient-poor soils (Bjørnstad 2012: 22; Frøseth 2006), and consequently seems to have been a reasonable adaptational strategy to local conditions at Spydeberg. Yet total dependency on this one species would also have reduced the farm's agricultural robustness over time.

Farming in eighteenth-century Østfold was labour-demanding and time-consuming. Ploughing and sowing started approximately twenty days after the fields had completely thawed, usually at the beginning of April. According to Wilse's (1991 [1779–1780]: 472–488) agricultural calendar, his workers began ploughing and sowing between mid-April and mid-May, and usually continued until the end of May. Harvesting began consistently around three months after the ploughing season had ended, but the timing was particularly sensitive to weather, and the work could last from around twenty to up to sixty days. Persistent or heavy precipitation could delay or disrupt the harvest, and grains left in the field after 20 September only rarely matured. Any remaining crops were cut green after 10 October and used for fodder.

Wilse's ploughing and harvesting dates give us an average cropping season of ~110–140 days between 1 May and 20 September, depending on the fields in question. Local topography, soil conditions and choice of grains, as well as the time of the ploughing and sowing, caused the individual fields to mature at different dates, creating a landscape with an intricate mix of shades of green and yellow. Wilse (1991 [1779–1780]: 177) found this phenomenon fascinating. He stated that local agriculture differed considerably in this respect from farming in the flat and more heavily cultivated countries further south, where the autumn grain colours were more even. However, these factors also made Norwegian agriculture more difficult to master (Wilse 1991 [1779–1780]: 162–163). Overall, the active cropping season for each field was significantly shorter than 140 days, and probably closer to 110, with the latter representing the statistical average of time between the beginning of ploughing and harvesting.

Analysing Crop Vulnerability

The impact of climate variability on crops can be roughly analysed by calculating the reduction in growing degree days (GDD). I have previously used GIS-based GDD modelling in a number of studies on the impact of sixth-century cooling on Scandinavian Iron Age societies (Gundersen 2021, 2025; Gundersen and van Dijk, forthcoming; van Dijk et al. 2023).[2] These studies demonstrated considerable regional variability in agricultural vulnerability to climate cooling in southern Norway. Thus, the sixth-century cooling should not be approached as a uniform crisis, but as something that affected Iron Age farmers in very different ways depending on local environmental conditions and farming practices. However, as farming practices and local climate variability are uncertain variables in a prehistoric setting, the models were based on idealised datasets that did not necessarily reflect actual historical impacts, but merely visualised different agricultural thresholds in selected landscapes. This paper utilises the same basic model as in the previous studies but improves the model output by integrating Wilse's meteorological data from the 1780s and his agricultural statistics for the 1770s (Pappert et al. 2021; Wilse 1991 [1779–1780]: 488). By using historical instead of idealised datasets, the results will provide a better starting point for discussing harvest failures in the past.

---

2. For other approaches to GDD and the 6th century cooling, see Stamnes (2016) and Arthur et al. (2024).

In simple terms, GDD is the sum of temperatures during the growing season. It is calculated by multiplying daily mean temperatures with the length of the growing season (i.e., the time of year with daily mean temperatures ≥5 degrees Celsius). During the reference period 1961–1990, the growing season ranged from <140 days in the major inland valleys in the east to >200 days along the coast (Map 2).

The minimum GDD required for grains to mature varies between species. Barley and rye are well adapted to northern climates, while wheat is temperature-demanding and thus vulnerable to summer cooling. Oats are also considered suitable for northern climates, but have a higher GDD requirement than barley and rye.

GDD also varies according to local environmental conditions. At higher latitudes, the longer hours of daylight during summer reduce GDD requirements. These requirements fall by 20 GDD per degree north of 60°N. Conversely, high precipitation levels require an increase in GDD (Frøseth 2004; Strand 1984: 27; Åssveen and Abrahamsen 1999). Rainfall exceeding 250mm from May to August increases requirements by 60–80 GDD per 100 millimetres rainfall for barley, 90–100 millimetres for oats and 100–110 millimetres for wheat. Thus, the wet climate makes cultivation more susceptible to cold weather on the western coast of Norway than in the dry regions of eastern Norway despite equal latitudes and temperatures. Therefore, agricultural vulnerability to climate change must be analysed at local levels, combining both social and environmental factors into the equation.

Local temperature measurements at modern weather stations in the Indre Østfold municipality for the 1961–1990 reference period (MET id 3750, 3710, 3460 and 3400) indicate a growing season of ~185 days and ~2,100–2,200 GDD (depending on altitude). For this area, situated at 59°N and with a total MJJA precipitation of 290mm (based on the 1961-1990 reference period), the minimum GDD requirements for modern oats are ~1,350–1,430, for barley ~1,290–1,370 and for wheat >~1,510. Thus, these requirements fall well within local GDD values.

Nevertheless, these twentieth-century values do not take into account eighteenth-century conditions. These include the time-consuming processes of preindustrial agriculture, and the ways that autumn precipitation affects the ripening of grains. Moreover, there have been systematic improvements and standardisation of crop varieties since the twentieth century. The improvements have reduced temperature requirements for crops, particularly wheat. The exact GDD requirements for grain types during the eighteenth century are not known, but there was evidently a great variety of locally developed subtypes adapted to local conditions. For instance, according to Wilse (1991 [1779–1780]: 189),

Integrating Agricultural Vulnerability and Climate Extremes

local wheat was considered among Spydeberg farmers as highly robust and able to mature under almost any circumstances, as long as the soil was properly cultivated – traits which contrast with the known temperature requirements for wheat today.

Early twentieth-century experimentation with traditional grains may help reduce these biases caused by modern standardisation. According to experimental agronomist Foss (1926), premodern barley types required between 1,200 and 1,350 GDD, oats 1,300–1,350 GDD and wheat a minimum of 1,550 GDD. His own experiments with winter rye resulted in mature grains even in conditions with as few as 1,050 GDD. Factoring in local precipitation values during 1961–1990 and latitude, barley would have needed ~1,240–1,390 GDD, oats ~1,350–1,400 GDD, rye ~1,060 GDD and wheat a minimum of ~1,600 GDD to mature. The following section will integrate these results with Wilse's farm statistics from the 1770s and weather data from the 1780s to analyse how climate variations might have affected local agriculture in Spydeberg during the late eighteenth century.

GDD Analysis

The year 1783 is the warmest and 1784 the coldest in Wilse's SMP records. Therefore, these two years have been selected for the analysis to visualise the likely minimum and maximum climate impact on agriculture in Indre Østfold during the 1780s. In 1783, the growing season lasted 196 days, which is slightly more than the 1961–1990 average of 185 days, while in 1784 it only lasted 159 days. The 1784 cold year is often attributed to the Laki fissure eruption (Zambri et al. 2019) and Wilse made several meteorological observations that were probably related to this event. However, further enquiries into the topic fall beyond the scope of this study.

Based on the statistical average length for the cropping season in 1769–1779, the middle dates for the ploughing and harvesting seasons (115 days between 13 May and 5 September) are set as constraints for the GDD accumulation for the years of 1783 and 1784 in Map 3 and Map 4. The risk for night frost increases significantly when daily temperatures fall to ≤10 degrees Celsius in autumn, but these frosts do little damage to the crops in spring (Strand 1984: 21–23). Wilse (1991 [1779–1780]: 176–177) described in similar ways how grains became particularly temperature-sensitive in autumn and susceptible to night frost. Only days with mean temperatures ≥5 degrees Celsius from May to July and ≥10 degrees Celsius in August and September are therefore included in the model.

The models are calibrated according to the minimum GDD requirements for oats, since oats were beyond comparison the most important crop in this region. The models are also combined with a soil classification dataset from the Norwegian Institute of Bioeconomy Research (NIBIO) to visualise the extent of the cultivated landscape today. The soil is mostly classified as having high potential for cereal cultivation; however, this evaluation is based on modern mechanised and fertilised agriculture, and thus the classification does not apply to conditions during the eighteenth century. The region is described by contemporary authors as marginal and with low soil quality (Fleischer 2003 [1743]; Sigholt 2003 [1743]; Wilse 1991 [1779–1780], 1992 [1793–1796]).

A few observations stand out. First, most farmland in the 1784 case shows a temperature sum slightly above the minimum requirements for oat cultivation. This means that outright failure is theoretically avoided.

Second, however, the temperature margins are rather low in 1784, including large areas with temperature margins of less than ten per cent. By contrast, in 1783, cropland remains well within the local temperature requirements. The contrast between these two years illustrates the vulnerability of premodern agriculture in Spydeberg to climate variability and seasonal changes. The MJJA temperature difference between 1783 and 1784 was only 1.4 degrees Celsius; yet the GDD models reveal that this difference could have caused very different outcomes for local agriculture. Turning to Wilse's farm statistics for the 1770s, we find that the yield ratio at Spydeberg varied considerably from year to year. For oats, this ratio ranged between 1:3.2 during 1774 and 1775 to 1:8.1 in 1770. His average yield ratio for oats was 1:3.9 during the so-called bad years of 1771–1775 and 1:5.6 for the rest of the decade, resulting in an overall average of 1:4.8. These yields were comparable to those of other areas of eighteenth-century Norway – that is, the best yields at his farm were comparable to yields in the best agricultural districts in Norway in the official protocols for 1723 and 1835, while the poorest yields at his farm were comparable to those of the worst districts (cf. Aschehoug 1890).[3] Similar numbers, ranging from 1:3.7 to 1:6.9,

---

3. On a national level, there is a substantial increase in mean yield ratio for oats from 1:3,1 in 1723 to 1:4,8 in 1835. Late 18th century topographers often reported higher yield ratios than listed in the 1723 protocols as well, leading some historians to conclude that the early protocols were probably based on net rather than gross productivity (Lunden 2002: 197). However, considering the large annual variability in Wilse's records, and the lack of consecutive official records from the 18th century, too much confidence should perhaps not be put in singular statements by topographers or the 1723 and 1835 cadastres. There is also the possibility that the ratios increased due to improved farming methods. Combined, the sources nonetheless indicate great regional variation with ratios ranging between 1:2 and 1:7, which resembles Wilse's own data for the 1770s.

Integrating Agricultural Vulnerability and Climate Extremes

have been obtained for Rogaland in 1775 (Lillehammer 1982: 121).

Third, the low-temperature margins obtained for 1784 indicate that local output was likely very low, and that crops would have been highly vulnerable to late summer precipitation. Harvesting would probably have stretched into autumn, thus increasing the risk that frosts or an early end to the growing season could ruin the crops. A similar situation developed in 1772, the single worst year in Wilse's observations, when a cold spring was followed by continuous summer rain and an early autumn. Wilse (1991 [1779–1780]: 176) observed that uncut mature oats rotted easily, meaning that farmers faced a delicate balancing act between leaving grains in the field long enough to mature and the risk of ruin if those crops were left uncut for too long.

Neither of these two scenarios is synonymous with a full-fledged disaster. Some yields can be expected even in the 1784 scenario. However, premodern agriculture in Indre Østfold appears to have been vulnerable to even minor temperature variations. The MJJA temperatures for 1784 were only 0.9 degrees Celsius below the 1961–1990 average, which would not be an extreme outlier in the context of Northern Hemisphere temperature variability over the last two millennia (cf. Büntgen et al. 2020). On the contrary, it is more likely to have been a recurrent situation.

Taken as a whole, the GDD models provide some evidence that 1783 was likely a good year for the farmers in Indre Østfold. The 1784 cooling probably had a negative impact on oat yields, but the harvest was likely not an outright failure. The comparison between these two years demonstrates that margins for agriculture were small and that food crises were likely to develop during severe cooling events and when unusual weather delayed the harvest.

These findings contrast with those of a previous GDD study in Sarpsborg, just south of Indre Østfold. The Sarpsborg study indicated that temperature margins for crops remained sufficient even during a 3-degree-Celsius cooling scenario (w.r.t. 1961–1990) (van Dijk et al. 2023). The methods used in these two studies are the same, but the previous model assumed more ideal conditions. It defined the cropping season as the number of days with mean temperatures ≥5 degrees Celsius and thus it did not consider the time-consuming labour associated with premodern farming. In this study, the model has been improved by adjusting the cropping season according to Wilse's farm statistics, which reduces the cropping season from 159 to 115 days for 1784. Nevertheless, it should be stressed that, even in its improved version, the model merely visualises thresholds for grain cultivation during different temperature scenarios. Certain weather phenomena such as the timing of precipitation and frosts that constitute crucial variables for crops have not been integrated into the model. It thus provides

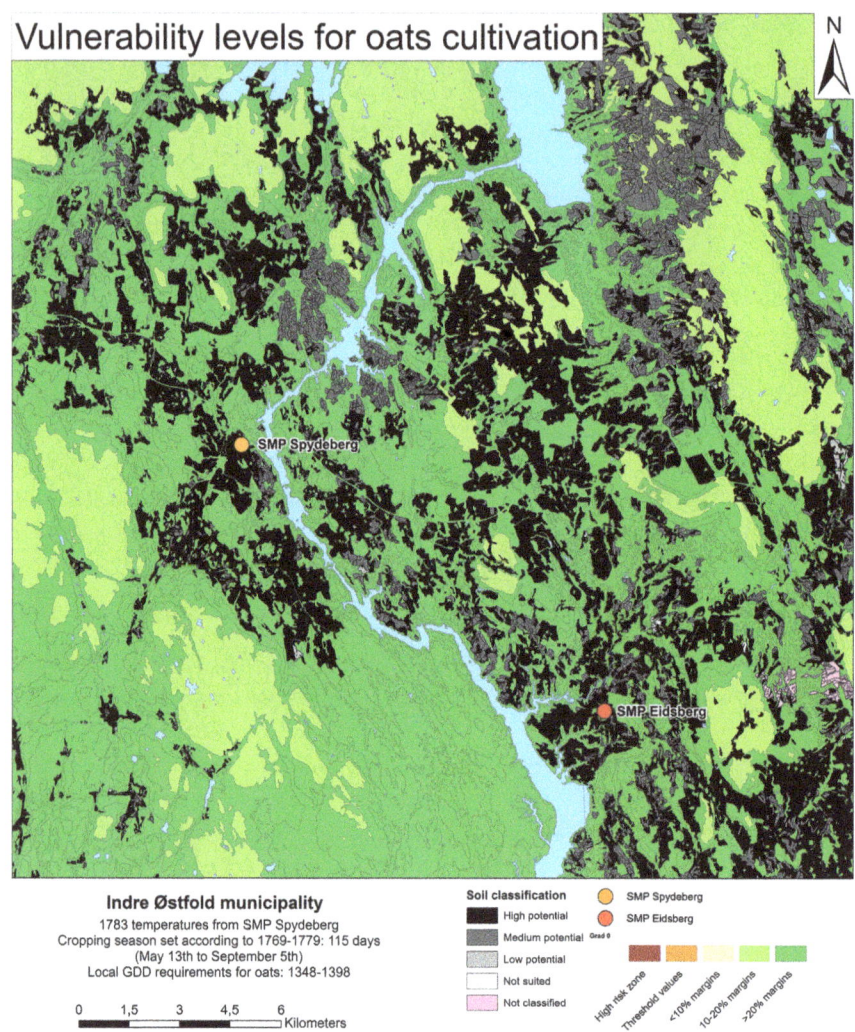

Map 3. GDD-model for Indre Østfold based on 1783 weather data from the SMP Spydeberg weather station.

some indication of how a farming economy might have been affected, but it can never fully replicate the actual historical consequences of climate variability.

Moreover, Wilse's farming principles might have differed from those of other farmers in the area. Wilse took a scientific approach to most of his activities, particularly agriculture, and he experimented and documented his methods in order to improve them. He refers to different practices among local farmers,

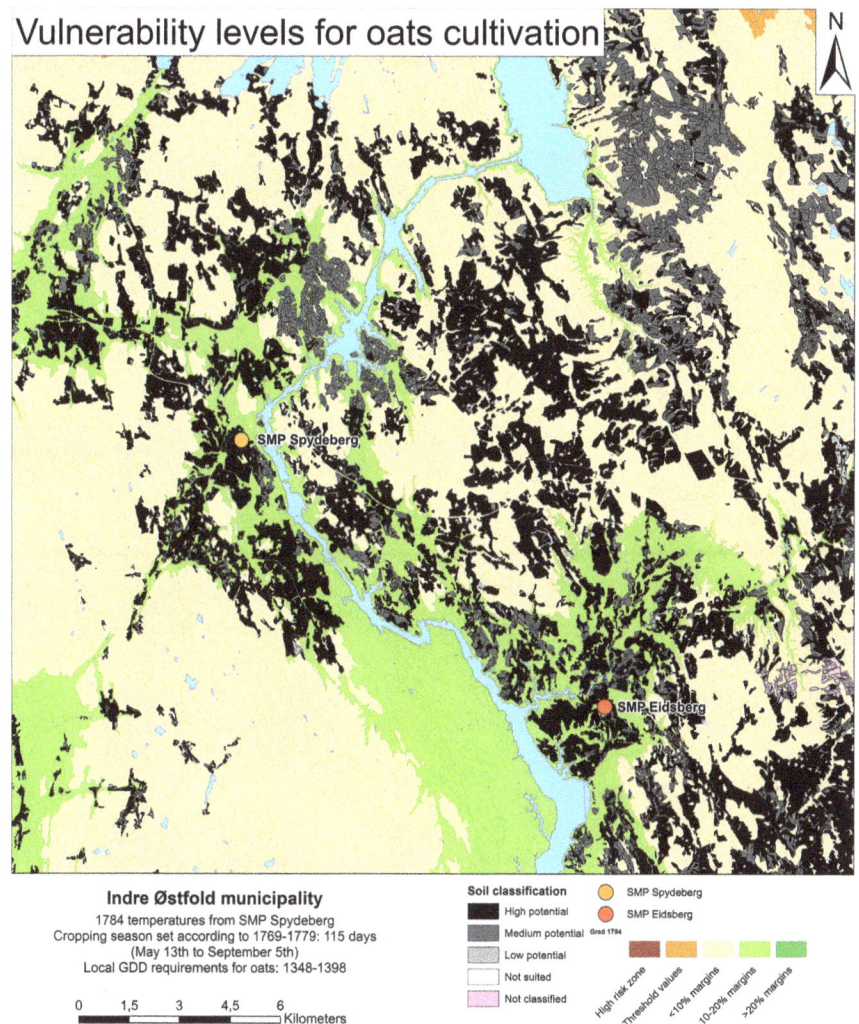

Map 4. GDD-model for Indre Østfold based on 1784 weather data from the SMP Spydeberg weather station.

and his agricultural calendar should be understood as an attempt to create a standard for how farming should be conducted in Spydeberg, rather than reflecting actual practices (i.e., Wilse 1991 [1779–1780]: 182). Although he respected experienced farmers and acknowledged their advice, he nonetheless flattered himself by stating that his oat crops performed better during the 1771–1775 crisis than many others in his parish (Wilse 1991 [1779–1780]: 195). In other words, a model based on singular accounts, such as Wilse's, does not reflect local variation in agricultural practices during the late eighteenth century. We also lack farm records from Østfold for the 1780s that could provide qualitative insights into the agricultural impacts of the cold summer of 1784. For lack of such data, the GDD model, combined with other historical sources, serves as a good starting point for discussing its social consequences.

Climate, Crop Failures and Crises in Eighteenth-Century Norway

Finally, Wilse's records provide useful insights into larger patterns of crop failures and crises across Norway during different decades of the eighteenth century. Norway was not self-sufficient in grains and was dependent on imports. From 1735 onwards, the southeastern parts of Norway, including Akershus, were also subject to a grain monopoly that prohibited imports from outside the borders of the dual monarchy of Denmark-Norway (Herstad 2000). However, in part due to slow communication, Danish grain imports regularly proved insufficient, especially when harvests were poor. Governmental responses were usually further delayed by the frozen sea in winter, which prevented shipments until spring. During crisis years, such as the early 1770s, royal decrees temporarily suspended the trade restrictions. In 1781 and 1782, crops again failed throughout Akershus, and prices soared (Sogner 1976). Timber exports from Norway also declined due to international conflict, leaving farmers with less income. While foreign grain imports from outside the union fell to a minimum in the latter half of the 1770s, they rose again in 1783, reaching 39 per cent of total grain imports in 1786 (statistical data in Herstad 2000: 378–381). At the same time, Danish imports fell by more than half from 1783 to 1784 and remained low for the next two years while grain prices kept increasing, thus contributing to social tensions (Fiskaa 2012; Løyland 2018). According to Sølvi Sogner (1976: 126), the early 1780s had many similarities with the early 1770s and could have developed into a similar demographic crisis.

The 1783 harvests were good, providing a temporary relief to the situation, but problems mounted once more in 1784. There are several contemporary accounts that support the results from the GDD analysis for Indre Østfold. Among these are weather observations by *amtmann* (magistrate) Christian Sommerfelt at Østre Toten, pastor Hans Strøm at Øvre Eiker, and pastor Axel Smith at Trysil (compiled by Wishman 2007: 8–30, 44). Smith, Strøm and Sommerfelt all reported on an unusually cold summer and poor harvests in 1784. At Østre Toten, it took unusually long for the grains to mature, while in Trysil the harvest failed completely, and the grains were cut green. According to Smith, there were only eleven weeks between the last thaw of spring and

Integrating Agricultural Vulnerability and Climate Extremes

first snow of autumn. In addition, pastor Hans Jacob Wille at Seljord reported on unstable weather in late July with floods and landslides.

Severe night frosts were reported in southern and eastern parts of the country from mid-August, and these destroyed much of the remaining crops (Wishman 2007: 22, 43). Wilse also recorded night frosts on the eighteenth and nineteenth of August in his report from Spydeberg to the SMP (Ephemerides 1787: 533). This means that, across much of Norway, the active cropping season for 1784 was probably shorter than indicated by the GDD model, thus resulting in smaller temperature margins and greater risk of poor yields. Famine, frost and poor harvests were also reported from Trøndelag and Central Sweden (Wishman 2007: 44).

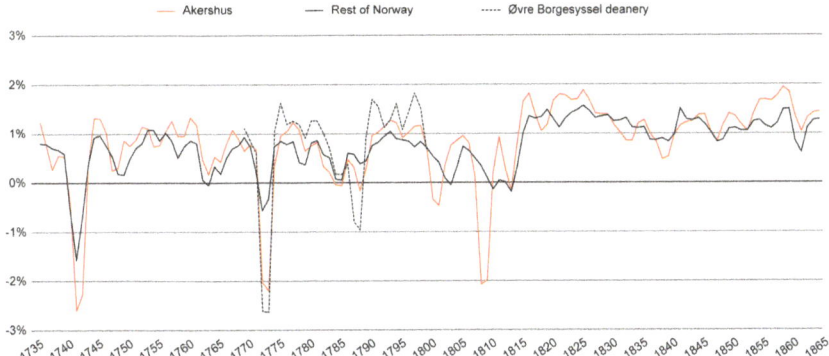

Figure 2. Population growth rates for Akershus County, Norway (excluding Akershus), and Øvre Borgesyssel deanery, based on demography studies by Drake (1969) and Sogner (1979). Drake's numbers for Akershus and Norway cover 1735–1865, while Sogner's numbers for Øvre Borgersyssel are restricted to 1769–1800.

Nevertheless, Norway averted a full-fledged crisis following the 1784 harvest failure as well. Norway experienced several demographic crises during the eighteenth and early nineteenth centuries, of which 1742–1743, 1773–1774 and 1809–1810 stand out as particularly severe (cf. Figure 2). There was an increase in the mortality rate during 1785, which resulted in a minor population drop in Akershus in 1786 (statistical data in Drake 1969; see also Sogner 1976). A similar development has been documented for Øvre Borgersyssel deanery, to which Spydeberg and Eidsberg belonged, with increased mortality rates for 1784 and 1785 (cf. Sogner 1979: 196). Unfortunately, we lack statistical data for Spydeberg for the 1780s. In his Spydeberg description, Wilse provides numbers organised in five-year intervals for births and deaths up to 1779 (Wilse 1991 [1779–1780]). His Eidsberg publication (Wilse 1992 [1793–1796]) provides overlapping data with Spydeberg for 1765 to 1779, and further up to 1794 (Figure 3). Apparently, the 1771–1775 crisis severely affected both communities. Wilse specifically mentions an extraordinary number of deaths in Spydeberg in 1773, which reflects the overall situation in Akershus (Sogner 1976: 116). The situation stabilised afterwards but was followed by a new increase in mortality in Eidsberg 1785–1789. The exact reasons are unknown. Wilse paid little interest to the matter in his Eidsberg publication, although it should be noted that this is a much smaller work than his more informative and comprehensive Spydeberg description.[4] Considering the result of the GDD analysis, and the various reports of poor harvests and hunger from middle, southern, and eastern Norway, it seems reasonable to suggest that the increased mortality rate in Eidsberg was to some degree related to the climate anomaly and cold summer weather that Wilse reported to the SMP. Wilse also reported on a cold summer in 1785, which likely added to an already difficult situation.

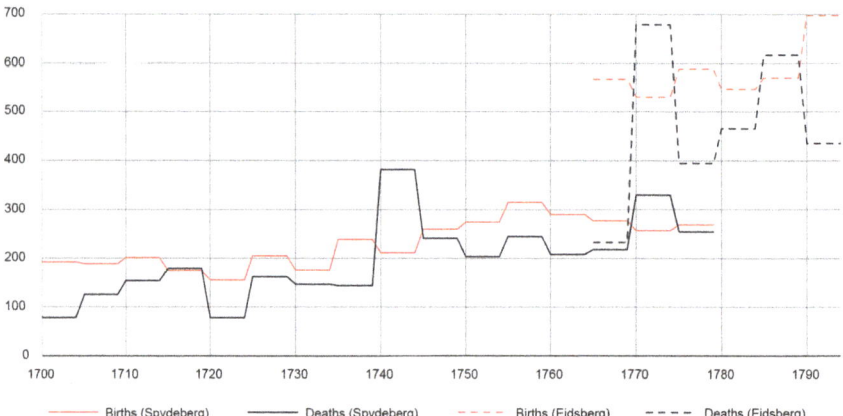

Figure 3. Number of births and deaths in Spydeberg (1696–1779) and Eidsberg (1765–1794) according to numbers provided by Wilse (1991 [1779–1780], 1992 [1793–1796]). Wilse organised his Spydeberg data in 5-year intervals. The Eidsberg numbers are here organised accordingly for better comparison between the two areas for the overlapping years.

---

4. This might have had personal reasons. In 1783, he started recording weather data three times daily for the SMP, which much have taken much of his time. He also worked on several other projects in parallel with the Eidsberg description (Stubberud 2016: 146). Moreover, he was struck by personal grief when he lost his wife in May 1783. In the church book, he recorded her passing with the phrase 'unforgettable lady'. According to Stubberud (2016: 74, 170), his wife's passing was a hard blow that left a long-lasting mark on him. His second wife died before the book was finished in 1796.

Integrating Agricultural Vulnerability and Climate Extremes

Akershus and Øvre Borgesyssel seem to have been particularly vulnerable to climate variability. The population growth rates for the crisis years of the eighteenth and early nineteenth centuries demonstrate a far more extreme situation in Akershus than in the rest of the country, including the three extreme situations during 1742–1743, 1772–1773, and 1809–1810. The numbers for Øvre Borgesyssel cover fewer years, and the 1740s and 1809–1810 are missing in this overview. However, the years 1772–1773 show an even higher population drop than in Akershus as well as a second significant decline during 1788–1789, which is far less discernible in the rest of the country. These discrepancies call for an explanation.

For the early 1740s, Astrid Løvlien (1977) has identified epidemics as the main cause for the high mortality rates in the west, while famine was more prominent in the east. However, there has been a tendency to consider either famine or epidemics as main causes, rather than considering the two as complementary perspectives on the same phenomenon (Dybdahl 2014). Accordingly, poor nutrition and starvation lead to poor public health, which increases the risk for diseases and epidemics that might spread beyond the original catchment area. For the 1740s, John Herstad argues that poor harvests were a significant factor in the east, which in combination with epidemics resulted in a 'combined crisis' that struck eastern Norway particularly hard. Sogner (1979) argues that coastal and near-border communities such as Øvre Borgesyssel were more exposed to epidemics. This was due to the presence of military encampments and more contact with the outside world. Moreover, she points out that these communities were often less self-sufficient in grains because livelihoods depended on lucrative timber exports. In this perspective, it was more profitable to invest in forestry and buy grains, which ultimately left populations dependent on food imports and vulnerable to trade fluctuations. These are sound arguments, and Wilse (1991 [1779–1780]: 205) himself complained about short-sighted and extensive deforestation that in the long run impoverished the farm economy. Accordingly, several farmers in the area were left unable to supply themselves even with firewood.

Additional contributing factors are to be found in contemporary topographical and historical written accounts. Historian Gerhard Schøning (2010 [1761]) specifically addressed the difficulties encountered during so-called agricultural 'bad years'. By systematically examining historical accounts of weather phenomena and poor harvests from prehistory to 1740, he identified three major phenomena that affected farmers: 'green years' (when grains failed to mature due to cold and wet conditions), 'shiny years' (droughts) and prolonged three-year incidents, of which the latter had particular potential to develop into full-fledged food crises, even in the best agricultural areas. He concluded that

Norway had become too dependent on grains and possessed too few granaries, making society highly susceptible to harvest failures. However, he also emphasised how marine fisheries often negated the worst consequences for the population.

Schøning's focus was Trøndelag in Central Norway, but other officials made similar observations about Østfold. In the early 1740s, the Danish-Norwegian government made a nationwide attempt to systematically document the economic, social and environmental conditions of the dual monarchy, which resulted in the 1743 survey. The survey addressed topics such as agricultural practices, diseases, environmental conditions and non-agrarian resource exploitation. The reports from Østfold are grim and clearly coloured by the recent crisis. They all addressed widespread harvest failures due to cold northern winds, unstable weather and cold summers (compiled in Røgeberg 2003). A telling example is pastor Johan Ludvig Christian Ludvigsen Paus who arrived in Eidsberg in 1741. In his report, he described harvest outputs down to a third compared to normal years as well as high death tolls due to plague-like diseases (Paus 2003 [1743]). According to Wilse (1992 [1793–1796]: 46), Paus and his wife died in 1744, and Paus left behind records full of complaints about starvation, misery and a financially ruined rectory: strong words, considering Wilse's own assessment of Eidsberg as a wealthier parish than Spydeberg. Paus and other local officials also mentioned high grain dependency and a very narrow farming strategy. Livestock was mostly kept for household needs because of long winters and limited fodder. Poor soils and little access to manure meant that farmers were totally dependent on oats, which, though well-adapted to such conditions, were less productive than other crops, as discussed in the previous sections. The surplus remained low even when harvests were good (cf. Røgeberg 2003).

The regional farming strategy left populations highly susceptible to weather conditions that damaged oats. These included late autumn wind and precipitation (Strand 1984: 64), as Wilse (1991 [1779–1780]: 176) also noticed. This created a major challenge if cold summers delayed the harvesting. Wilse (1991 [1779–1780]: 182) furthermore observed that oats were more vulnerable to the autumn cold than barley. Other species have their weaknesses as well, but the point is that a more diversified agricultural strategy would have made farming more resilient towards shifting weather phenomena and climatic extremes. Restricted access to manure limited farmers' options and locked them into a narrow farming strategy. Moreover, in contrast to Schøning's (2010 [1761]) study of Middle Norway, terrestrial wildlife and fish seem to have been scarce and insufficient to prevent starvation when harvests failed, and *amtmann* Baltzar Sechmann Fleischer (2003 [1743]) also complained about a general lack of interest among the populace in investing more in marine fisheries. Higher grain dependency, and less access to commercial fishing, is also considered by

Integrating Agricultural Vulnerability and Climate Extremes

Herstad (2000: 315–319) to have contributed to the higher mortality rates in the east than in the rest of the country.

Wilse's own accounts indicate that much remained the same more than a generation later (Wilse 1991 [1779–1780]: 161–171, 182, 248–252). Lack of summer pastures and limited winter fodder meant that the farmers could only keep livestock in low numbers. Probably for the same reason, farming was still heavily dependent on oats (Wilse 1991 [1779–1780]: 182). Surplus production was still low; no local granaries had been built; fishing, hunting and trapping were common but small-scale, and did not make up an important part of the local diet or economy (Wilse 1991 [1779–1780]: 264, 275, 305–307, 323–324; 1992 [1793–1796]: 77; 1993: 118). Diets remained primarily grain-based, usually in the form of porridge.[5] The fields were fallowed every third and manured every twelfth year. Crop rotations were practised, sometimes combined with grazing. However, planting of oats still dominated, meaning crop rotation was less effective than in areas with a more diversified strategy.

Apparently, key risk factors identified for this region during the early 1740s crisis were still present in Indre Østfold towards the end of the century. Poor environmental preconditions for cultivation and husbandry, as well as limited access to and exploitation of non-agrarian resources, meant that the local populace was locked in a farming strategy that made them highly vulnerable to climate anomalies or weather extremes. Moreover, as Wilse (1991 [1779–1780]: 362) observed, the population had been growing, with an increase of 25 per cent during the past eighty years, immigration and emigration excluded. Soaring grain prices and falling timber prices during the 1780s made the situation even worse. Despite increased mortality rates in the late 1780s, a population crisis seems to have been avoided, probably due to factors situated outside the local community. Sogner (1976) discusses important measures employed in the early 1780s by the *stiftamtmann* (county governor) for Akershus, Albert Philip Levetzau. These included his active use of royal granaries in the fortress cities to help vulnerable social classes, which might have countered a possible outbreak and spread of epidemics as well.

This last example serves as a reminder that quantifiable data and GDD modelling must be contextualised and examined alongside qualitative historical records. Nevertheless, the model illustrates the potential for integrating agricultural and meteorological data in the analysis of climate and weather impacts on a preindustrial farming society. In this way, such modelling can help bridge between different historical records and substantiate the claim that climatic

---

5. Wilse appears quite frustrated when commenting on local diets, stating '…soon there will be nothing but porridge and porridge again all year round'. (Wilse 1991 [1779–1780]: 306, own translation).

anomalies and long-term climate change had the potential to severely affect societies and foster social change.

Conclusion

To critically investigate harvest failures and societal consequences in eighteenth-century Norway, this study has deployed a growing degree days (GDD) model that integrates contemporary meteorological and agricultural data by pastor Jakob Nicolaj Wilse. The model result suggests that preindustrial farming in Indre Østfold, South-eastern Norway, had small temperature margins to rely on, which made farmers vulnerable to climate fluctuations, such as the 1784 cold spell. The model is discussed against contemporary topographical accounts to explain the particularly high mortality rates in this area during the crisis years of the early 1740s, early 1770s and mid-1780s. Apparently, poor soils and limited access to pastures locked farming in a narrow strategy heavily dependent on oats. Combined with little access to non-agrarian resources, these factors made the community highly vulnerable to climatic shocks.

The study provides new insights on the role of climate variability for agricultural outputs in preindustrial Norway, and thus its potential role for society in broader terms. In a more general sense, it also demonstrates how integrated socionatural approaches can be used to analyse climate and society interactions in the past. However, as this study also demonstrates, causality is not necessarily linear. The complexity of social systems, and how humans choose to respond to changing socionatural conditions, makes any discussion of cause-and-effect a difficult task, especially in those cases where the empirical records are fragmented or lacking. However, integrated analytical approaches can help to identify major risk drivers and substantiate whether it had the potential to influence the course of events. Well documented and situated at the margins of agriculture, eighteenth-century Norway has a unique potential for methodological advancements in this respect. The historical records from this period are rich in both quantitative and qualitative terms and even include contemporary weather observations by early scientists and farmers. To improve our understanding of disasters and social impact, equal interest should be paid to the cases where crises do seem to have developed, as the cases where they did not. More specifically, the contrasting social trajectories in the aftermath of the climate anomalies of the early 1740s, early 1770s and mid-1780s call for more research initiatives.

The late eighteenth and early nineteenth centuries are a transformative period in Norwegian history that culminates with the end of the Danish-Norwegian realm, a constitutional assembly and a union between Norway and Sweden in 1814. The process cannot be separated from a growing political movement for self-determination rooted in – amongst other things – dissatisfaction with the Danish government's inability to procure food for a starving population during the crisis years, especially the Napoleonic wars (Glenthøj and Ottosen 2014). Harvest failures and ecological shocks are an important part of this picture. Although there is a growing international body of climate studies in history and archaeology (Huhtamaa and Ljungqvist 2021; White et al. 2025), climatic stress is conspicuously absent in the Norwegian historical narratives, especially when it comes to the social significance of the LIA (see Seland, this volume). For instance, prevailing analyses of the so-called Lofthus rebellion in Southern Norway, 1786–1787, pay little attention to how poor harvests in the early 1780s exacerbated an already difficult situation among the farmers (Fiskaa 2012; Pedersen and Munch-Møller 2012; Rislå 2023; Sandvik and Dørum 2012; Sunde 2020). This is despite the fact that these hardships served not just to legitimise the struggle but explicitly influenced their decision to act (cf. Løyland 2018). As discussed in this paper, the 1784 cold spell is likely to have caused widespread crop failures throughout the country. Considering the prevailing cold temperatures in 1785, the situation likely endured for some time. This is not to say that the 1786–1787 rebellion was 'caused' by climatic factors, as other important political and socio-economic trajectories were clearly present as well, but merely point out that the role played by the 1784–1785 anomaly in the timing, legitimisation and geographical scope of these events deserves to be properly scrutinised.

Acknowledgements

This work is funded by the FRIPRO project CLIMCULT at the University of Oslo (Research Council of Norway, grant 315441). The author would particularly like to thank Dominik Collet, Sam White and Matias Kallevik for their constructive and helpful comments.

Bibliography

Arthur, F., K. Hatlestad, K.-J. Lindholm et al. 2024. 'The impact of volcanism on Scandinavian climate and human societies during the Holocene: Insights into the Fimbulwinter eruptions (536/540 AD)'. *The Holocene* 34 (5): 619–633. https://doi.org/10.1177/09596836231225718.

Aschehoug, T.H. 1890. *Statistiske studier over folkemængde og jordbrug: i Norges landdistrikter i det syttende og attende aarhundrede / af T.H. Aschehoug*. Kristiania: H. Aschehoug. https://babel.hathitrust.org/cgi/pt?id=hvd.32044059012435&seq=5.

Behre, K.-E. 1981. 'The interpretation of anthropogenic indicators in pollen diagrams'. *Pollen et Spores* 23 (2): 225–245.

Bjørnstad, Å. (2012). *Vårt daglege brød: kornets kulturhistorie*. 2nd ed. Oslo: Vidarforlaget.

Büntgen, U., D. Arseneault, É. Boucher et al. 2020. 'Prominent role of volcanism in Common Era climate variability and human history'. *Dendrochronologia* 64: 125757. https://doi.org/10.1016/j.dendro.2020.125757.

Collet, D. 2019. *Die doppelte Katastrophe: Klima und Kultur in der europäischen Hungerkrise 1770–1772*. Göttingen: Vandenhoeck & Ruprecht. https://doi.org/10.13109/9783666355929.

Drake, M. 1969. *Population and Society in Norway, 1735–1865*. Cambridge: Cambridge University Press.

Dybdahl, A. 2014. 'Sult eller sykdom? Hva var årsaken til den demografiske krisen i Norge først på 1740-tallet?' *Michael. Tidsskrift for samfunnsmedisin og medisinsk historie* 11 (1): 9–27.

Dybdahl, A. 2016. *Klima, uår og kriser i Norge gjennom de siste 1000 år*. Oslo: Cappelen Damm Akademisk.

Ephemerides. 1783–1794. *Ephemerides: Observationes anni 1781–1791*. Manheimii: Societatis Meteorologicae Palatinae.

Ephemerides. 1787. *Ephemerides Societatis Meteorologicae Palatinae. Observationes Anni 1785*. Manheimii: Societas Meteorologica Palatina. https://www.digitale-sammlungen.de/en/view/bsb10723491.

Eriksen, A. 2007. *Topografenes verden: fornminner og fortidsforståelse*. Oslo: Pax. https://www.nb.no/items/URN:NBN:no-nb_digibok_2013091905070.

Eriksen, A. 2020. 'History, exemplarity and improvements: 18th century ideas about man-made climate change'. *Culture Unbound* 11 (3–4): 353–368. https://doi.org/10.3384/cu.2000.1525.1909302.

Federhofer, M.-T. 2002. 'Værtegn. Om Jacob Nicolaj Wilses (1735–1801) meteorologiske notasjonssystem'. *Nordlit* 11: 91–102. https://doi.org/10.7557/13.2070.

Fiskaa, I. 2012. 'Lofthusreisinga i Agder og Telemark 1786–87'. In K. Dørum and H. Sandvik (eds), *Opptøyer i Norge 1750–1850*. Oslo: Scandinavian Academic Press. pp. 103–155.

Fleischer, B.S. 2003 [1743]. 'Fredrikstad kjøpstad og Smålenes amt'. In K.M. Røgeberg (ed.) *Norge i 1743: innberetninger som svar på 43 spørsmål fra Danske Kanselli: 1: Akershus stift: Østfold, Akershus*. Vol. 1. Oslo: Solum Forlag. pp. 169–185.

Foss, H. 1926. *Beretning fra Statens forsøksstasjon for fjellbygdene 1925. Ottende arbeidsår. Forsøk med rug og hvete i fjellbygdene*. Oslo: Grøndahl & Søns Boktrykkeri.

Frøseth, R.B. 2004. 'Korn'. In G.L. Serikstad (ed.) *Økologisk handbok. Matvekster*. GAN Forlag. pp. 167–187. https://docplayer.me/1628789-Okologisk-handbok-matvekster-gan-forlag-as.html.

Frøseth, R.B. 2006. 'Økologisk kornproduksjon: Arts- og sortsvalg'. *Bioforsk TEMA* 1 (36): 1–2. https://nibio.brage.unit.no/nibio-xmlui/bitstream/handle/11250/2505428/Bioforsk-TEMA-2006-01-36.pdf?sequence=1&isAllowed=y.

Glenthøj, R. and M.N. Ottosen. 2014. *1814. Krig, nederlag, frihet. Danmark-Norge under Napoleonskrigene*. Oslo: Spartacus.

Gundersen, I.M. 2021. *Iron Age Vulnerability. The Fimbulwinter Hypothesis and the Archaeology of the Inlands of Eastern Norway*. Ph.D. thesis. Oslo: University of Oslo. https://10.5281/zenodo.5782896.

Gundersen, I.M. 2025. 'Global challenges, local impacts. Contextualising the 6th-century climate extreme and Iron Age farming societies'. In M. Østmo, A. Tvedte-Kristoffersen and M. Moen (eds), *Telling Different Stories: Combining Perspectives on the Viking Age*. *Viking*, Special Volume 3: 163–190. https://journals.uio.no/vikingspecialvolumes/article/view/12203/10278.

Gundersen, I.M. and E. van Dijk. Forthcoming. 'Interlocking climate and society: Climate and harvest failure in sixth-century Scandinavia'. In A. Franklin-Lyons and T. Newfield (eds), *Contextualizing Medieval Food Shortages: Causes, Definitions, and Historiography*. Amsterdam: Amsterdam University Press.

Hanssen-Bauer, I., E.J. Førland, I. Haddeland et al. 2017. *Climate in Norway 2100 – A Knowledge Base for Climate Adaptation*. Vol. 1/2017. Oslo: The Norwegian Centre for Climate Services (NCCS). https://www.miljodirektoratet.no/globalassets/publikasjoner/m741/m741.pdf.

Herstad, J. 2000. *I helstatens grep: kornmonopolet 1735–88*. Vol. 8. Oslo: Tano Aschehoug. https://www.nb.no/items/URN:NBN:no-nb_digibok_2008110300060?page=0.

Huhtamaa, H. and F.C. Ljungqvist. 2021. 'Climate in Nordic historical research – a research review and future perspectives'. *Scandinavian Journal of History* 46 (5): 665–695. https://doi.org/10.1080/03468755.2021.1929455.

Lillehammer, A. 1982. 'Rug – bygg – havre. Ei nyoppdaga kjelde om korndyrkinga i Rogaland frå 1775'. In A. Lillehammer (ed.) *Faggrenser brytes. Artiklar tileigna Odmund Møllerop 7. desember 1982*. Vol. Ams-Skrifter 9. Stavanger: Arkeologisk museum i Stavanger. pp. 115–122. https://www.nb.no/items/URN:NBN:no-nb_digibok_2017100405159?page=119.

Ljungqvist, F.C., A. Seim and D. Collet. 2023. 'Famines in medieval and early modern Europe – Connecting climate and society'. *WIREs Climate Change* e859: 1–12. https://doi.org/10.1002/wcc.859.

Lunden, K. 2002. *Frå svartedauden til 17. mai: 1350–1814*. Oslo: Samlaget.

Lundstad, E., S. Norrgård and A.E.J. Ogilvie. 2025. 'The development of meteorological institutions and early instrumental climate data in the Nordic countries', this volume. https://doi.org/10.63308/63881023874820.ch01.

Løvlien, A. 1977. *Dødelighetskrisa på 1740-tallet: En sammenligning mellom to norske landsdeler*. MA thesis. Bergen: University of Bergen. https://www.nb.no/items/a0792eaaa89feeccb443edaa5ce6ed64r.

Løyland, M. (ed.) 2018. *Lofthusoppreisten. Rettsmateriale frå kommisjon og høgsterett 1789–99*. Vol. 4. Oslo: Riksarkivet.

Maliks, J. and S. Johansen. 'Schønings klimateorier: studier i naturens endringer og menneskets livsvilkår'. In P.-O.B. Rasch (ed.) *Norsk vitenskap i støpeskjeen*. Oslo. Universitetsforlaget. pp. 132–150. https://doi.org/10.18261/9788215072432-24-07.

Pappert, D., Y. Brugnara, S. Jourdain et al. 2021. 'Unlocking weather observations from the Societas Meteorologica Palatina (1781–1792)'. *Climate of the Past* 17 (6): 2361–2379. https://doi.org/10.5194/cp-17-2361-2021.

Paus, J.L.C. 2003 [1743]. 'Eidsberg prestegjeld'. In K.M. Røgeberg (ed.) *Norge i 1743. Innberetninger som svar på 43 spørsmål fra Danske Kanselli. 1: Akershus stift og amt: Østfold, Akershus*. Oslo: Solum Forlag. pp. 295–297.

Pedersen, A. and M.G. Munch-Møller. 2012. 'Protester i nødstid'. In K. Dørum and H. Sandvik (eds), *Opptøyer i Norge 1750–1850*. Oslo: Scandinavian Academic Press. pp. 253–258.

Pfister, C., R. Brázdil, J. Luterbacher, A.E.J. Ogilvie and S. White. 2018. 'Early modern Europe'. In S. White, C. Pfister and F. Mauelshagen (eds), *The Palgrave Handbook of Climate History*. London: Palgrave Macmillan. pp. 265–295. https://doi.org/10.1057/978-1-137-43020-5_23

Rislå, Ø.K. 2023. *Et opprør i opplysningens tid. Christian Lofthus og hans bevegelse i en transnasjonal kontekst*. MA thesis. Kristiansand: Universitetet i Agder. https://uia.brage.unit.no/uia-xmlui/bitstream/handle/11250/3075738/no.uia%3Ainspera%3A143763668%3A37319536.pdf?sequence=1.

Ryan, S.M. 2023. 'Norwegian climatology, the Republic of Letters and the Nordic Enlightenment'. *Annals of Science* 80 (4): 303–336. https://doi.org/10.1080/00033790.2023.2209095.

Røgeberg, K.M. (ed.) 2003. *Norge i 1743. Innberetninger som svar på 43 spørsmål fra Danske Kanselli. 1: Akershus stift og amt: Østfold, Akershus*. Vol. 1. Oslo: Solum forlag.

Sandvik, H. and K. Dørum. 2012. 'Skatteopptøyer og bondebegevelser'. In K. Dørum and H. Sandvik (eds), *Opptøyer i Norge 1750–1850*. Oslo: Scandinavian Academic Press. pp. 39–44.

Schøning, G. 2010 [1761]. 'Kort beretning om ein del uår og misvekst, særleg i Trondhjems stift i Noreg'. Trans. by T. Moen, in S. Johansen (ed.) *Grønår og skinår. Klimahistorie for Trøndelag gjennom 1000 år*. Trondheim: NTNU og Tapir Akademisk Forlag. pp. 21–41. https://www.nb.no/items/0b88d711a78b86879821b245519956ef.

Seland, E.H. 2025. 'Climate narratives in Norwegian public histories', this volume. https://doi.org/10.63308/63881023874820.ch11.

Sigholt, J.T. 2003 [1743]. 'Rakkestad fogderi og sorenskriveri'. In K.M. Røgeberg (ed.) *Norge i 1743: innberetninger som svar på 43 spørsmål fra Danske Kanselli: 1: Akershus stift: Østfold, Akershus*. Vol. 1. Oslo: Solum Forlag. pp. 247–261.

Sogner, S. 1976. 'A demographic crisis averted?' *Scandinavian Economic History Review* 24 (2): 114–128. https://doi.org/10.1080/03585522.1976.10407847.

Sogner, S. 1979. *Folkevekst og flytting: en historisk-demografisk studie i 1700-årenes Øst-Norge*. Oslo: Universitetsforlaget.

Stamnes, A.A. 2016. 'Effect of temperature change on Iron Age cereal production and settlement patterns in Mid-Norway'. In F. Iversen and H. Petersson (eds), *The Agrarian Life of the North 2000 BC–AD 1000: Studies in Rural Settlement and Farming in Norway*. Oslo: Cappelen Damm Academic. pp. 27–40. https://library.oapen.org/handle/20.500.12657/31162.

Strand, E. 1984. *Korn og korndyrking*. Oslo: Landbruksforlaget. https://www.nb.no/items/URN:NBN:no-nb_digibok_2012011808175?page=0.

Stubberud, T. 2016. *J.N. Wilse. En opplysningsmann*. Rakkestad: Valdisholm Forlag.

Sunde, K.A. 2020. *Om Lofthusopprøret, og hvordan bøndenes klager vant gjenklang i kommisjonen*. MA thesis. Oslo: University of Oslo. https://www.duo.uio.no/handle/10852/84693.

van Dijk, E., I.M. Gundersen, A. de Bode et al. 2023. 'Climatic and societal impacts in Scandinavia following the 536 and 540 CE volcanic double event'. *Climate of the Past* **19** (2): 357–398. https://doi.org/10.5194/cp-19-357-2023.

Wanner, H., C. Pfister and R. Neukom. 2022. 'The variable European Little Ice Age'. *Quaternary Science Reviews* **287**: 107531. https://doi.org/10.1016/j.quascirev.2022.107531.

White, S., D. Collet, A. Alcoberro et al. 2025. 'Climate, peace, and conflict – past and present: Bridging insights from historical sciences and contemporary research'. *Ambio*. https://doi.org/10.1007/s13280-024-02109-1.

Wilse, J.N. 1778. *Meteorographia compendiosa, eller en kort Maade, ved beqvemme Tegn, saa nøye og udførlig, som man vil, at optegne alle slags Veyr, Veyrskikker og Luftsyner, det er: alle slags meteorologiske Iagttagelser, og deraf giøre daglige og aarlige Lister til Sammenligning af adskillige Tiders og Steders Veyrlig og Klima med Kobber*. Kiøbenhavn: Det Kongelige Universitet Bogtrykkerie, Det Kongelige Danske Videnskabernes Selskab. https://www.nb.no/items/URN:NBN:no-nb_digibok_2017090826009.

Wilse, J.N. 1790–1798. *Reise-Iagttagelser i nogle af de nordiske Lande, med Hensigt til Folkenes og Landenes Kundskab, først bestemte som Bidrag til det Tydske Bernoulliske Verk*. Vol. 1–5. Kiøbenhavn: Paa S. Poulsens Forlag.

Wilse, J.N. 1991 [1779–1780]. *Physisk, oeconomisk og statistisk beskrivelse over Spydeberg præstegield og egn i Aggershuus-stift udi Norge*. Christiania: C.S. Schwach.

Wilse, J.N. 1992 [1793–1796]. *Topographisk Beskrivelse af Edsberg Præstegjeld*. Rakkestad: Valdisholm Forlag.

Wilse, J.N. 1993. *Reiser i Østfold på 1700-tallet*. Ed. by T. Stubberud. Rakkestad: Valdisholm Forlag.

Winkler, P. 2023. 'The early meteorological network of the Societas Meteorologica Palatina (1781–1792): Foundation, organization, and reception'. *History of Geo- and Space Sciences* **14** (2): 93–120. https://doi.org/10.5194/hgss-14-93-2023.

Wishman, E.H. 2007. *Vær og klima over indre strøk av Østlandet 1781-1790. Et bidrag til Norges klimahistorie basert på fire emetsmenns meteorologiske og klimarelaterte observasjoner og andre nordiske kilder*. Vol. 46. Stavanger: Arkeologisk museum i Stavanger.

Ystgaard, I. and R. Sauvage. 2025. 'A series of unfortunate events: Two central Norwegian settlements facing the climatic downturn after AD 536–540', this volume. https://doi.org/10.63308/63881023874820.ch03.

Zambri, B., A. Robock, M.J. Mills and A. Schmidt. 2019. 'Modeling the 1783–1784 Laki eruption in Iceland: 1. Aerosol evolution and global stratospheric circulation impacts'. *Journal of Geophysical Research: Atmospheres* 124 (13): 6750–6769. https://doi.org/10.1029/2018JD029553.

Åssveen, M. and U. Abrahamsen. 1999. 'Varmesum for sorter og arter av korn'. *Grønn forskning* 2: 55–59.

The Author

Ingar Mørkestøl Gundersen is a Postdoctoral Research Fellow in the Department of Archaeology, Conservation, and History at the University of Oslo. He has published on a range of topics on climate history and the Scandinavian Iron Age. Gundersen received his Ph.D. in 2022 with the thesis 'Iron Age Vulnerability', which investigated the archaeological evidence for a sixth-century climate crisis in eastern Norway. His doctoral research was part of the VIKINGS project (Volcanic Eruptions and their Impacts on Climate, Environment, and Viking Society in 500–1250 CE). Together with Dr Manon Bajard, he received the Inter Circle U. prize 2022 for outstanding examples of cross-disciplinary research. He is currently part of two research projects on the Nordic Little Ice Age (ClimateCultures, University of Oslo and *The Nordic Little Ice Age (1300–1900) Lessons from Past Climate Change (NORLIA)* at the Centre for Advanced Study at the Norwegian Academy of Science and Letters.

Chapter 10

'AN ICE BREAKUP AS IN THE GOOD OLD DAYS'. ICE JAMS IN THE AURA RIVER, TURKU SOUTHWEST FINLAND, 1739–2025

Stefan Norrgård

Abstract

This study investigates the historical occurrence and impact of ice jams in the Aura River, Turku, Finland, from 1739 to 2025. Ice jams, which are common in regions with prolonged river ice cover, can cause significant water level rises and subsequent flooding. The study analyses historical documents, including newspaper articles and weather journals, to reconstruct past ice jam events and their effects. Key findings highlight the role of natural and anthropogenic obstacles, such as bridges and skating rinks, in initiating ice jams. The study further examines the methods employed to mitigate these events. An ice jam index is developed to categorise the severity of events over time. The results indicate a clear change towards fewer severe ice jam events in the twentieth century. This research contributes to understanding hydrological extremes and their historical context in Finnish rivers.

Introduction

On 16 April 1919, on the Aura River in Turku, an ice jam had formed against the Aurasilta Bridge, the second of the two bridges in the city centre. The breakup began in the afternoon and, as ice kept pushing against the pillars, the jam worsened, blocking the flow of water as the day progressed. The water levels rose an estimated 280 centimetres, which was enough to push blocks of ice up on the stone quays. Afterwards, as the ice jam let go, the surge damaged four ships moored in the river. The local newspaper, *Åbo Underrättelser*, labelled it 'An ice breakup as in the good old days' (ÅU 18 April 1919). The editor estimated that it was the first time in more than ten years that the spring ice breakup had been exceptionally dynamic. A review of the local newspapers shows that

doi: 10.63308/63881023874820.ch10

the last comparable ice jam event occurred sixteen years earlier. In 1903, the jam lasted for an estimated five hours, and the tail of the ice jam extended 500 metres upstream from the Aurasilta Bridge. At its climax, the water level rose to an estimated 365 centimetres. As the ice jam let go, the surge damaged several ships moored in the river over winter (Norrgård 2020). But what about the occurrence of ice jams before 1903 and after 1917?

The purpose of the present study is twofold, and the paper is likewise divided into two parts. The first part uses historical material to identify and categorise ice jam events to create an ice jam index for the city of Turku, Finland, between 1739 and 2025. Based on the same material, in the second part, the purpose is to discuss what contemporary observers considered the main reason for the recurring ice jams. Additionally, the article highlights past attempts to mitigate impact and prevent ice jams from occurring. Thus, this study contributes to the discussion of what causes ice jams, especially in past societies and urban environments. It shows how past societies have discussed, combated and reflected on the impacts and causes behind ice jams. The general purpose is to leverage historical documents to create research opportunities. As such, this study exemplifies the use of historical documents to identify hydrological extremes in Finnish rivers.

The main motivation for this study is that, as noted by Fukś (2023), awareness of river ice research is low and river ice topics were not addressed in the IPCC's cryosphere chapter in 2021. Moreover, there is a research gap on historical flood intensity and ice jams in Finland. Although some studies have dealt with future flood scenarios in connection with climate warming (e.g., Verta and Triipponen 2011; Lindenschmidt et al. 2022) no long-term chronologies or historical investigations exist (e.g., Blöschl et al. 2020). Except for some historical case studies from spring ice jams in the Tornio River (Saarnisto and Helama 2017) and Aura River (Norrgård 2020), or ice jam events during mid-winter breakups (Norrgård 2022b), research on ice jams is limited to few technical papers in conference proceedings (e.g., Ahopelto et al. 2015; Aaltonen and Huokuna 2017). Most research on ice jams comes from North America and especially Canada (Rohr 2022; Rokaya et al. 2018).

Ice Breakups and Ice Jams

Ice jams of various intensities are regular and probable phenomena in cold regions where rivers annually freeze over and have a cover of ice for one month or longer. The relationship between ice jams and climate is complex and influenced by various factors. For instance, trends in the timing of the ice-off event in the Aura River show a considerably higher correlation with February temperature compared to precipitation over the 1961–2020 period (Norrgård and Helama 2019; 2022). Ice jams, however, depend on a multitude of factors at the time of occurrence, for example, the thickness and strength of the ice, precipitation, discharge rates, temperatures, snow melt and ice decay and solar radiation (e.g., Beltaos and Prowse 2001).

Ice jams are unpredictable and can cause extreme floods, which is why they are serious threats to riverine societies (Rokaya et al. 2018b). In attempts to predict ice jams, Guo (2002) suggested that ice jam occurrence could be determined based on the relationship between river discharge and ice thickness. However, ice jam research has been limited by the shortage of long-term data (Zangb et al. 2022; Beltaos 1997).

Whether ice jams occur in autumn, winter or spring, they are caused by miscellaneous constrictions in the river channel. The obstacles can be either natural (e.g., stones, stationary ice, river narrowing) or anthropogenic (e.g., bridges or other stationary objects) that prevent the dislodgement of the ice on its path towards the sea. Bridges are known to impede the passage of ice during breakup, particularly in a narrow section of the river (Beltaos et al. 2006).

A jam can stay in place for minutes, hours or days, depending on the size of the river and the location of the jam. When jamming has occurred, ice blocks moving down the river can cause the jam to start growing in both height and length. The tail of an ice jam can be anything from a few metres to several kilometres long. The jam usually causes water levels to rise behind the blockage. When the height of the jam has reached a certain peak, water and ice start spreading to the sides as in 1919. This continues until the jam breaks, which creates a forward-going surge. The surge causes the water level upstream to drop quickly whereas it may cause further ice jams or flooding downstream (Madaeni et al. 2020; Lindenschmidt et al. 2018; Beltaos and Prowse 2001).

The Aura River and Turku

The Aura River starts in Oripää and is seventy kilometres long with a drainage basin of 874 square kilometres and enters the Archipelago Sea, a sub-basin of the Baltic Sea, in Turku. In the 1700s, Turku had a population of a few thousand and by the end of the 1800s about 30,000. More recently, the city has grown rapidly and in 2024 it had over 206,000 inhabitants.

The first description of an ice jam is from April 1649 (Hausen 1882), but it was not until the 1700s that ice breakups became a real nuisance. In the mid-1700s, erosion caused by ice breakups threatened to destroy one of the main roads and, because the wooden pilings that had protected the embankments since the 1600s required much maintenance, it was decided that the city centre had to be protected with stone quays (Norrgård 2022a). Hereafter, stone quays have always been a prominent feature of the city centre and a reference point for the height of the water level during ice jams.

For a long period of time, Turku had only one bridge in the city centre. This bridge, however, was destroyed in the great city fire in 1827. In 1831, it was replaced by two bridges: Aurasilta Bridge and Tuomiokirkkosilta Bridge. They were built south and north of the first bridge, respectively. These were the only bridges in the city centre until 1939. Map 1 shows the main bridges in Turku and the years they were built. Before the third bridge was built in 1939, different harbours, such as the fishing harbour, commodity and passenger harbours, were situated in close vicinity to the bridges.

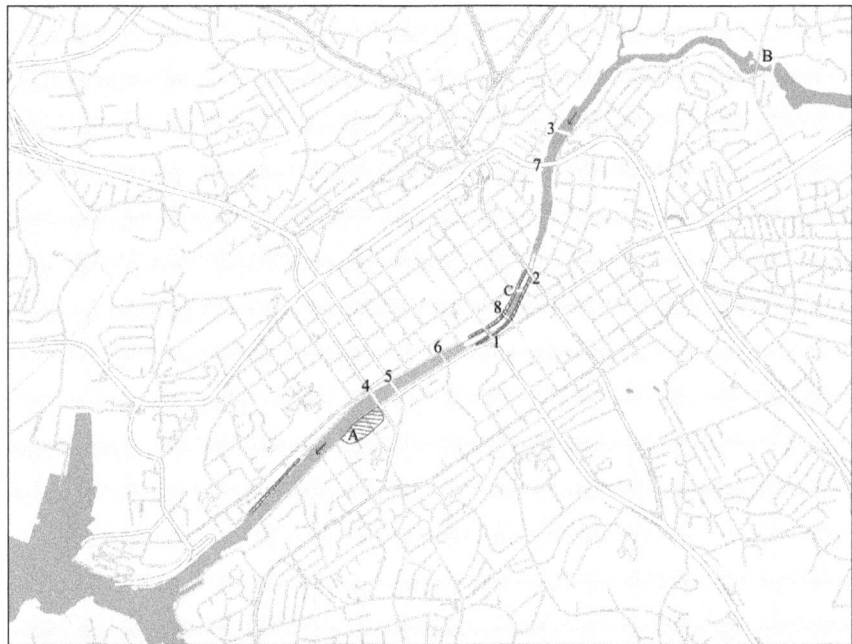

Map 1. City map of Turku. The numbers indicate when each bridge was first opened for traffic. Some bridges have been renewed over the years but the location has not changed. 1. Aura Bridge (1831); 2. Tuomiokirkkosilta (1831); 3. Railway bridge (1899); 4. Martinsilta Bridge (1940); 5. Myllysilta Bridge (1974); 6. Teatterisilta Bridge (1997); 7. Tuomaansilta Bridge (2000); 8. Library Bridge (2013). On the river, the white areas mark the location of the skating rinks. The striped areas indicate flooded areas in 1837, 1838, 1853 and 1857. (A) marks the location of the sign showing the water level in 1853 and 1857. (B) marks the Halinen Dike and the location of the sign showing water level heights in 1837 and 1882. Base Map by phettula on Esri (with data from Lantmäteriet, National Land Survey of Finland, Esri, TomTom, Garmin, Foursquare, GeoTechnologies, Inc, METI/NASA and USGS).

'An Ice Breakup as in The Good Old Days'

In spring, essentially no ships or boats could enter the harbours before the river was free of ice. The economic benefit of knowing when the river was ice-free was therefore why the professor of medicine at the Royal Academy of Turku, Johan Leche (1704–1764), started documenting ice breakup observations in 1749 (Norrgård and Helama 2019; 2022). He later categorised the intensity of his observations over the period 1749–1763 and his interest in ice breakups led to inquiries about earlier events. As a result, he described the ice jam flood in 1744, during which the basements in houses near the river were flooded and the jam threatened to destroy the bridge (Leche 1763). Professor Leche's categorisation and descriptions are valuable because he documented all types of breakups – from extreme to thermal breakups driven by temperature increase – and impacts that are identified in later records.

Material Employed in the Present Study

For most of the city's history, the ice breakup signalled the arrival of spring and the start of a new economic cycle. The ice jam event in 1837, described as the first extreme event in a long time, was a milestone event because it occurred on the threshold of the maritime boom and industrialisation that followed. As the economic importance of the river grew, local newspapers, available in both Swedish and Finnish, provided detailed descriptions of ice breakups and ice jams. Not limited by space and always competing for readers, lengthy and detailed descriptions were favoured during extreme events while slow and uneventful ice breakups generated short and disappointing reflections.

In the pre-1837 period, there are less than a handful of newspapers and the descriptions of ice jam events are sparse and vague. For this early period, the categorisation relies on descriptions found in the meteorological journals kept at the Royal Academy (Holopainen et al. 2023) and a diary kept by Johan Winter (1788–1872) between 1812 and 1828. Only the most extreme ice jam events and the least intensive ice breakups, i.e. thermal breakups, were considered reliable for the pre-1837 period. Thermal breakups are the opposite to extreme ice jam events and easily identified because it was noted that the ice breakup was slow and uneventful, which was, to some extent, a disappointment.

The ice jam index, the attempts to mitigate impacts, and the parts dealing with the causes of recurring ice jams are based on newspaper descriptions starting from 1771 (Table 1). The most reliable descriptions are from 1837–1950. During this period, the articles often include variables permitting a detailed study of each ice jam event. They might describe the weather leading up to the event; observations of water levels and flow intensities; minute description of how the event progressed; sporadic ice thickness descriptions; preventative

Figure 1a. The mid-winter ice jam on 27 February 1887. The photo is taken from the Tuomiokirkkosilta Bridge. It is the only picture taken when an ice jam had reached its climax in the Aura River in Turku. The events in 1837 and 1857 were claimed to be similar. Source: Ole Aune, Åbo Akademi University Picture Collections.

measures before, during and after the event; descriptions of the debris in the ice, and finally, the havoc caused by the surge as the ice jam broke. After the Second World War, the descriptions become vague and increasingly variable. The change is partly caused by professionalisation of journalism, which meant that the story-telling aspect vanished, which often included the journalists' own vivid descriptions and personal experiences. Partly the change also occurred because the harbours moved towards the mouth of the river and partly because the modernisation of society meant that the city and its citizens became less dependent on the ice on the Aura River.

Pictorial evidence is scarce and the only photo that shows the height of the water level during an ice jam in the 1800s is from a partial but extreme mid-winter ice breakup on 27 February 1882 (Norrgård 2022b). The descrip-

'An Ice Breakup as in The Good Old Days'

Figure 1b. A photo taken from the same place as in 1887, but with the purpose of showing the height of the stone quays from the water level between the bridges. The photo was taken on 27 February 2025. Source: Stefan Norrgård.

tions from this ice jam event are similar to extreme spring ice breakups and it has therefore been used as a reference point when comparing to past events. Figure 1a shows the height of the water level, as seen from Tuomiokirkkosilta Bridge, on 27 February 1882. The ice jam was at least 500 metres long and the most disastrous mid-winter ice breakup in the history of Turku. For comparison, Figure 1b shows the same place on 27 February 2025. The latter photo includes the height of the different tiers of the quays. In the 1990s and 2000s, the newspapers occasionally published photos of the ice breakup while in progress. Some videos of contemporary events are few. Videos of the ice breakups in 2010 and 2011 are found on YouTube, and even though they were eventful, they are not comparable to past events.

Stefan Norrgård

Table 1. The newspapers and years used to create the ice jam index.

Newspaper	Years
Aura	1880–1896
Helsingin Sanomat (HS)	1994
Sanomia Turusta	1850–1903
Tidningar Utgifne af et Sällskap I Åbo	1771–1778,1782–1785
Turkulainen	1915–1917
Turun Lehti	1882–1919
Turun Sanomat (TS)	1905–2020
Tähti	1863–1867
Uusi Aura	1897–1929
Åbo Allmänna Tidning	1810–1819
Åbo Morgonblad	1821
Åbo Nya tidningar	1789
Åbo Posten	1873–1883
Åbo Tidning	1800–1809
Åbo Tidning	1882–1906
Åbo Tidningar (Åtar)	1791–1799
Åbo Tidningar (Åtar)	1820–1861
Åbo Underrättelser (ÅU)	1824–2020

The Aura River Ice Jam Index

Indices are frequently used for historical flood reconstructions and the most common is a three-level categorisation. Kiss (2019) has an overview of different types of methods used in previous studies. In accordance with this, the water level height, which also indicates the magnitude of the ice jam, was divided into three categories: (3) extreme, (2) strong and (1) mild to weak (Figure 2). The index includes 209 observations and covers 73 per cent of the 1739–2025 period. Zero (0) denotes insufficient or nodescriptions. For the 1700s and 1800s, there are no observations but, in the late 1900s and early 2000s, the lack of descriptions mainly suggests that the ice breakup was mild to weak. A category (2) event, like that in 1994, would have made the news.

Leche's (1763) categorisation between 1749 and 1762 is included in the index. He divided the intensity into three categories (strong, mild and weak) but the only extreme ice jam event seems to have occurred in 1744. Leche would, presumably because he lived an estimated forty metres from the river (Norrgård 2024), have included his own descriptions of a category (3) event if he had witnessed one. Thus, in the index, 'strong' ice breakups were included as category (2) events. 'Mild' and 'weak' events were categorised as (1) events as he noted that the ice melted because of increasing temperatures and the water level was low (Holopainen et al. 2023). Observations of the ice breakups between 2018 and 2025 were made by the author as part of an ongoing research study to better understand the breakup process and the reliability of qualitative descriptions.

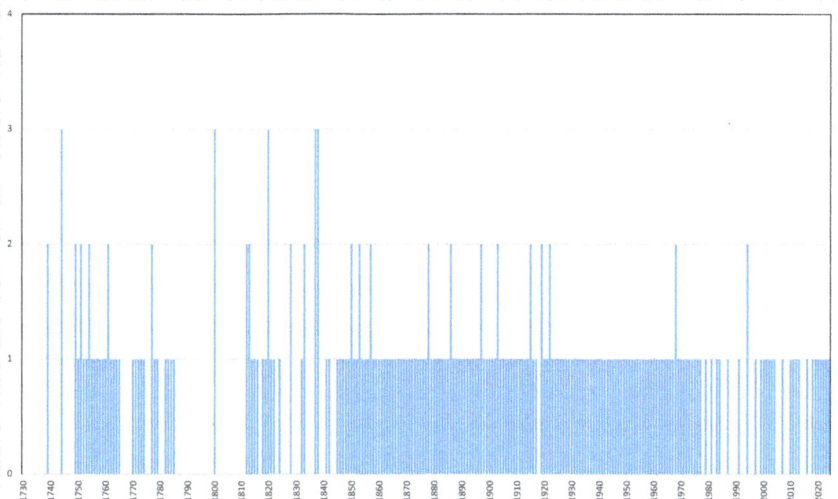

Figure 2. The Aura River ice jam index, 1739–2025, in Turku. See text for description.

Category (3) includes events when the water level was high enough to flood houses near the river, such as in 1744, 1800, 1837 and 1838. The event in 1800 was exceptional because ice was reported to have entered some houses through the windows. The mid-winter ice breakup in 1882 would fall under this category, but it is worth noting that the river did not become ice-free until spring.

Category (2) contains ice breakups with water levels between 120 and 300 centimetres. During some years, ice blocks covered the stone quays or the nearby street after the water level had subsided, but the newspapers fail to explain where. Part of the challenge to determining the exact location is that the streets parallel to the river bear the same name, from the city centre to the industrial area, on both sides. Currently, the height of the stone quays between the Aurasilta Bridge and Tuomiokirkkosilta Bridge would require that the water levels exceeded 250 centimetres, depending on the thickness and the amount of ice, before the ice would rise onto the embankments. The only years when the water level was estimated as clearly over 300 centimetres were in 1837 (360 centimetres), 1838 (375 centimetres) and 1903 (365 centimetres), and these refer to the ice jam in front of the Aurasilta Bridge. Before the third bridge was built in 1939, some of the ice jams at the bridge were followed by smaller and short-lived ice jams caused by stationary stronger ice. The surge or a second or third ice jam sometimes flooded areas closer to the mouth of the river where the landscape flattens out and the width of the river doubles from circa 50 to 100 metres compared to the city centre. Nonetheless, floods were uncommon because, as the river widens, discharge rates fell quickly, which prevented flooding. The exceptions were, for example, 1853 and 1857. These floods were commemorated with a metal plaque attached to a house on Savenvalajankatu an estimated fifty metres from the river and 250 metres below the current location of the Martinsilta Bridge (see Map 1). The plaque indicated the height of the water level above the street in 1853 (45 centimetres) and 1857 (60 centimetres). This house has been demolished and the exact place for these measurements therefore remains unknown. Stone quays were progressively built on this section of the river in the late 1800s, therefore preventing floods.

Category (1) events denote all other breakups, and it is the most diverse category. The water level varied from low to normal or slightly higher than normal. Some of these events caused damage because of high flow rates and smaller and short-lived ice jams. In 1946, following a small ice jam at Martinsilta Bridge, the 42-metre museum ship *Sigyn* was ripped off its moorings with the surge (ÅU 2 April 1945). The harbour diaries, for example, also mention smaller ice jams in 1960, 1962, 1963 and 1966. Category (1) comprises descriptions like 'exceptionally low water levels', 'the water level is as low as during the summer months' and claims that 'the ice melted in its place'. These descriptions imply that the breakup was thermal, i.e., the ice breakup was driven by increasing temperatures and not discharge and water levels.

It should be duly noted that the ice jam index does not imply that an ice breakup without high water levels would have been uneventful, only that the water level and possible damage were less severe. Of the 22 category (2) and (3) events, nineteen of the events occurred between 3–28 April. Only two occurred in late March (1903, 1968) and one in May (1812). Eleven events between 1837 and 1919 seem to have been initiated, as described by the newspapers, by intense rains a day or two before the ice jam occurred. The lack of descriptions does not indicate a lack of precipitation. This ambiguity is a limiting factor when relying only on newspapers and observations made by journalists. Nonetheless, category (1) events (87.5 per cent) clearly dominate.

'An Ice Breakup as in The Good Old Days'

'Watch the Niagara Falls, Go to Halinen!'

The Halinen Dike (see Map 1), six kilometres from the mouth of the Aura River, plays an important part in the history of the breakup process and ice jams in Turku. In the 1700s, the dike was built of wood and because it was repeatedly destroyed during ice breakups, it was rebuilt using stone in 1824. It was partly destroyed in the ice breakup in 1837, but the following version survived until a new and higher dike was built on top of the old dike in 1972.

The dike separated the lower reaches of the river from the upper reaches and divided the ice breakup process into two stages. In the first stage, the ice below the Halinen dike broke up and the river became ice-free. Ice jams initiated in this order, as in 1886, 1897 and 1994, were less severe. In the second stage, on average one to three days after the lower reaches of the river were ice-free, as snowmelt increased further inland and the water levels rose, the ice behind the dike rose over the edge and rapidly traversed the city. These blocks of ice were the so-called *Old men from Halinen* (in Swedish: *Hallisgubbar*; in Finnish: *Halisten ukot*) (henceforth the Halinen ice) and essentially referred to all ice from upstream. The Halinen ice was notorious for rapidly traversing the city, owing to the increased water level and flow rates needed to lift the ice up and above the dike. The Halinen ice is mentioned in the historical records for the first time in 1697 (Norrgård 2022a).

There were two exceptions to this two-stage ice breakup scenario. First, during drier years, the Halinen ice would rot behind the dike, which was considered a great disappointment. In 1759, for example, as ice rotted behind the dike, Leche wrote, 'No ice breakup this spring' (Holopainen et al. 2023). Second, driven by rains, the Halinen ice and the ice below the dike started to move at the same time. These ice breakups initiated worst-case scenarios. This was most likely the case in 1800, and with some certainty in 1813, but definitely in 1837. The dike was partially destroyed and, as an ice jam formed against the piers of the Aurasilta Bridge, the water level reached a maximum height of around 360 centimetres, flooding streets and basements. As the ice jam let go, the surge took two wooden ships with it and the streets were flooded and covered in ice blocks between 74 and 104 centimetres thick. It took hundreds of men several days of work to push the ice back into the river (Åtar 26 April 1837). The following year, on 24 April, this scenario repeated itself and the water level was recorded as half a foot higher (Åtar 28 April 1838)

The arrival of the Halinen ice was often closely monitored by the media. It was also a local tradition to go and watch the water cascade over the dike in the 1900s. For example, *Turun Sanomat* wrote on 11 April 1969, 'Watch the Niagara Falls, go to Halinen!' Despite the comparison to Niagara, the water level had not risen enough to push the 45-centimetre-thick ice over the dike.

In the past, during the most extreme years, the height of the water level was etched into a house situated below the dike. The etchings marked out the highest water level at 522 centimetres from 1837 and below that the water level of 1882 (Rinne 1905). The house has been torn down and, after the dike was rebuilt in 1972, the water level regulating mechanism, or flood gates, should keep the Halinen ice from going over the edge of the dike. Still, the Halinen Ice managed to crawl over the dike and destroy the bridge immediately below in 1976 (TS 20 April 1976).

What Caused the Ice Jams?

The newspapers highlight three possible causes believed to initiate ice jams: the bridges, the skating rinks and the lack of dredging. Distinguishing one cause from the other was probably challenging because they all referred to the same location in the city centre and the narrowest parts of the river. This was, however, never considered in the newspapers.

In the 1700s, descriptions imply that the ice jams occurred at the bridge. Its construction is not well documented, but being built of wood, withstanding the force of the ice, the bridge required much maintenance. For most of the period, it seems to have had one pier, designed as a wedge towards the stream and covered in metal. The wedge is visible on some city maps and the pier appears in some paintings in the latter part of the 1700s. The size of the pier provides a potential explanation for the identified ice jams until the bridge burnt down in 1827.

Of the two bridges that replaced the destroyed bridge in 1831, the Aurasilta Bridge was initially called the Stone Bridge. Its name derived from the two stone piers that were considered to cause ice jams. At first sight, it would seem the most natural explanation because all significant ice jams occurred in front of it. Minor ice jams occurred in front of the Tuomiokirkkosilta Bridge, which had two sleek wooden piers, but ice jams never seem to have threatened the structure of the bridge in the same manner as the Aurasilta Bridge. At the end of the 1800s, the bridges were in such a poor condition that they had to be renewed. In 1899, the Tuomiokirkkosilta Bridge was made into an arch bridge to allow the ice to flow freely under it. Ironically, the only significant spring ice jam at the bridge, caused by stationary ice, occurred in 1915.

The Aurasilta Bridge was renewed in 1907. The old piers were barely standing and replaced with one pier in the middle of the river, including a wedge against the current to withstand the force of the ice. The current bridge still has the same pier. The height of the bridge was lowered from 725 to 450 centimetres and the abutments moved five metres into the river. Lowering the

'An Ice Breakup as in The Good Old Days'

height was an unexpected decision. However, this implies that the ice jams before 1907 were more intense than the ice jams after. Although an arch bridge was considered the best solution, the soil was deemed too unstable to bear its weight. The main result of rebuilding the first bridge into an arch bridge was that the entire force of the ice was directed towards the pillars of the Aurasilta Bridge. Despite the changes and the fears that arose, category two ice jams decreased markedly. However, another change occurred simultaneously, and this could provide a better explanation for the decrease in events.

During the latter half of the 1800s, skating on the river grew in popularity. The history of skating on the river ice is still unwritten, but the first maintained skating rink between the bridges was arranged in 1875. Maintaining the skating rinks enabled the ice to grow thicker and stronger over the course of the skating season, which provided the ingredients necessary to create an ice jam against the piers of the Aurasilta Bridge. In 1878, one skating rink was located below the Aurasilta Bridge, and the newspapers noted that it was the stronger ice below the bridge that seemed to have caused the ice jam. (ÅP 10 April 1878) There was, however, also a skating rink above it that was not discussed. In 1886, when the next significant ice jam occurred, the city had given permission to arrange four skating rinks on the ice (see Map 1). The first sawing of the skating rinks, to give room for the ice to move, was before the ice jam in 1897. Both skating rinks between the bridges were sawed into smaller pieces before the ice started breaking up. Despite this, it was these ice blocks that, in combination with the piers of the Aurasilta Bridge, were considered to have caused the jam. (ÅT 16 April 1897)

In 1900, the city started requiring that those maintaining the skating rinks had to saw them if the conditions were such that there was a chance of an ice jam occurring. At this stage, however, lighted artificial skating rinks were already competing with those on the river. Skating rinks were still arranged, and ash was sometimes spread on the ice to speed up the melting, but they seemed to have decreased in popularity. In 1915, while romanticising the idea of skating on the river, the city arranged a skating rink in the river and again, an ice jam occurred. This was the first time since 1903, and it happened again in 1922. This implies that the skating rinks played a part in creating ice jams, but, although skating rinks were still arranged in the river until 1939, there were no significant ice jams. This, however, could depend on the warmer winters in the 1930s (Tuomenvirta 1995), which made it difficult to maintain the skating rinks. In the post-war period, skating rinks have not been arranged on the river, even though attempts to revive the tradition were made in the 1970s.

In the late 1880s, the lack of dredging between the bridges was discussed when the riverbed was visible during drier summers. In 1888, while repairing the

Tuomiokirkkosilta Bridge, the workers noted that the river had frozen almost throughout. There was only a trail of water under the sixty-centimetre-thick ice. This could, arguably, have increased the risk of ice jams. It was speculated that soil was transported from upstream via erosion, and this fell to and accumulated on the riverbed when an ice jam formed against the Aurasilta Bridge.

Preventing and Managing Ice Jams

Since the 1700s, different methods have been utilised to mitigate or prevent ice jams from occurring. Most attempts were made in retrospect and because of an ice jam the previous year. In 1739, for instance, an ice jam nearly destroyed the bridge and the next year there was a suggestion to saw the ice to prevent a similar scenario. Whether the sawing was executed is uncertain. Evidence of the first sawing of the ice is from 1745, and again it was initiated because of an extreme ice jam the preceding year. In 1786, there was a general recommendation that the ice should be sawed to avoid ice jam floods (Norrgård 2022a). Such recommendations were, however, easily forgotten if an extreme ice jam event was followed by a series of weaker ice breakups.

In 1753, in a thesis on how to plan a city, the suggestion was that the citizens should be prepared to protect critical structures during ice breakups. Using any instrument longer than their arms, they should push the ice away from bridges, mills and other structures to prevent these from being destroyed (Norrgård 2022a). Eventually, this suggestion resulted in the deaths of two men, who were lowered onto the ice jam from the Aurasilta Bridge in 1837. As they tried to push the ice blocks away from the bridge, the jam suddenly let go and while the first man was swept away under the ice, the other was crushed against the piers. Despite the fatalities, throughout the 1800s and 1900s, men armoured with axes and wooden poles, standing on the ice or in boats, or being lowered from the bridge, were often used to combat ice jams.

In the 1800s, alongside sawing of the ice, icebreakers were used almost annually between 1873 and 1939 (Norrgård and Helama 2019). The use of ice breakers did not prevent ice jams from occurring but during some years they were utilised to break stronger ice in front of an existing ice jam, as for instance in 1922. As already noted, it was customary to spread ash on the ice between the bridges in an attempt to weaken it or then again to saw the ice where the ice skating rinks had been maintained over winter.

Another method was dynamite, used for the first time in 1897. Several charges were placed on both sides of the river in the midst of the ice jam in front of the Aurasilta Bridge, but the attempts were futile (ÅU 16 April 1897). Dynamite was used again in 1926 and 1946. On 25 March 1953 when repairing the Tuomiokirkkosilta Bridge, 'hundreds' of charges were used according to the newspapers. Unsuccessfully, at first, because the ice was fifty centimetres thick in the middle of the river and seventy centimetres near the riverbanks (ÅU 27 March 1953). Four decades later, in 1994, when an ice jam was threatening the Aurasilta Bridge, dynamite was considered too dangerous to use in the city centre (TS 5 April 1994). In the last forty years, technical solutions and heavy machinery have replaced dynamite and manpower. In 2010, to protect the old Myllysilta Bridge, a crack was sawed in the sixty- to ninety-centimetre-thick ice, after which air pressure was used to create a current to weaken the ice from below (TS 19 March 2010). The next year, to protect the construction site of the new Myllysilta Bridge, excavators worked day and night to crush the ice into smaller pieces and push it under the bridge (YouTube 2011).

Climatic Reflections

Category (3) events have not occurred in over 180 years and category (2) events became increasingly sporadic in the 1900s. It is challenging to establish a climatic cause for the decline because of the complicated influences of factors such as ice thickness, snow cover, precipitation and solar radiation (Beltaos et al. 2006).

Intense rains clearly initiated eleven ice jams, a majority occurring in April and, thus, earlier ice breakups could provide a plausible explanation for a decrease in ice jams. Figure 3 shows the ice-off dates for the Aura River between 1949 and 2025. Ice-off events before April were extremely rare in the 1800s (1), rare in the 1900s (18), but dominate in the 2000s (13). Moreover, while there is a clear trend towards earlier ice breakups, for the first time since 1749, the river did not freeze up completely in 2008 and 2020 (Norrgård and Helama 2022). Of the 31 ice breakups between 1995 and 2025, i.e. after the ice jam in 1994, eighteen have occurred before April. In 2023 (observations by author), the river did not freeze completely until 7 March and eighteen days later it was ice-free. In 2024, the river froze in early February, but the weather was challenging and it was difficult to establish a clear freeze-up date. On 13 February, the ice was between twenty and thirty centimetres thick. The ice started breaking up in early March and continued until the river was ice-free on 25 March. In 2025, the river did not freeze until 17 February and was ice-free twelve days later. Later freeze-ups and earlier breakups mean that there is not enough time for the ice to grow thick and strong. In other words, the climatic conditions needed to provide ice cover, and the conditions required to initiate an ice jam, have become infrequent. Moreover, a study of the impacts of precipitation on ice cover (Fujisaki-Manome et al. 2020) showed that winter precipitation reduced ice thickness. Monthly mean temperature and precipitation data from a model

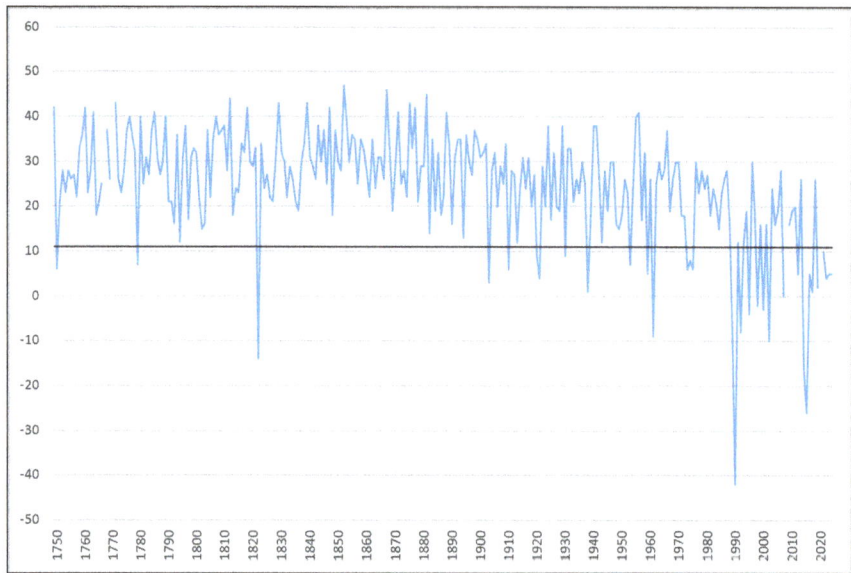

Figure 3. The ice-off dates in the Aura River, 1749–2025. The figure shows when the river was ice-free in relation to the spring equinox (commonly 20 or 21 March), which is indicated by zero (0). A negative value indicates that the river was ice-free before spring equinox. To better show changes in the timing, the black line indicates 31 March. For methods and in-depth analyses for the 1749–2020 period, see Norrgård and Helama (2019; 2020).

created by the Finnish Meteorological Institute (FMI) for the 1960–2020 period (Aalto et al. 2016; 2013), showed that the January–February period has become wetter and warmer when compared to the twenty years before (1968–1993) and after (1995–2020) the ice jam in 1994. Thus, increased temperatures and precipitation, both affecting the thickness, duration and strength of the ice cover, would reduce the probability of ice jams, or make them less severe, as noted by Newton et al. (2024).

Comparing the 2020s to the conditions in 1994 shows a stunning difference. Over the 1969–2020 period, February 1994 was the driest month of all and the third coldest. The mean temperature for the entire period was −5.12 degrees but in February 1994 it dropped to −13.4 degrees. March was only slightly colder than average but the wettest month over the entire 1960–2020 period. The mean precipitation was more than twice the average of 35.54 millimetres. In other words, an extended period of unusually low temperatures with almost no precipitation let the ice grow thick and strong. This was followed by extreme amounts of precipitation that created excellent conditions for the ice jam on 3 April. Comparing 1994 to 1968, when temperature and precipitation for both February and March were slightly above or below average, shows how intricate ice jam studies are. In the case of the Aura River, more in-depth case studies and analyses are needed.

The Aurasilta Bridge and the skating rinks provide a plausible explanation for the ice jams between the 1880s and 1920. However, the frequency of ice jam events would probably have been higher in the 1900s if all climatic variables had remained unchanged since the 1800s. This hypothesis is based on two dominating factors. First, the number of bridges in the city centre has increased from two to six since 1939 (see Map 1). Second, the mechanical stress of the ice cover during winter decreased after the harbours and shipyard, sugar and cotton industries moved away from the city centre in the second half of the 1900s. Consequently, there are no economic incentives to speed up the ice breakup process. Despite these changes, ice jams have become increasingly rare.

Of the 100 observations between 1839 and 1939, there were 88 category (1) events and for 42 years the water level was described as low, exceptionally low or equivalent. The study therefore shows that category (1) events have always been a prominent feature of the ice breakups. The frequency of category (1) events explains why the Halinen ice became notorious: it traversed the city with high intensity. The Halinen ice enabled extreme ice jams but was not a prerequisite for causing them. Introducing the water level regulation mechanism at the Halinen dike lowered the probability of category (3) events.

Photos from 1994 show no sign of ice above the second tier (see [Figure 1b](#)), suggesting that the water level rose only to an estimated 250 centimetres between the bridges. At its peak, one newspaper reported there was ice on the Aurasilta Bridge at 1.30 AM (ÅU 6 April 1994). This suggests that the jam was high in the middle of the river. However, the Halinen ice never arrived and therefore the ice never rose above the second tier onto the embankments as in 1919. In 1919, the surge damaged four ships (ÅU 18 April 1919); in 1994, it dragged a ship moored below Aurasilta Bridge off its moorings and pushed it into Myllysilta Bridge several hundred metres downstream. In *Turun Sanomat*, the 1994 event was later called the 'the greatest ice breakup of all times' (TS 11 April 1994).

Conclusions

It is in human nature to try to control the ice breakups and minimise their impacts. In Turku, the main goals have been to protect the bridges and the ships moored in the river over the winter. However, the city is situated at the mouth of the Aura River and trying to control the ice breakup, by sawing the ice or using dynamite and icebreakers, and hoping it would change the outcome, seems not to have had the desired effect. Except for the use of icebreakers, there were no annual traditions. Most attempts to control the ice breakup depended on the ice breakup the preceding year and the severity of the winter, i.e. the thickness of the ice. Today category (3) events are part of the climatic history of Turku. A lack of observations from the 1700s makes the extreme ice jam events in the 1800s stand out. However, the index shows a clear change in the frequency and intensity since the 1920s.

The study shows that historical documents can be used to investigate past ice jam events. Local knowledge is usually essential when studying ice jams and, for this purpose, newspapers are exemplary sources. Besides describing the timing of the ice breakup, newspapers further depict the attempts to prevent, mitigate and lessen impact before, during and after an ice jam event.

In this study, the focus was on ice jams in Turku, but the river is long, and ice jams have occurred also in other places. In 1839, for example, there was no information from Turku, but 35 kilometres upstream, in Pöytyä, an ice jam flood destroyed six bridges, several barns, several smaller rafts, mills and dikes. Future studies could extend the chronology to the entire river, whereas in-depth case studies could establish the impact of climate warming and possible changes in discharge rates, temperature and precipitation during the winter months in Turku. This could help to understand the occurrence and probability of future ice jams.

Bibliography

Newspapers

Turun Sanomat: (TS) 19 April 1922; 5 April 1994; 19 March 2010.
Åbo Posten: 10 April 1878.
Åbo Tidningar: (Åtar) 26 April 1837; 28 April 1838.
Åbo Tidning: 16 April 1897.
Åbo Underrättelser (ÅU) 29 April 1853; 5 April 1886; 16 April 1897; 18 April 1919; 3 April 1945; 27 March 1953; 6 April 1994.

Research articles and secondary literature

Aalto, J., P. Pirinen, J. Heikkinen and A. Venäläinen. 2013. 'Spatial interpolation of monthly climate data for Finland: comparing the performance of kriging and generalized additive models'. *Theoretical and Applied Climatology* **112**: 99–11. https://doi.org/10.1007/s00704-012-0716-9.

Aalto, J., P. Pirinen and K. Jylhä. 2016. 'New gridded daily climatology of Finland: Permutation-based uncertainty estimates and temporal trends in climate'. *Journal of Geophysical Research: Atmospheres* **121** (8): 3807–3823. https://doi.org/10.1002/2015JD024651.

Aaltonen, J. and M. Huokuna. 2017. 'Flood mapping of river ice breakup jams in River Kyrönjoki delta'. *CGU HS Committee on River Ice Process and the Environment. 19th Workshop on the Hydraulic of Ice Covered Rivers.* Whitehorse, Yukon, Canada, 9–17 July. https://cripe.ca/docs/aaltonen-huokuna-2017-pdf?wpdmdl=1549&refresh=67aff380612a01739584384.

Ahopelto, L., M. Huokuna, J. Aaltonen and J.J. Koskela. 2015. 'Flood frequencies in places prone to ice jams, case city of Tornio'. *CGU HS Committee on River ice Process and the Environment. 18th Workshop on the Hydraulic of Ice Covered Rivers.* Quebec City, QC, Canada, 18–20 August.

Beltaos, S. and T.D. Prowse. 2001. 'Climate impacts on extreme ice-jam events in Canadian rivers'. *Hydrological Sciences* **46** (1): 157–181. https://doi.org/10.1080/02626660109492807.

Beltaos, S., M. Lindon, B.C. Burrel and D. Sullivan. 2006. 'Formation of breakup ice jam at bridges'. *Journal of Hydraulic Engineering* **132** (11): 1229–1236. https://doi.org/10.1061/(ASCE)0733-9429(2006)132:11(1229).

Blöschl, G., A. Kiss, A. Viglione et al. 2020. 'Current European Flood-rich period exceptional compared with past 500 years'. *Nature* **583**: 560–566. https://doi.org/10.1038/s41586-020-2478-3.

Fujisaka-Manome, A., E.J. Anderson, J.A. Kessler, P.Y. Chu and A.D. Gronewold. 2020. 'Simulating impacts of precipitation on ice cover and surface water temperature across large lakes'. *Journal of Geophysical Research: Oceans* **125** (5): e2019JC015950. https://doi.org/10.1029/2019JC015950.

Fukś, M. 2023. 'Changes in river ice cover in the context of climate change and dam impacts: a review'. *Aquatic Sciences* **83**: 113. https://doi.org/10.1007/s00027-023-01011-4.

Guo, Q. 2002. 'Applicability of criterion for onset of river ice breakup'. *Journal of Hydraulic Engineering* **128** (11): 1023–1026. https://doi.org/10.1061/(ASCE)0733-9429(2002)128:11(1023).

Hausen, R. (ed.) 1882. *Diarium Gyllenianum eller Petrus Magni Gyllenii Dagbrok 1622–1667.* Helsingfors: J. Simelii arfvingars tryckeri.

Holopainen, J., S. Helama and S. Holopainen. 2023. 'Kuninkaallisen tiedeakatemian arkistoon sijoitetut Turun vuosien 1748–1800 päivittäiset säähavainnot historiallisena ja elektronisena informaationa' [The observations from Turku 1748-1800, in the Royal Academy of Sciences, as an historical and electronic record]. *Auraica* **14** (1): 39–57. https://doi.org/10.33520/aur.142174.

Kiss, A. 2019. *Floods and Long-Term Water-Level Changes in Medieval Hungary.* Cham: Springer.

Leche, J. 1763. 'Utdrag af 12 års Meteorologiska Observationer, gjorda i Åbo: Sjette och Sista Stycket'. *Kongl. Vetenskapsacademiens Handlingar* (Jul. Aug. Sept.): 257–268.

Lindenschmidt, K.E., M. Huokuna, B.C. Burrell and S. Beltaos. 2020. 'Lessons learned from past ice-jam floods concerning the challenge of flood mapping'. *International Journal of River Basin Management* 16 (4): 457–468. https://doi.org/10.1080/15715124.2018.1439496.

Lindenschmidt, K.E. et al. 2022. 'Assessing and mitigating ice-jam flood hazards and risk: A european perspective'. *Water* 15 (1): 1–23. https://doi.org/10.3390/w15010076.

Madaeni, F., R. Lhissou, K. Chokmanni, S. Raymond and Y. Gauthier. 2020. 'Ice jam formation, breakup and prediction methods based on hydroclimatic data using artificial intelligence: A review'. *Cold Regions Science and Technology* 174: 103032. https://doi.org/10.1016/j.coldregions.2020.103032.

Newton, B., S. Beltaos and B.C. Burrell. 2024. 'Ice regimes, ice jams, and a changing hydroclimate, Saint John (Wolastoq) River, New Brunswick, Canada'. *Natural Hazards* 120: 12613–12642. https://doi.org/10.007/s11069-024-06736-5.

Norrgård, S. 2020. 'The Aura River ice jam in Turku, March 1903'. *Arcadia* 10. https://doi.org/105282/rcc/8982.

Norrgård, S. and S. Helama. 2019. 'Historical trends in spring ice breakup for the Aura River in Southwest Finland, AD 1749–2018'. *The Holocene* 29 (6): 953–963. https://doi.org/10.1177/0959683619831429.

Norrgård, S. and S. Helama. 2020. 'Tricentennnial trends in spring ice break-ups on three rivers in northern Europe'. *The Cryosphere* 16 (7): 2881–2898. https://doi.org/10.5194/tc-16-2881-2022.

Norrgård, S. 2022a. 'Klimatanpassning i Åbo 1700–1827' [Climate adaptation in Turku 1700–1827]. *Historiska och Litteraturhistoriska Studier* 97: 71–99. https://doi.org/10.1030667/hls.112271.

Norrgård, S. 2022b. 'Tre klimathistoriska perspektiv på islossningen i Aura å' [Three historical perspectives on ice breakups in the Aura River]. In N.E. Villstrand and K. Westerlund (eds). *Is – på olika vis*. Åbo: Sjöhistoriska institutet vid Åbo Akademi. pp. 47–64.

Norrgård, S. 2024. 'En märklig mans liv och arbete. Johan Leche (1704–1764) och meteorologins historia i Åbo, Finland och Sverige under 1700-talet' [A peculiar man's life and work. Johan Leche (1704–1764) and the history of meteorology in Turku, Finland and Sweden during the 18th century]. *Auraica* 15 (1): 57–87. https://doi.org/10.33520/aur.157036.

Prowse, T. and S. Beltaos. 2002. 'Climatic control of river-ice hydrology: A review'. *Hydrological Processes* 16 (4): 805–822. https://doi.org/10.1002/hyp.369.

Rinne, J. 1905. *Turusta Halisten Koskelle* [From Turku to the Halinen rapids]. Helsinki.

Rokaya, P., S. Budhathoki and K.-E. Lindenschmidt. 2018a. 'Ice-jam flood research: A scoping review'. *Natural Hazards* 94: 1439–1457. https://doi.org/10.1007/s11069-018-3455-0.

Rokaya, P., S. Budhathoki and K.-E. Lindenschmidt. 2018b. 'Trends in the timing and magnitude of ice-jam floods in Canada'. *Scientific Reports* **8**: 5834. https://doi.org/10.1038.s411598-81-24057-z.

Saarnisto, M. and S. Helama. 2017. 'Miljön och omgivningarna kring Särkilax i Torne Älvdal från 1400- till 1600-tal' [The environment and the surroundings in Särkilax in the Torne River valley]. In S. Torikka (ed.) *1617 Övertorneå storsocken under en dramatisk tid.* Väyläkirja: EU. pp. 437–455.

Salinas, J.L., A. Kiss, A. Viglione, R. Viertl and G. Blöschl. 2016. 'A fuzzy Bayesian approach to flood frequency estimation with imprecise historical information'. *Water Resources Research* **52**: 6730–6750. https://doi.org/10.1002/2016WR019177.

Tuomenvirta, H. 1995. 'The warmest decade in Finland – the 1930s'. In P. Heikinheimo (ed.) *International Conference on Past, Present and Future Climate. Proceedings of the SILMU conference held in Helsinki, Finland, 22–25 August 1995.* Helsinki: Painatuskeskus. pp. 258–261.

Veijalainen, N. 2011. *Estimation of Climate Change Impact on Hydrology and Floods in Finland.* Ph.D. thesis. Aalto: Aalto University. https://urn.fi/URN:ISBN:978-952-60-4614-3.

Verta, O.-M. and J.-P. Triipponen. 2011. 'The Kokemäenjoki River Basin flood risk management plan – A national pilot from Finland in accordance with the EU flood directive'. *Multidisciplinary Hydrology Research* **60** (1): 84–90. https://doi.org/10.1002/ird.668.

Zhanbg, F., M. Elshamy and K.-E. Lindesnschmidt. 2022. 'Climate change impacts on ice jam behaviour in an inland delta: A new ice jam projection framework'. *Climatic Change* **171**: 13. https://doi.org/10.1007/s10584-022-03312-3.

YouTube videos

Aurajoen jäiden lähtö keväällä 2011 [Aura River Spring ice breakup 2011]. 2011. https://youtu.be/0zhtEs4i02c?si=ilNWiPZn2dE4hrhN.

Ice breakup – islossning – Jäiden lähtö. 2010. https://youtu.be/akluvQS-bnY?si=thEwdYyCOFAWMDrQ.

The Author

Stefan Norrgård is a senior researcher and climate historian at the Department of History at Åbo Akademi University in Turku, Finland. Subsequent to reconstructing climate in West Africa during the 1700s, his research interests have centred on riverine ice breakups in Finland. He has reconstructed spring ice breakups for both the Aura River (Turku) and the Kokemäki River (Pori) between the 1700s and 2000s. He has several publications on ice breakups but his research field also covers historical climate adaptation processes and meteorological observations in Finland and Sweden in the 1700s. His ongoing research project, founded by the Kone Foundation, investigates climate, culture and society in Finland in the 1700s.

NARRATING CLIMATE HISTORIES

Chapter 11

CLIMATE NARRATIVES IN NORWEGIAN PUBLIC HISTORIES

Eivind Heldaas Seland

Abstract

From the mid-nineteenth to the early twenty-first century, a number of multi-volume histories of Norway were published by major trade presses. These were written mostly by groups of professors at Norwegian Universities taking charge of one volume/historical period each, and were aimed at the general public. Sales were high, and the impact on public perception of history can likely only be compared with that of the most successful school textbooks. This chapter examines the role of climate and climate change as an agent of historical change in these works, concentrating on prehistory as well as the periods corresponding to what are today known as the Late Antique / Early Medieval Little Ice Age, the Medieval Warm Period and the Little Ice Age. While these terms rarely appear in the examined works, the potential role of climate and climate change in bringing about historical change is discussed to varying degree. These accounts are viewed through the lens of the narrative theory, enabling us to classify narratives of identity, decline and growth propelled by climate and climate change, as well as to pursue the limited, but growing awareness of environmental history in mainstream Norwegian historiography.

Introduction

As historians, we emphasise to our students that history is not identical with the past. While the past was real and it is possible to get secure knowledge about many aspects of it, it is gone, and what we are left with are data, sources and signs that allow us to create representations, some prefer to say narratives, about that past. These representations are intended to function in the present (Veyne 1971). Which narratives of the past historians have found interesting, important and useful has changed considerably over time. The study of this is called historiography.

doi: 10.63308/63881023874820.ch11

Even if history, much less historiography, is not identical with the past, they are worthwhile pursuits, because they say something about how people used to think about societal change, causality and human agency at different times. History, in a wide sense of the word, and with all its shortcomings, remains, after all, the only basis for making informed choices about the present and predictions for the future (Adamson, Hannaford and Rohland 2018; Guldi and Armitage 2014; Van de Noort 2011). Knowledge about how scholars perceived the role of climate in societal change and processes of causality in the past informs us about how we arrived where we are and what options we currently have (Seland 2017; Seland and Kleiven 2023: 170–74).

This paper addresses the roles ascribed to climate and climate change in survey works of Norwegian history published over the course of the twentieth century, here called 'public histories'. In the nineteenth century, when history was first established as an academic discipline, this happened in close symbiosis with the nation-state. In Norway, as in other countries influenced by the German university system, the nation was the obvious unit of analysis, and the history of this nation from its origin to the present day became the dominating grand narrative of history, (Burrow 2009: 455–66). A task of primary importance for the first generations of government-paid professional historians was thus to research, write and disseminate national history to as large a public as possible. A central tool in this was schools, where instruction in religious and classical history and literature was largely replaced with Norwegian and World history in the wake of school and curriculum reforms in 1869 and 1896 (Thue 2019). In creating a public sphere of history, large, multivolume national histories became the preferred medium of professional historians in Norway, inspired by developments in German, Danish and Swedish historiography (Heiret, Ryymin and Skålevåg 2013: 8–9). Most of these works were written by university professors. The first ones were single-authored, but over time each volume came to be written by a separate author, an expert on a particular historical period, and sometimes by two authors writing separate parts of the same volume. The works were commissioned by commercial publishers and aimed at a general, although educated readership. None of the public histories cite references to secondary literature beyond general bibliographies in the most recent cases, and credit or criticism for earlier scholarship is rarely given by name. Nevertheless, these works were also venues for the dissemination if not exactly of research, then surely the scholarly state of the art, making them relevant also as a source of information for the history of scholarship. The potential caveat that these works might not have been representative of their time, but rather reflecting the interests and idiosyncrasies of their respective authors, is unlikely. The public histories were vetted by teams of series editors and academic consultants. They were reviewed, mostly favourably, in academic journals and the press. Authors were expected to be the best in their fields and the books were supposed to represent a changing state of the art. Otherwise, there would not have been a need for new series every second decade or so; and discussion and polemics about causation and interpretation belonged to the genre. Even if authors disagreed with a view, they would sometimes include it in order to polemicise against it, as we shall see below.

The public histories continued to carry considerable prestige within the historical discipline through the twentieth century (Rykkelid 2021: 19–26), but the genre disintegrated with changing reading habits and book markets around the turn of the millennium, as might be illustrated by the steeply declining number of volumes in the latest works included in this survey (Table 1).

Table 1. Norwegian public histories, 1851–2011.

Peter A. Munch, *Det norske Folks Historie* (1851–1863), 8 vols
Ernst Sars, *Udsigt over den norske historie* (1873–1891), 4 vols
Norges historie fremstillet for det norske folk (1909–17), 13 vols
Det Norske folks liv og historie (1930–38), 11 vols
Thorleif Dahl (ed.) *Vårt folks historie* (1961–64), 9 vols
Knut Mykland (ed.) *Cappelens norgeshistorie* (1976–80), 15 vols
Knut Helle (ed.) *Aschehougs norgeshistorie* (1994–98), 12 vols
Samlagets norgeshistorie (1999), 6 vols
Hans Jacob Orning (ed.) *Norvegr* (2011), 4 vols

Methods

My research question is simply what role climate plays as an agent of historical change in public histories of Norway published between 1909 and 2011. The two works included in Table 1, published in the nineteenth century (nos. 1 and 2), were instrumental in constituting the genre as it would also manifest itself in later series (Melve 2013: 36–37), but have been included here mainly for purposes of comprehensiveness. They were written before scientific knowledge of climate change had started to make an impression on the historical discipline and before the start of modern archaeology. These works contain discussions of the polar regions in Greco-Roman literature, including Pytheas' voyage to Thule c. 325 BCE, and elements of environmental determinism in shaping the character of the ancient inhabitants of Scandinavia that completely disappeared in later texts. While this is interesting in its own right (see Lid 2022), climate is perceived as essentially static, and these early works are thus not useful for the present purpose.

There are two recent reviews of Nordic and Norwegian historical climate scholarship, (Huhtamaa and Ljungqvist 2021; Pedersen 2016). While both very useful, their emphasis is on mapping the field and on research rather than on the genre-specific narrative analysis aimed for here. In his MA thesis, Johannes Rykkelid investigated how climate is conceptualised in two of the works discussed here (nos. 6 and 7). His work has a wider scope and addresses a smaller corpus than the present study. Bibliometric and narrative analyses are well-established approaches to the study of contemporary climate change and discourse (e.g., Fløttum and Gjerstad 2016; Fu and Waltman 2022; Moezzi, Janda and Rotmann 2017), and more limited work has been conducted in other historical and archaeological fields (Marx, Haunschild and Bornmann 2018; Marx 2017; Seland 2017, 2020).

Rather than reading 82 volumes of Norwegian history cover to cover, I have used the digitalised first editions in the Norwegian National Library, making case-insensitive and truncated searches for '*klima*' (climate), '*varme*' (heat), '*kulde*' (cold), '*temperatur*' (temperature), '*istid*' (ice age), '*isbre*' (glacier), '*dyrking*' (cultivation), '*nedbør*' (precipitation), '*tørke*' (drought) and '*flom*' (flood) in order identify relevant passages. As it turns out, '*klima*' works well as a search term, and the other terms turn up in conjunction with this in relevant contexts. Words describing temperature are mainly used to describe human temper or attitude, especially in the early works. Moreover, I limited my search to the volumes covering prehistory, in most of these works defined as the pre-Viking age, the Middle Ages and the Early Modern Period, in order to catch any discussions of what palaeoclimatologists describe as the Deglaciation, the Holocene Thermal Maximum, the Roman Warm Period, the Late Antique / Early Medieval Little Ice Age, the Medieval Warm Period and the Little Ice Age.

To appreciate the role ascribed to climate and climate change, the reading of the passages thus identified was informed by narrative analysis. Historians have been aware of the importance of narratives in historical prose since the 1970s. Paul Veyne described how historical narratives are constructed by organising events and actions into plots, thus establishing chronology and notions of causality (Veyne 1971). Phillipe Carrard argued that narratives and ideas of causation are present in all historical texts, even those that explicitly reject narrative history, such as the so-called New History or structural history to which much of twentieth-century environmental history belongs (Carrard 1992). Such narratives can be identified, still according to Carrard, by studying the poetics of history – that is, the conventions and characteristics of historical texts. Others have developed these insights into a typology of historical narratives, including narratives of progress, decline, continuity and change, that may serve to explain, justify or criticise, not only the past, but also the present of the historian and their implied reader (Heiret, Ryymin and Skålevåg 2013; Rüsen 2005).

Results

The early literature, pre-World War Two, continues older traditions of discussing reports of the northern regions in Classical literature with emphasis on Pytheas' voyage to Thule. Unlike the nineteenth-century works and the earlier topographic traditions, these references are seen as purely descriptive accounts of prevailing climatic conditions, not as having explanatory power with regard to the nature and qualities of northerners (Bugge 1912: 64–70; Lid 2022; Shetelig 1930: 96–98). These accounts all but disappear in the post-WWII works, likely due to the increasing doubt displayed in Classical scholarship about the veracity of such accounts, emphasising their literary rather than their documentary nature (cf. Romm 1992).

By the outset of the twentieth century, however, scientific knowledge about climate change, especially the existence of the ice ages, began to make its inroads in historiography. This was about time. Norwegian geologist Jens Esmark had realised the existence of the last glacial period as early as the 1820s, and it was soon after documented and described in detail by Louis Agassiz (Agassiz 1840; Hestmark 2018). Ellsworth Huntington published his *The Pulse of Asia* in 1907, which became a foundational text of environmental determinism that was widely read and acclaimed at the time (Huntington and Crumley 2013). Professor of history at the University of Oslo (then Kristiania), Medievalist Alexander Bugge (1870–1929), wrote the three first volumes of the first Norwegian public history published after Norwegian independence. His treatment of climate is all gathered in the first volume, describing the pre-800 CE period. He describes there how deglaciation made settlement possible, how the land level has been rising, and how the postglacial climate was milder than in his own age, by two degrees centigrade as he claims without citing sources. He also states that the animals and plants still current in his time reached present-day Norway in this period parallel with human immigration (Bugge 1912: 14–15). In this he foreshadows the climate history periods we today describe as the deglaciation and the Holocene Thermal Maximum. Moreover, Bugge describes Roman Iron Age and Viking Age climates as colder than this, and more in line with that prevailing in his present time (1912: 85). Thus, he reveals no awareness of the Roman or Medieval warm periods or the Late Antique / Early Medieval Little Ice Age. Interestingly, however, he also ascribes humans a potential to influence climate and cites Italy and the Netherlands as examples of places where climate and vegetation have changed in a benign direction since prehistory due to human agency, unlike in his native Norway, which remained little influenced by human activity (1912: 241). In this he is almost certainly channelling the Norwegian eighteenth-century historian Gerhard Schöning (1722–1780), who compared ancient Roman authors' descriptions of wild and cold Western European landscapes with the cultural landscapes of his own age (Eriksen 2019).

Norske Folks Liv og Historie was published in eight volumes between 1930 and 1938. The first volume covered the period up to 1000 CE, and was written by Haakon Shetelig (1877–1955), a professor at Bergen Museum from 1914, and an important figure in early Norwegian archaeology. At Bergen Museum Shetelig worked alongside pioneers in climate research including Vilhelm Bjerknes (1862–1951). Shetelig would certainly have been aware of Huntington's work, which by then had been expanded with the notoriously influential *Climate and Civilization* (Huntington 1915). While climate does not play a major role in Shetelig's volume, he is explicit that it is not a stable factor, but has changed multiple times through Norwegian prehistory (Shetelig 1930: 10–11). The Holocene Thermal Maximum is recognisable in his treatment of the Neolithic and Bronze Ages, which he sees as more benevolent towards agriculture than his own time (1930: 29, 53, 63). Similarly, he describes a humid and cold period in the Early Iron Age, a period with a paucity of archaeological finds, and improved climatic conditions in the high Medieval period, when Greenland was settled by the Norse (1930: 94–95, 365). Thus, to Shetelig, climate is seen not only as changing, as recognised already by Bugge, but also as a potential agent of historical change. For the first time, a causal correlation is proposed, viz. between climate change and changing agricultural output.

The notion of a Medieval Warm Period appears here for the first time in the surveyed corpus of public histories, but it was not completely novel. Benjamin Franklin had floated the idea in his conversations and correspondence with the Swedish explorer and naturalist Pehr Kalm in the mid-eighteenth century (Jonsson 2015: 117–19). While it is hard to know if Shetelig was familiar with Kalm's work he would have been familiar with the debate that had taken place in more recent times between Norwegian medievalist Edvard Bull and the Icelandic geologist Thorvaldur Thoroddsen about whether the Icelandic climate was colder in the Late Middle Ages than in the period of Norse settlement of the island in the ninth and tenth centuries (Bull 1913, 1915; Pedersen 2016; Thoroddsen 1914).

In Volume 3, professor of agricultural history Sigvald Hasund (1868–1959) wrote that documentary sources indicate a late medieval climate deterioration starting in the fourteenth century. He also describes more regional problems in Iceland connected with volcanism and increased sea ice (1934: 148, 195–96). This is no doubt a direct echo of the Bull–Thoroddsen debate. While acknowledging that the climate was not stable, and that conditions for agriculture in Iceland were indeed marginal, Hasund nevertheless argues the case of societal resilience, stating that bad years have occurred in all periods of history. Other volumes in the work do not mention climate at all. Interestingly this also pertains to Edvard Bull's second volume on the period 1000–1280 (Bull 1931), notwithstanding his earlier interest in the subject (Bull 1913, 1915; Huhtamaa and Ljungqvist 2021; Ogilvie 1984; Pedersen 2016).

The first three volumes out of nine in *Vårt folks historie*, published 1961–1964, were written by Anders Hagen (1921–2005, Volume 1) and Charles Joys (1910–1994, Volumes 1–3). Hagen was senior keeper ('*førstekonservator*') at Oslo University Collection of Antiquities when his volume was written and later went on to serve as director of Bergen Museum for almost three decades. He was also a pioneer in Norwegian Cultural heritage protection and Nature conservation (Solberg 2005). Similar biographical information about Joys is lacking, but he seems to have been teaching mainly primary and secondary school, and most of his publications address Norwegian medieval church history. In their co-authored Volume 1, the Holocene Thermal Maximum is clearly marked out, with the detail that glaciers had completely melted by c. 3000 BCE and the claim that summers were more benign than those prevailing in their own time (Hagen and Joys 1962: 21–24, 66). In Volume 2, Joys highlights that Norse communities in Iceland and Greenland were especially vulnerable to climate change due to geographical marginality (Joys 1962: 189, 200). This echoes the earlier claims by Bull and Hasund. Joys, however, also identifies the onset of Late Medieval climate deterioration in 1270, which is remarkably in line with recent scholarship. This is seen as a prime explanation for the so-called Late Medieval Crisis (Joys 1963, 371–72). Climate or climate change is not mentioned in Volume 4 by Reidar Marmøy (1963), covering the period up to 1660 and thus the coldest part of the Little Ice Age.

The first work of public history to take a more scientific approach to the issue was *Cappelens Norgeshistorie*, published 1976–1980. The first volume, covering the pre-Viking period until the eighth century CE, was written by the young archaeologists Bente Magnus (b. 1939) and Bjørn Myhre (1938–2015). Myhre was at Bergen Museum at the time and later at the University of Oslo. Magnus later worked as an independent archaeologist in Sweden. Climate is cast as the main agent of change throughout the volume, which also features timelines and discussions of the scientific evidence (Magnus and Myhre 1976). The historical volumes that follow this up are less explicit, but also more detailed and nuanced than earlier works in identifying for the first time in this corpus, a mild period in the high Middle Ages in combination with the onset of colder conditions in the late thirteenth century (Gunnes 1976: 108, 112; Lunden

1976: 226). Both developments are used to explain extent of settlement and volume of agricultural production. Climate deterioration in the Late Medieval period is discussed in a separate sub-chapter in Volume 4, where it is ascribed considerable explanatory value (Imsen and Sandnes 1977: 148–52, 158–59, 166, 182). After Volume 4, however, climate is only mentioned in static terms, thus missing out entirely on the coldest period of the Little Ice Age.

The last major series on Norwegian history, *Aschehougs Norgeshistorie*, was published in 1994–1998. Volume 1, by archaeologist Arnvid Lillehammer (b. 1943), follows the cue of Magnus and Myre, in discussing evidence of climate change in prehistory in detail and by casting it as a major agent of societal change (Lillehammer 1994). In the two volumes covering the high Medieval period, however, only lip-service is paid to benign climate in the North Atlantic and a possible climate deterioration from the late thirteenth century (Helle 1995; Krag 1995). Volume 4, dealing with the Late Medieval Period, was written by the director of Norsk Folkemuseum, Halvar Bjørkvik (1924–2021); in contrast, he devotes eleven pages to the topic and gives it major explanatory weight (Bjørkvik 1996: 32–43). While Volume 5 is the first in this corpus to mention the Little Ice Age by that name, this only happens in passing and, perhaps symptomatically, as something that takes place somewhere else (Rian 1995: 112, 165, 222). The variation in the coverage between the different volumes is striking. Climate takes the lead in processes of change in the prehistoric period and received thorough coverage in the volume on the Late Medieval Period, but is all but ignored in the other three volumes discussed here. The notion of a Medieval Warm Period proposed by Hubert Lamb was perhaps not widely supported yet (Lamb 1965), but had, as shown above, been discussed among historians by the 1970s and was foreshadowed by the Bull-Thoroddsen debate in the early twentieth century. The existence and reality of an Early Modern Little Ice Age was, however, well established by this time (Grove 1988; Lamb 1982; Figure 1). That some authors engaged with climate change while others all but ignored it seems to indicate that individual author interest and opinion took prevalence over available evidence and knowledge in the coverage of climate history.

After the turn of the millennium, there was no longer a market for multi-volume, illustrated public histories of the kind that had been published throughout the preceding century. Two later series may, however, illustrate more recent developments. *Norsk Historie* from Samlaget, published in 1999 in six volumes, appears to have been directed at the textbook market rather than the general public, with fewer images, keywords in the margins, and explicit discussions of scholarship, although without proper references. The series leaves out the archaeological part of the Norwegian past and starts with a volume on Medieval history until 1300 CE by Jón Viðar Sigurðsson, a historian at the University of

Climate Narratives in Norwegian Public Histories

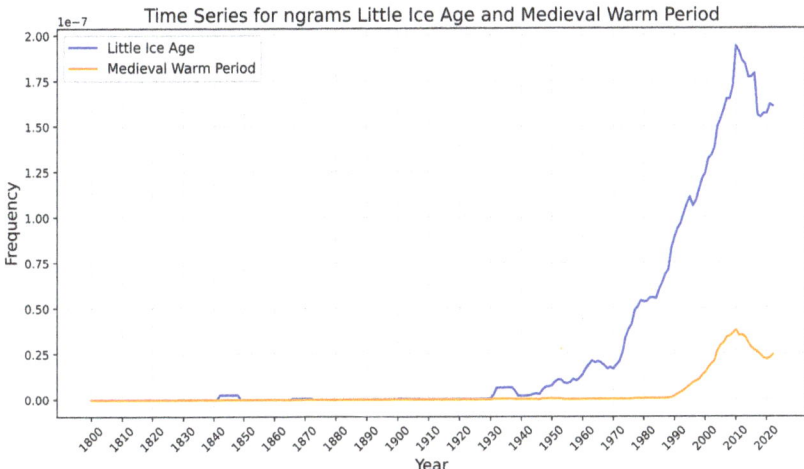

Figure 1. Use of the terms 'Little Ice Age' and 'Medieval Warm Period' in English language books. Source: Google N-Grams (https://books.google.com/ngrams).

Oslo. Despite extensive treatment of agriculture, climate and climate change are not discussed at all (Sigurdsson 1999). Volume 2, by historians Geir Atle Ersland, University of Bergen, and Hilde Sandvik, University of Oslo, covers the period 1300–1625. It discusses the Late Medieval Crisis and rejects climate change as a significant explanation (Ersland and Sandvik 1999: 47–48). It completely fails to mention the Little Ice Age. The last work of this kind, *Norvegr*, was published in 2011 in four relatively slim volumes. The author of Volume 1, Hans Jacob Orning, University of Oslo, is the first historian to integrate climate throughout his narrative, which covers the period from the deglaciation until 1400 (Orning 2011). Volume 2 mentions climate only in static terms and also ignores the existence of the Little Ice Age, which can hardly be considered anything short of remarkable at this late point in time (Njåstad 2011).

Discussion

As we have seen, climate and climate change have been present from the start in the grand narratives of Norwegian history, but mainly in the archaeological past, to a smaller extent with regard to the Late Medieval crisis, and hardly at all after the Late Medieval Period. It has been considered a major agent of change in prehistory and has for the most part been mentioned, accepted or rejected as one in Medieval history. Authors writing about the Early Modern Period have all but ignored the topic. With regard to climate history, the period terms of the discipline are rarely used in these more general works. We may nevertheless recognise periods such as the Holocene Optimum and the first onset of the Little Ice Age if a twelfth- to thirteenth-century date is accepted for the onset of colder conditions. The Roman and Medieval warm periods are sometimes alluded to without being named, while the Late Antique Little Ice Age and the Early Modern parts of the Little Ice Ages remain all but absent.

Archaeologists and historians writing about prehistory and thus engaging with archaeological scholarship have clearly been more open to and informed about the role of climate change than historians. This goes back to Shetelig's 1931 volume, and climate became fully integrated in narratives from the 1970s, which also explicitly engaged with palaeoclimatological scholarship. Historians dealing with the High Medieval and Early Modern periods have been more hesitant, a few of them even outspokenly sceptical or dismissive, and have, with one notable exception (Orning 2011), tended to relegate discussions of climate from their main narratives into separate paragraphs or sections. That authors dealing with the Late Medieval period have been more interested in climate change could perhaps be explained by the fact that scholarship on this period of Norwegian history has been dominated by experts on agricultural history, including Sigvald Hasund and Kåre Lunden mentioned above, and that a main historical narrative of this period has been that of decline. Works where coverage of climate varies between different volumes, and individual volumes that outright dismiss or ignore the relevance of climate change indicate that this reflects decisions made by authors and editors rather than lack of familiarity with climate history. In general, however, climate change has been considered important for the Late Medieval Crisis and somewhat significant to the Viking and High Medieval expansion, but conspicuously irrelevant with regard to the Early Modern Period.

In terms of narrative analysis, climate serves as a static backdrop throughout this historiography. Norwegian climate was marginal for agriculture, and premodern existence was always precarious. Most works address climate, at least in passing, in order to explain and underline this narrative of marginality. Climate change, however, also takes on roles of events that propel the historical narrative forward, especially in prehistory and in the Late Medieval period, and thus become key elements of the plot as conceived in Veyne's historical epistemology (Veyne 1971). With regard to Carrard's notion of 'poetics of history' (Carrard 1992), climate is seen as either a benign and stabilising factor, for instance with regard to the Neolithic and early Bronze Age periods, or more often as a threatening, hostile, disruptive, unpredictable, and malign factor. Narratives may also be classified with those of growth and decline taking prominence, but stories of constant vulnerability and resilience in face of a hostile and unpredictable climate are also present in the material. While there is an overtly deterministic vein in much of the archaeological scholarship, an outspoken resistance towards deterministic explanations is evident in many of the historical volumes.

In conclusion, climate and climate change have been a part of Norwegian historiography since the start of the tradition, but mainly with respect to the earliest parts of history. The lack of awareness of or openness to scholarship on the Little Ice Age is glaring, and it is tempting to ascribe this to the methodological nationalism that critics have claimed used to characterise the discipline in the period under discussion (Tvedt 2013), which must have influenced the education, interest and reading of the individual authors.

Acknowledgements

Thanks are due to Leidulf Melve and the anonymous reviewer for their valuable input, and to Tomáš Glomb for his help with Figure 1.

Bibliography

Adamson, G.C.D., M.J. Hannaford and E.J. Rohland. 2018. 'Re-thinking the present: The role of a historical focus in climate change adaptation research'. *Global Environmental Change*, 48: 195–205.

Agassiz, L. 1840. *Études sur les glaciers*. Neuchatel: Jent et Gassmann.

Bjørkvik, H. 1996. *Folketap og sammenbrudd: 1350–1520*. Vol. 4. Oslo: Aschehoug.

Bugge, A. 1912. *Tidsrummet indtil ca 800*. Kristiania: Aschehoug.

Bull, E. 1913. 'Klima og historie'. *Samtiden* 24: 200–208.

Bull, E. 1915. 'Islands klima i oldtiden'. *Geografisk Tidsskrift* 23.

Bull, E. 1931. *Fra omkring 1000 til 1280. Det norske folks liv og historie gjennem tidene*. Oslo: Aschehoug.

Burrow, J. W. 2009. *A History of Histories: Epics, Chronicles, Romances and Inquiries from Herodotus and Thucydides to the Twentieth Century*. London: Penguin.

Carrard, P. 1992. *Poetics of the New History: French Historical Discourse from Braudel to Chartier*. Baltimore, MD: Johns Hopkins University Press.

Eriksen, A. 2019. 'History, exemplarity and improvements: 18th century ideas about man-made climate change'. *Culture Unbound: Journal of Current Cultural Research* 11 (3–4): 353–368. https://doi.org/10.3384/cu.2000.1525.1909302.

Ersland, G.A. and H. Sandvik. 1999. *Norsk historie 1300–1625: eit rike tek form*. Oslo: Samlaget.

Fløttum, K. and Ø Gjerstad. 2016. 'Narratives in climate change discourse'. *Wiley Interdisciplinary Reviews: Climate Change* 8: e429. https://doi.org/10.1002/wcc.429.

Fu, H.-Z. and L. Waltman. 2022. 'A large-scale bibliometric analysis of global climate change research between 2001 and 2018'. *Climatic Change* **170** (3–4): 36. https://doi.org/10.1007/s10584-022-03324-z.

Grove, J.M. 1988. *The Little Ice Age*. London: Methuen.

Guldi, J. and D. Armitage. 2014. *The History Manifesto*. Cambridge: Cambridge University Press.

Gunnes, E. 1976. *Rikssamling og kristning 800–1177*. Oslo: Cappelen.

Hagen, A. and C. Joys. 1962. *Forhistorisk tid og vikingtid*. Oslo: Aschehoug.

Hasund, S. 1934. *Tidsrummet 1280 til omkring 1500. Det norske folks liv og historie gjennem tidene*. Oslo: Aschehoug.

Heiret, J., T. Ryymin and S.A. Skålevåg. 2013. *Fortalt fortid: norsk historieskriving etter 1970*. Oslo: Pax.

Helle, K. 1995. *Under kirke og kongemakt: 1130–1350*. Oslo: Aschehoug.

Hestmark, G. 2018. 'Jens Esmark's mountain glacier traverse 1823 – the key to his discovery of Ice Ages'. *Boreas* **47** (1): 1–10.

Huhtamaa, H. and F.C. Ljungqvist. 2021. 'Climate in Nordic historical research – a research review and future perspectives'. *Scandinavian Journal of History* **46** (5): 665–95. https://doi.org/10.1080/03468755.2021.1929455.

Huntington, E. 1915. *Civilization and Climate*. New Haven, CT: Yale University Press.

Huntington, E. and C. Crumley. 2013. 'The pulse of Asia (1907)'. In L. Robin, S. Sörlin and P. Warde (eds), *The Future of Nature*. New Haven, CT: Yale University Press. pp. 121–33.

Imsen, S. and J. Sandnes. 1977. *Norges historie. Bind 4. Avfolkning og union: 1319–1448*. Oslo: Cappelen.

Jonsson, F.A. 2015. 'Climate change and the retreat of the Atlantic: The cameralist context of Pehr Kalm's voyage to North America, 1748–51'. *The William and Mary Quarterly* **72** (1): 99–126.

Joys, C. 1962. *Hellig Olavs arv*. Oslo: Aschehoug.

Joys, C. 1963. *Fra storhetstid til unionstid*. Oslo: Aschehoug.

Krag, C. 1995. *Vikingtid og rikssamling: 800–1130*. Oslo: Aschehoug.

Lamb, H.H. 1965. 'The early medieval warm epoch and its sequel'. *Palaeogeography, Palaeoclimatology, Palaeoecology* 1: 13–37.

Lamb, H.H. 1982. *Climate, History and the Modern World*. New York: Routledge.

Lid, H.A. 2022. 'Dyrking av nordmenns karaktertrekk blant norske patrioter ca. 1770–1814'. Master's thesis. University of Agder. https://uia.brage.unit.no/uia-xmlui/handle/11250/3003209.

Lillehammer, A. 1994. *Aschehougs Norgeshistorie. Bind 1. Fra jeger til bonde: inntil 800 e.Kr*. Oslo: Aschehoug.

Lunden, K. 1976. *Norge under Sverreætten, 1177–1319: høymiddelalder*. Oslo: Cappelen.

Climate Narratives in Norwegian Public Histories

Magnus, B. and Myhre, B. 1976. *Norges historie. Bind 1. Forhistorien: fra jegergrupper til høvdingsamfunn.* Oslo: Cappelen.

Marmøy, R. 1963. *Vårt folks historie, IV: Gjennom bølgedalen.* Oslo: Aschehoug.

Marx, W., R. Haunschild and L. Bornmann. 2017. 'The role of climate in the collapse of the Maya civilization: A bibliometric analysis of the scientific discourse'. *Climate* 5 (4): 88.

Marx, W. 2018. 'Climate and the decline and fall of the Western Roman Empire: A bibliometric view on an interdisciplinary approach to answer a most classic historical question'. *Climate* 6 (4): 90.

Melve, L. 2013. 'Mellomalderen i dei nasjonale forteljingane'. In J. Heiret, T. Ryymin and S.A. Skålevåg (eds), *Fortalt fortid: norsk historieskriving etter 1970.* Oslo: Pax. pp. 35–60.

Moezzi, M., K.B. Janda and S. Rotmann. 2017. 'Using stories, narratives, and storytelling in energy and climate change research'. *Energy Research & Social Science* 31: 1–10.

Njåstad, M. 2011. *Norvegr. Norges historie. Bind II: 1400–1840.* Oslo: Aschehoug.

Ogilvie, A.E.J. 1984. 'The past climate and sea-ice record from Iceland, part 1: Data to A.D. 1780'. *Climatic Change* 6 (2): 131–52. https://doi.org/10.1007/BF00144609.

Orning, H.J. 2011. *Norvegr. Norges historie. Bind I: Frem til 1400.* Oslo: Aschehoug.

Pedersen, E.S. 2016. 'Klimaets plass i norsk historie: En forskningshistorisk oversikt'. *AmS-Varia* 58: 61–76.

Rian, Ø. 1995. *Aschehougs Norgeshistorie. Bind 5. Den nye begynnelsen: 1520–1660.* Oslo: Aschehoug.

Romm, James S. 1992. *The Edges of the Earth in Ancient Thought: Geography, Exploration, and Fiction.* Princeton, NJ: Princeton University Press.

Rykkelid, J.H. 2021. '"I vårt barske og lunefulle klima": Om fremstillingen av klima i norgeshistoriene utgitt i etterkrigstiden (1945–89)'. Master's thesis. University of Oslo. https://www.duo.uio.no/handle/10852/91283.

Rüsen, J. 2005. *History: Narration, Interpretation, Orientation.* New York: Berghahn Books.

Seland, E.H. 2017. 'Tøffe valg med store konsekvenser. Historiske fortellinger i norsk klimapolitikk'. In T. Ryymin (ed.) *Historie og Politikk – Historiebruk i norsk politikkutforming etter 1945.* Oslo: Universitetsforlaget. pp. 221–49.

Seland, E.H. 2020. 'Climate change in urban biographies: Stage, event, agent'. *Journal of Urban Archaeology* 2: 187–96.

Seland, E.H. and K.F. Kleiven. 2023. *En kort introduksjon til klimahistorie: menneskene og klimaet etter siste istid.* Oslo: Cappelen Damm akademisk.

Shetelig, H. 1930. *Fra oldtiden til omkring 1000 e.Kr. Det norske folks liv og historie gjennem tidene.* Oslo: Aschehoug.

Sigurdsson, J.V. 1999. *Norsk historie 800–1300: frå høvdingmakt til konge- og kyrkjemakt.* Oslo: Samlaget.

Solberg, B. 2005. 'Anders Hagen'. In *Norsk biografisk leksikon.* Oslo: Store Norske Leksikon. https://nbl.snl.no/Anders_Hagen.

Thoroddsen, Th. 1914. 'Islands klima i Oldtiden'. *Geografisk Tidsskrift* 22.

Thue, F.W. 2019. 'Den historiske allmenndannelse: Historiefaget i høyere/videregående skole, 1869–2019'. *Historisk tidsskrift* **98** (2): 167–90. https://doi.org/10.18261/issn.1504-2944-2019-02-04.

Tvedt, T. 2013. 'Om metodologisk nasjonalisme og den kommunikative situasjonen – en kritikk og et alternativ'. *Historisk tidsskrift* **91** (4): 490–510. https://doi.org/10.18261/ISSN1504-2944-2012-04-02.

Van de Noort, R. 2011. 'Conceptualising climate change archaeology'. *Antiquity* 85 (329): 1039–48.

Veyne, P. 1971. *Comment on ecrit l'histoire: essai d'epistemologie*. Paris: Du Seuil.

The Author

Eivind Heldaas Seland is a Professor of Ancient History and Premodern Global History at the University of Bergen, Norway. His research interests focus on how scholars utilise historical climate data and climate change to explain societal transformations.

Chapter 12

GLACIER POETRY IN NORWEGIAN LITERARY HISTORIOGRAPHY

Kristine Kleveland

Abstract

Perceptions of glaciers through poetry are significant to Nordic climate history. This paper examines Norwegian literary critical discussions of glacier poetry. Eighteen glacier poems are mentioned in fifteen works of literary history. However, the glacier motifs themselves are seldom given much attention by the critics and glacier poetry has not explicitly been identified as a distinct poetic tradition. The glacier motifs in the eighteen poems and discussions of them show a couple of tendencies in uses and understandings of glaciers. These are explored through close readings of four poems, written by Andreas Munch, Arnulf Øverland, Gunvor Hofmo and Kristofer Uppdal. These poems suggest trends in how glacier poetry is indirectly canonised, often emphasising patriotism and national romanticism, identity-building and fortifying backgrounds, and themes of solitude and isolation, thereby reinforcing ideas about the 'people of the north' and the unique role glaciers play in their cultural narrative.

Introduction

Norwegian geography, demography and literary tradition should all suggest that poetry that in some way portrays glaciers would have been written, published and read. Yet, a canonisation of glacier poetry does not exist in Norwegian scholarly tradition. In this paper, it is argued that glacier poetry, or glaciers as motifs in Norwegian poetry, have been overlooked and that Norwegian glacier poetry is a poetic tradition that has yet to be explicitly stated. There is no established tradition of Norwegian glacier poetry in the literary histories, and few of the glacier poems that exist have been given attention in these literary histories. Nevertheless, glacier poetry exists in most literary histories through mentions of glacier poems, though often without an explicit focus on their glacial aspects.

doi: 10.63308/63881023874820.ch12

Kristine Kleveland

Norwegian glacier poetry is an essential part of Nordic climate history. Poetry may not allow a statistical analysis of climate change. Nevertheless, climate is embedded in glacier poetry, and poetry may allow us to consider changes in the view on, understanding of, and relation to glaciers. Glacial poetry gives a unique insight into human understanding of, and relation to, glaciers and the surrounding landscapes.

In this paper, eighteen glacier poems, mentioned across fifteen Norwegian literary histories, are objects of study. The poems 'Brudefærden' by Andreas Munch, 'Vort land' by Arnulf Øverland, 'Vinden har bare' by Gunvor Hofmo and 'Isberget' by Kristofer Uppdal are read closely to illustrate a few tendencies in the literary histories' indirect canonisation of glacier poetry. In general, when literary histories mention glacier poems, the glacier motifs are often read as having had a fortifying and identity-building effect on the Norwegian people, entrenching ideas about the 'people of the north'. Partly overlapping with the identity-building readings is a tendency to read glacier motifs through a patriotic lens. A third tendency is to thematise solitude and isolation, utilising the possible isolating and conserving qualities of ice.

Methods and Findings

This exploration combines distant reading and close reading. It draws on literary studies traditions but also has points of contact with the field of history. 'Narrative history' is a method historians use to unveil a certain aspect or lens of history. In this case, the narrating is done through what Franco Moretti calls 'distant reading' (2013), which makes it possible to work with a corpus of this size. Eighteen of the most influential literary histories from 1896 to 2018 were chosen and systematically searched through for words that indicate cold phenomena. These searches included search words such as '*jøkul**', '**fonn**', '*snø*' and '*tind**'. They also include '*isbre**', '*isbræ**' and '*glac**', but they do not include variations of the Danish word '*gletsjer*'. The searches are not exhaustive as not every word and every possible spelling of it is covered. There are also some flaws in the machine reading programmes' reading of the texts. Nevertheless, the results show tendencies in the focus and writings of Norwegian literary history and are enough to draw some general conclusions on the representation of glacier poetry in Norwegian literary histories.

Glacier Poetry in Norwegian Literary Historiography

The investigation of these eighteen literary histories shows eighteen glacier poems.[1] In the further exploration of the glacier motif's effects in a selection of these, close readings are used, in addition to a discussion of the literary histories' presentation of the poems.

Table 1. Glacier poems in Literary Histories

Literary histories	Glacier poems mentioned
Jæger, *Illustreret norsk literaturhistorie* (1896)	Sylvester Sivertson (1833) 'Norafjeld med Jøkkel blaa'
	Jonas Lie (1867) 'Dikt fra halvhundreaarsfesten den 17de mai 1864'
Bing, *Norsk litteraturhistorie* (1904)	Henrik Wergeland (1844) 'Hardanger'
	Johan Sebastian Welhaven (1848) 'I Uveiret'
Nærup, *Illustreret norsk litteraturhistorie: sidste tidsrum 1890–1904* (1905)	Arne Garborg (1895) 'Mot Soleglad'
Hambro, *Kortfattet norsk litteraturhistorie: med illustrasjoner* (1912)	Sylvester Sivertson (1833) 'Norafjæld med Jøkel blaa'
	Bjørnstjerne Bjørnson (1859) 'Der ligger et land mod den evige sne'
Elster, *Illustrert norsk litteraturhistorie* (1934-35)	Sylvester Sivertson (1833) 'Norafjell med Jøkel blaa'
	Andreas Munch (1850) 'Brudefærden'
Liestøl and Stang, *Norges litteraturhistorie* (1938)	Bjørnstjerne Bjørnson (1858) 'Over de høie fjælle'
Bull, Paasche, Winsnes and Houm, *Norsk litteraturhistorie* (1955–63)	Henrik Wergeland (1833) 'Paa Skakastølstind'
	Andreas Munch (1850) 'Brudefærden'
	Arne Garborg (1895) 'Mot Soleglad'
	Kristofer Uppdal (1920) 'Isberget'
	Gunnar Reiss-Andersen (1946) 'Tre ganger Norge'
	Tore Ørjasæter (1953) 'Reiser seg fritt'
Dahl, *Norges litteratur* (1981–89)	Johan Sebastian Welhaven (1934) *Norges Dæmring*
	Andreas Munch (1850) 'Brudefærden'

---

1. Of the eighteen works that were searched through, glacier poems were found in most. There are, though, no mentions in either *Norsk kvinnelitteraturhistorie, Norsk barnelitteraturhistorie* or *Skriftbilder: samisk litteraturhistorie*. The reason for this lack of focus on glacial motifs specifically in these three literary minorities, literature by women, by Sami and for children, would be interesting to discuss further but is just briefly considered in the analysis of 'Vinden har bare' by Hofmo.

Kristine Kleveland

Literary histories	Glacier poems mentioned
Beyer and Beyer, *Norsk litteraturhistorie* (1996)	Bjørnstjerne Bjørnson (1859) 'Der ligger et land mod den evige sne'
	Kristofer Uppdal (1920) 'Isberget'
Fidjestøl, Kirkegaard, Aarnes, Aarseth, Longum and Stegane, *Norsk litteratur i tusen år: teksthistoriske linjer* (1996)	Bjørnstjerne Bjørnson (1859) 'Der ligger et land mod den evige sne'
	Arnulf Øverland (1929) 'Vort land'
	Rolf Jacobsen (1951) 'Nordfjord'
Rottem, *Norges litteraturhistorie: etterkrigslitteraturen* (1996-98)	Gunvor Hofmo (1978) 'Vinden har bare'
Havnevik, *Dikt i Norge: lyrikkhistorie 200–2000* (2002)	Bjørnstjerne Bjørnson (1859) 'Der ligger et land mod den evige sne'
	Arne Garborg (1895) 'Mot Soleglad'
	Kristofer Uppdal (1920) 'Isberget'
	Arnulf Øverland (1929) 'Vort land'
Hagen, Haaberg, Sejersted, Selboe and Aaslestad, *Den norske litterære kanon* (2007–09)	Bjørnstjerne Bjørnson (1858) 'Over de høie fjælle'
	Bjørnstjerne Bjørnson (1859) 'Der ligger et land mod den evige sne'
	Kristofer Uppdal (1920) 'Isberget'
	Olav H. Hauge (1951) 'Jøkulen og treet'
Andersen, *Norsk litteraturhistorie* (2012)	Arne Garborg (1895) 'Mot Soleglad'
	Kristofer Uppdal (1920) 'Isberget'
Sørbø, *Nynorsk litteraturhistorie* (2018)	Arnljot Eggen (1951) 'Eld og is'

Kristofer Uppdals' well-known poem 'Isberget' is mentioned the most, six times, with Bjørnstjerne Bjørnson's 'Der ligger et Land mod den evige Sne' as first runner-up with five mentions. Arne Garborg's 'Mot Soleglad' from *Haugtussa* is referred to four times while Sylvester Sivertsons' 'Norafjeld med jøkel blaa' and Andreas Munch's 'Brudefærden' are mentioned three times each. The rest is discussed, cited or quoted in one or two of the literary histories.

Through the searches for glacier poetry, cold elements appeared in descriptions of personal characteristics or interpersonal relationships, and to describe moods in scenes or literary works. Where glaciers and ice are explicitly referred to in the literature, they are treated as background motifs and are not given a central role in the readings. In Andersen's reading of Garborg's 'Mot soleglad', for instance, he mentions the glacier in relation to lighting and mood in the poem (Andersen 2012: 267). Ice is understood as an environmental circumstance that has hardened and shaped the people in the north in the war poems by Reiss-Andersen (Bull et al. 1955, vol. 6: 300). Uppdal's 'Isberget' stands out as

Glacier Poetry in Norwegian Literary Historiography

the only example where the iceberg itself is the driving force and main motif of the poem. The iceberg is, though treated as an iceberg, also read as a symbol of something other, something more than 'just ice' (Havnevik 2002: 280; Hagen et al. 2007: 41; Andersen 2012: 361).

The earliest of the glacier poems discussed in the literary histories are read as patriotic and part of a National Romantic literary tradition. Wergeland is a dominating figure in this context and is mentioned as the author of two glacier poems. Even the poems not written by him are often compared to Wergeland's writing. The glaciers in romantic poems from the first half of the nineteenth century are not given much attention in the literary histories' readings. They seem to be a part of the landscape or background, whereas these landscapes are repeatedly read as something particularly Norwegian. Often, the glaciers serve as a contrast to ripened fruit and blooming fields, as in 'Hardanger' by Wergeland and 'Brudefærden' by Munch. In a more abstract sense of background, glaciers are seen as obstacles or as harsh living conditions for individuals or specifically Norwegians, having a fortifying effect, as in *Norges Dæmring* by Welhaven and 'Over de høie fjelde' by Bjørnson. There lies a patriotism in the readings of these lyric glaciers. The perceptions of glaciers in Norwegian poetry from 1833 till the late 1800s are quite similar, though with different emphases. Of the nineteenth-century poets, Wergeland, Sivertson, Welhaven, Munch, Bjørnson and Garborg are mentioned in relation to glacier poems several times throughout the literary histories, and Wergeland, Welhaven and Bjørnson are represented with two glacier poems each. The literary histories' readings of these poems are quite similar, from Jæger (1896) to Sørbø (2018). Given that the writing of literary history itself has undergone extensive changes throughout this timespan, this consistent reading may indicate two things. One is the tendency of literary histories to reproduce certain tropes and readings. The other is the possibility that these poems are patriotic and national romantic to the extent that they demand the same reading, regardless of literary theoretical view.

The selection of glacier poems from the twentieth century discussed in the literary histories is more varied than that from the nineteenth century. Only Uppdal's 'Isberget' and Øverland's 'Vort land' are discussed more than once ('Isberget' by Bull et al. 1961, vol. 5; Beyer and Beyer 1996; Havnevik 2002; Hagen et al. 2007; Andersen 2012; and 'Vort land' by Fidjestøl et al. 1998; Havnevik 2002). The larger volume of glacier poems from the twentieth century could indicate a larger number of glacier poems written in this period but could also indicate less of a tendency to canonise or to reproduce the ideas of former literary histories. The glacier poems from the 1920s to the 1940s continue the patriotic tendency but differ as the two world wars are backdrops. The glaciers in Øverland and Reiss-Andersen's poems are part of hardening and fortifying natural conditions that have led the people of the north to grow strong and resilient. There lies a brutality in Uppdal's 'Isberget' as well, but this poem stands out from the other glacier poems with the iceberg as the lyric subject. The brutality comes from the unambiguous glacier motif. The brutality of the iceberg is in some way similar in 'Isberget' to the glacier in Wergeland's 'Paa Skakastølstind', but, while Uppdal's iceberg is the driving force with agency and will, Wergeland's glacial landscape seems to be bound by the power of gods and destiny. The twentieth-century glacier poems discussed in the literary histories seem to be more open to interpretation, and the glaciers seem to have multiple functions, more so than the earlier ones that are categorised as patriotic and national romantic. Glaciers are more ambiguous motifs in the twentieth century, showing both brutality and vulnerability, closeness and distance, vitality and death. The reasons for this ambiguity may be a combination of three factors: that literary history writing has changed, that the genre of lyric has changed and that the view and understanding of nature have changed.

A tendency that seems consistent throughout the period is a reading of glacier poetry as thematising loneliness and melancholy. Liestøl and Stang see Bjørnson's 'Over de høie fjelde' in relation to the episodes of overwhelming anxiety that Bjørnson was prone to (1938: 210). Bull et al. see Wergeland's 'Paa Skakastølstind' in relation to Werglenad's beliefs, often defined by darkness, 'mørket som vil snevre lys-kretsen inn.' (Bull et al. 1955–63, vol. 3: 213). The same authors see Ørjasæter's 'Reiser seg fritt' as an illustration of a difficult time in his life, 'den skjebnetunge tiden, angsten og uroen' (Bull et al. 1955–63, vol. 5: 637). Rottem sees Hofmo's 'Vinden har bare' as a calling for a non-responsive god (1996: 329) and Hagen, Haarberg, Sejersted, Selboe and Aaslestad sees Hauge's 'Jøkulen og treet' as thematising relational distance between man and woman (Hagen et al. 2007: 196). Sørbø sees Eggen's *Eld og is* as being about loneliness and isolation (Sørbø 2018: 343). The readings of glacier poems as expressions of loneliness and melancholy are in some cases examples of historical-biographical readings, and not always well rooted in the lyric texts. Other times these readings are examples of a more text-oriented reading, referring to the elements of the lyric texts rather than external factors.

Analysis

The three main tendencies described above are examined further through close readings of a few selected poems. The close readings are done with attention to the readings from the literary histories in order to see not only how the glaciers are perceived in the lyric texts, but how these perceptions have been understood as literary motifs. The national romantic context of the glaciers is represented by

Glacier Poetry in Norwegian Literary Historiography

Munch's 'Brudefærden'. The continuation of Norwegian patriotism in a post-world war context is represented by Øverland's 'Vort land'. The isolating effect of glaciers is represented by Hofmo's 'Vinden har bare'. In addition to these, the most frequently mentioned poem, Uppdal's 'Isberget', is discussed alongside the literary histories' readings of it. The former readings of 'Isberget' stand out from the other readings of glacier poetry and do not fall in line with any of the mentioned tendencies. 'Isberget', the poem that features glaciers most explicitly is also the one that is most frequently referred to in the literary histories. This calls for a close reading.

'Brudefærden' ('The wedding procession'), 1850

Andreas Munch's poem is an ekphrasis referring to Adolf Tidemann and Hans Gude's painting 'Brudeferd i Hardanger'. In the painting, the glacier is a central part of the background. Similarly, the poem depicts the bridal couple on the fjord surrounded by majestic and lush nature. The glacier encapsulates the picture, being mentioned both in the first and the fourth stanza:

> Der aander en tindrende Sommerluft
> Varmt over Hardangerfjords Vande,
> Hvor høit mod Himlen i blaalig Duft
> De mægtige Fjelde stande.
>
> Det skinner fra Bræ, det grønnes fra Li,
> Sit Helligdagsskrud står Egnen klædt i –
> Thi se! over grønklare Bølge
> Hjemglider et Brudefølge.
>
> ...
>
> Der blaaner fra Kløft, der skinner fra Bræ,
> Der dufter fra blomstrende Abildtræ –
> Ærverdig står Kirken paa Tangen
> Og signer med Klokkeklangen.

> There breathes a sparkling summer air
> Warm over the waters of the Hardangerfjord,
> Where high towards the sky in scent of blue
> The mighty mountains stand.
>
> It gleams from glaciers, it greens from the slope,
> The region is dressed in its holiday attire –
> For see! over the green-clear waves
> A bridal procession glides home.

...

Blue emerges from the clefts, it gleams from the glaciers,
Fragrance emits from blooming apple trees –
Reverently the church stands on the headland
And blesses with the sound of bells.[2]

The two appearances of the glaciers are the same phrase, 'Det skinner fra bræ' ('It shines from the glacier'). In both cases, the glacier appears in interaction with flourishing nature. As the glacier shines, the slopes turn green, and the scent of blossoming apple trees emanates. A similar interaction takes place in Wergeland's poem 'Hardanger', which also refers to the nature of Hardanger.

...
hvor ved den krystalne Bræ
blomstrer snehvidt Abildtræ,
medens i en Snefonns Spoer
vilde Rose lystigt groer,

...
where the crystalline glacier
snow-white apple trees blossom,
while in a snowdrift's trail
wild roses joyfully grow.

The snow-white apple tree blossoms as the glacier sparks like crystals in Wergeland's poem. Both poems show the glaciers as a part of Norwegian nature, not in contrast to the lushness of apple trees, but as equally central in the romantic picture of the nation.

Like the picture it is written about, Munch's poem contains several national romantic elements: people dressed in national costumes in a wooden boat, by a wooden church. The dressed-up people are reflected in the surrounding nature: the woods answer with cheering celebration when shots are fired. Likewise, nature is reflected by the people in it: the bride is as lovely as the fjord. The poem consolidates romantic tropes in referring to the painting:

I dette bævende, flygtige Nu,
Før Draaben av Aaren er trillet,
Har Kunsten fæstet med kjærlig Hu
Det hele, straalende Billed.

---

2. All translations from the Norwegian originals are by the author

Glacier Poetry in Norwegian Literary Historiography

Og løfter det stolt for Verden frem,
At Alle kan kjende vort herlige Hjem
Og vide de Eventyr klare,
Som Norges Fjorde bevare.

In this trembling, fleeting Now,
Before the drop from the vein has rolled,
Art has captured with love
The whole, radiant Image.

And proudly presents it to the world,
That all may know our magnificent Home
And be aware of the adventures
That Norway's fjords preserve.

Munch praises nature as well as art. A summarising translation of the last stanza follows: Art captures, with loving care, the radiant scene and presents it so that everyone can recognise our glorious home and know the fairytale that the fjords of Norway preserve.

Elster is the first of three to discuss Munch's 'Brudefærden' in the literary histories (1935). In *Illustrert norsk Litteraturhistorie* he refers to 'Brudefærden' as an example of a 'successful and melodious tableau poem', and places it as central to the national romantic breakthrough (Elster 1935: 160). In their work *Norsk litteraturhistorie*, Bull et al. quote the part of 'Brudefærden' where the glacier is first mentioned, but do not comment further on it (Bull et al. 1955–63, vol. 3: 441). They go on to discuss *Den Eensomme*, (*The Lonely*), where the subject is bound to climb peaks covered in snow, 'med Fjeld efter Fjeld, med Bræ efter Bræ' ('Mountain top after mountain top and glacier after glacier') (Munch 1846: 81). The mountain tops and glaciers are not given much attention but are seen as obstacles or adversity. Dahl quotes the second appearance of the glacier in 'Brudefærden' in his literary history *Norges litteratur. Tid og Tekst*. Dahl sees the poem as 'en eneste lang hyllest til norsk natur' (one massive praise of Norway's nature) (Dahl 1981, vol. 1: 126). He carries forward the national romantic reading of the poem without going deeper into the matter. 'Brudefærden' thus remains a national romantic ekphrasis with a glacier as part of the tableau in the literary histories. All three literary histories that highlight Munch's poem refer to the glacial part, but none of them explicitly point to the shining glaciers.

Kristine Kleveland

'Vort land' ('Our land'), 1929

Arnulf Øverland's poem 'Vort land' is filled with ice and coolness, although it does not explicitly contain glaciers. While the first half of the poem is filled with the waves of the ocean and summer nights, the second half is frozen solid:

Og sent på efteråret
indunder fjeldets horn
har vore fædre skåret
det dyrebare korn.
Om noget frøs, om sneen faldt,
så måtte ingen klage,
det kunde frosset alt.

På hvite vinterveier
går folk med øks og sag
indover skog og heier
ved første gry av dag.
Du kjendte kanske også dem,
som blev i elveflommen,
før tømmeret kom frem.

De stille stjerner stiger
i luftens frosne væld
og våker over riker
av sne og hvite fjeld.
Det er vort land, vort fædreland,
og det har altid tilhørt
den norske arbeidsmand!

And late in autumn
beneath the mountain's horns
our forefathers have cut
the precious grain.
If any froze, if snow fell,
then no one should complain,
all of it could have frozen.

On white winter roads
people go with axe and saw
into forests and heaths
at the break of day.

Glacier Poetry in Norwegian Literary Historiography

You might have known them too
who stayed in the river's flood,
before the timber came through.

The silent stars rise
in the frosty might of the air
and watch over realms
of snow and white mountains.
This is our land, our fatherland,
and it has always belonged to
the Norwegian working man!

The frost humbles the people as it sets the premises for their food source and livelihood. The frost takes some of the crops but could always have taken more. Therefore, there is nothing to complain about: 'Om noget frøs, om sneen faldt, / så måtte ingen klage, / det kunde frosset, alt.'. Low temperatures and harsh weather humble and harden the people and are the basis for human life. Still, the ancestors have harvested and learned to live in these harsh conditions. While the stars rise in the freezing air and keep watch over the kingdom of snow-covered mountain peaks, 'våker over riker / av sne og hvite fjeld', it is stated that the land has always belonged to the Norwegian working man, 'det har altid tilhørt / den norske arbeidsmand!'. The patriotic poem is both inclusive and exclusive, defining a clear 'we' with the 'our' and essentially allocating the land of Norway to the Norwegian working man.

Like in 'Brudefærden', 'Vort land' contains mirroring between the people and their surroundings. The first half of 'Vort land' is warm, soft and with open landscapes like the sea and wetlands. These landscapes are mirrored in affects: worried mothers and thoughtful sisters writing letters for their brothers out at sea. The second half of the poem is not only notably colder but also harder, sharper and more masculine. The landscapes consist of mountains and woods. Fathers cut the crops, and workers collect lumber with axes and saws, everything in cold conditions.

'Vort land' is discussed in two literary histories, both quoting the last stanza of the poem. In *Norsk litteratur i tusen år: Teksthistoriske linjer*, Fidjestøl et al. see the poem as worker-patriotism (1998: 491–492). It builds on and contributes to the myth of the cold north, where the climate has hardened and cultivated the people. The cold has made Norwegians, and especially the workers, smart and sturdy. They use the winters for logging and preparations for timber floating. Havnevik also writes about 'Vort land' in *Dikt i Norge: lyrikkhistorie 200–2000*. He sees the poem as a national anthem that mainly speaks to workers, fishermen, farmers and lumberjacks (Havnevik 2002: 292). Neither of the 'Vort land' discussions emphasise the ice and snow itself but focus on the coldness' effect on the people of the cold north.

'Vinden har bare' ('The wind only has'), 1978

The only glacier poem written by a female author that is mentioned in the Norwegian literary histories is 'Vinden har bare' by Gunvor Hofmo. The poem is only discussed in one of the literary histories. The reasons for this striking predominance of male glacier poetry authors could be several. Firstly, female authors have long been underrepresented in literary history. However, it is striking that there are no references to glacier poems in either *Norsk kvinnelitteraturhistorie* by Engelstad et al. (1988–90), *Norsk kvinnelitteraturhistorie* by Gaski (1998) or *Norsk barnelitteraturhistorie* by Birkeland, Risa and Vold (2018). The absence of glacier poetry in the literary histories specifically concerning female authors, Sami and children's literature could point to an understanding of glaciers as a masculine motif. This understanding does not reflect the reality of published glacier poems. There are several glacier poems written by female authors, and glacier works by Hulda Garborg, Signe Seim, Guri Botheim, Nanna Schwengaard and Synnøve Persen are some that could be included in the literary histories. Hofmo's lyric iceberg in 'Vinden har bare' differs from the ice in the previously discussed poems in that it is tied to inner life.

> Vinden har bare
> ett ærend ikveld:
> å gjenkjenne
> din sjels rørelser
> som spinner seg
> rundt et punkt
> av kosmisk syn.
> Og det er det
> store blikk
> fra månen som troner
> på hvelvingen
> og stjernene
> som slår kulden
> inn i din angst
> for hver ting
> som står så
> bitterlig alene
> i rummet.
> Det kantrer et isfjell
> av Sorg

Glacier Poetry in Norwegian Literary Historiography

for hver blomst
som står der.
Tankens selhunder
får ikke feste
og faller ned
fra isens loddrette
sider.

Inntil den store
ilden kommer
som Gud venter med
til Natten er inne
og de store landskap
og byen brenner.

The wind has only
one errand tonight:
to recognise
the movements of your soul
spinning itself
around a point
of cosmic vision.

And it is
the great gaze
from the moon that thrones
the vault
and the stars
that strikes cold
into your anxiety
for everything
standing so
bitterly alone
in space.
An iceberg of Sorrow
capsizes
for every flower
that stands there.
The seals of thought
find no grip
and fall down
from the ice's vertical
sides.

Kristine Kleveland

Until the great
fire comes
which God awaits
until Night has arrived
and the great landscapes
and the city burn.

The seals of thought fall off icebergs of sorrow and start beating the cold into the angst of the lyric subject, you. Human beings are not only reflected in their surroundings, but the phenomenology takes place within the lyric subject and works within it. The poem is packed with natural phenomena such as the wind, the moon and stars. Concerning the lyric subject, these phenomena have abilities such as recognition, sight and the ability to beat cold. The cold landscape of icebergs is alternated with flames in the last stanza, where God is waiting for the night to set fire to the large landscapes and the city. The icebergs are shifting landscapes, the unpredictable ground that capsizes and sends the seal dogs overboard. As symbols, they could point to sudden capsizes in life, or the turns of emotions that could throw off reason and thought.

'Vinden har bare' is only mentioned once throughout the literary histories, by Rottem in *Norges litteraturhistorie: etterkrigslitteraturen*. He uses 'Vinden har bare' to illustrate what he calls the problem with the state of nature and state of mind in relation (Rottem 1996, vol. 1: 329). He refers to the first few lines of the poem: 'Vinden har bare / ett ærend ikveld: / å gjenkjenne / din sjels rørelser' ('The wind has only / one errand tonight: / to recognise / the movements of your soul') (Hofmo 1978). When the poet seeks grounding for thought in a nature scene, the 'thought's seals' descend, Rottem writes and quotes the second half of the poem, where the seals fall down from an iceberg of sorrow (Rottem 1996, vol. 1: 329). Hofmo's poem is one the glacier poems that are seen as thematising loneliness, melancholy and solitude. Olav H. Hauge's 'Jøkulen og treet' is another one. The understandings of these poems as expressions of loneliness can be seen as historical-biographical to the extent that literary history authors read more into it than what the poem itself requires. Nevertheless, the isolating effect of actual ice can make it seem to invite such interpretations.

'Isberget' ('The iceberg'), 1920

As the title suggests, the poem has an iceberg as the main motif and the lyric subject. The iceberg tells the story of its journey from being calved by its mother, the ice-blue glacier, to drifting south, incorporating smaller icebergs and scraping along the ocean floor, to capsize and intentionally pose a threat to those who come near. The first stanza goes as follows:

Glacier Poetry in Norwegian Literary Historiography

Mi mor, den isblaa breden, kalva meg ei natt
i stjerneskin under polarhimlens frosne blaa,
eg vart ei klote, eit isberg – ruvelegt aa sjaa!
Vart født i barn-rider, hardare enn dødens strid.
Breden skreik i, og slepte meg, og havet svara med skratt.
Var radt som jordkula klovna i den ilske-rid.
Eg lyfte meg i undring, mi mor ha' kalva meg,
eg var eit isberg, og stirde sør, min faders veg,
hans lodne kulde-røyk strauk meg, tok meg med i si natt.

My mother, the ice-blue glacier, calved me one night
in starlight under the frozen blue of the polar sky,
I became a globe, an iceberg—towering to see!
Was born in child-pains, harder than the struggle of death.
The glacier screamed and released me, and the sea answered with laughter.
It was as if the earth had split in the fierce pain.
I rose in wonder, my mother had calved me,
I was an iceberg, gazed southward, my father's way,
his shaggy cold smoke brushed me, took me along in his night.

Rather than illustrating a tendency in the literary histories' discussions of glacier poetry, the discussion of 'Isberget' by Kristofer Uppdal differs from the discussions of other glacier poems. Discussions of 'Isberget' are more in both quantity and volume than those of other glacier poems. These discussions pay most attention to the ice, as the poem itself demands. Bull et al. are the first to discuss 'Isberget' in *Norsk litteraturhistorie* (1961). They use the poem to illustrate the first part of Uppdal's literary output. This part is characterised by his praise of the strength of personality, and where the power of the spirit is revealed primarily in those who can bear their loneliness and their destiny (Bull et al. 1955–63, vol. 5: 570). In *Dikt i Norge: lyrikkhistorie 200–2000* Havnevik writes more thoroughly about 'Isberget', and singles it out as Uppdal's greatest and most significant poem (Havnevik 2002: 279). What Uppdal does in his poetry, Havnevik claims, is to amplify the simplicities of nature and human life, and turn an iceberg into The Iceberg, 'Isberget' (Havnevik 2002: 279). He claims that the iceberg in the poem is more of an iceberg than other icebergs but, in the following, he treats the iceberg as a symbol rather than an actual iceberg. The symbol in question is unlike the traditional symbol with a specific meaning. Here, being an example of early modernism, the symbol is open for interpretation and does not itself point to its meaning. 'No one tells you what The iceberg actually is', says Havnevik and implies that it must be more than 'just' an iceberg (Havnevik 2002: 280).

Hagen, Haaberg, Sejerstad, Selboe and Aaslestad's discussion of 'Isberget' in *Den norske litterære kanon* is rather elaborative as well. The authors state that the poem is one of the undeniably sublime texts in the modern lyrical canon (Hagen et al. 2007: 41). They point to the epic simplicity of the poem at the same time as it is packed with symbolic meaning (Ibid.). A couple of the stanzas are quoted before they refer to the iceberg as a cool and monstrous form of nature and foreign intelligence, described as rugged and repulsive (Hagen et al. 2007: 42, 43).

Andersen has written the most recent large Norwegian literary history, *Norsk litteraturhistorie*, which confirms the canonisation of two well-known glacier poems: Garborg's 'Mot soleglad' and Uppdal's 'Isberget'. Andersen reproduces the understanding of Uppdal's iceberg as a symbol without the poem bearing the meaning of the symbol. He states that the reader must make up their own interpretation of what the iceberg is: 'Det står ikke noe sted i diktet at det skal leses som et bilde på det menneskelige jeg. En slik tolkning må leseren selv stå for.' (Andersen 2012: 361). He then refers to the author Tarjei Vessas and his suggestion that the iceberg simply is an iceberg. Andersen does not agree but states that the allegory is an original one and that the poem is expressionistic (Andersen 2012: 361). In *Nynorsk litteraturhistorie*, Sørbø's history of literature in Nynorsk, he briefly mentions 'Isberget', but does not discuss the content further.

Because the subject I in 'Isberget' is an iceberg, the discussions give more attention to this iceberg than the ice in other glacier poems. Uppdal's glacier poem has been widely discussed, and there are different opinions on what kind of interpretation the poem demands. However, there seems to be a consensus that the iceberg must be something more than just an iceberg, or that it asks to be seen as something more. The extensive room for interpretation in the poem can be understood as demonstrating the complexity and richness of the iceberg as a figure or symbol.

Conclusion

The results of searches for glacier poems in the Norwegian literary histories show that glaciers are central elements in the literary history, although they are often implicit and in the background and surrounding landscapes. The searches resulted in a list of eighteen glacier poems, written and discussed throughout the nineteenth and twentieth centuries. The close reading of the selection of poems spans several directions when it comes to the lyric effects of the glacier motifs. The glaciers are portrayed in positive ways: as a source of fertile landscapes and swelling fruits as seen in both 'Brudefærden' and 'Hardanger'. It is part of the tableau that shows the greatest and most festive parts of Norwegian nature and culture, as seen in the ekphrasis 'Brudefærden', and in the painting it refers to. The glaciers can also have an 'parenting' effect on the local population, as seen in 'Vort land'. This effect, however, crosses over to rather negative portrayals of glaciers. The ice is sometimes seen as cynical and cold, able to destroy crops and undermine the livelihood, as seen in 'Vort land'. In Uppdal's 'Isberget' the iceberg is also described as cynical and with an explicit intention to destroy, absorb and to pose a threat to its surroundings.

The distance from being part of the background and surrounding, swelling, nature in 'Brudefærden' (1850), to being a fast-moving, sublime and actively threatening iceberg in 'Isberget' (1920) is striking. The distance in time is also significant, over seventy years. Like everything else, these two poems are products of their times and say something about the views, values and understandings they originate from. Changes in climate have led to large shifts in all of these, views, values and understanding in the last decades. In the timespan between these two poems, factors such as an increase in tourism and therefore tourist-related accidents could be a bigger cause of change in the portrayal of glaciers. General urbanisation can explain a greater distance from the motif. Additionally, the shipwreck of the RMS *Titanic* in 1912 surely had an impact on the becoming of 'Isberget'. The changing views of and uses and effects of glacier motifs continue. A quite different side of the glacier is portrayed in Hofmo's 'Vinden har bare' (1978). The icebergs in this poem are dense and more abstract, being phenomenology of emotion, icebergs of sorrow.

The glacier motifs in some of the poems mentioned are reflective of the human beings in them. The reflecting effect of ice and the former tendency to use it reflectively in poetry could explain the common interpretation that the iceberg in 'Isberget' too must reflect something other than itself, for instance, a floating symbol of human behaviour. The reflecting effect of glaciers in poetry is not too imaginatively distant from the actual effects of glaciers. Glaciers are the parts of the earth with some of the highest albedo effects, at least when looking at the parts covered in white, clean snow. In addition to being reflective, ice is see-through, and thus contains an interesting oxymoron. In Hofmo's 'Vinden har bare', the ice does not reflect human emotions, but plays them out as part of the body, making the body too see-through like ice.

Although there is a lot to be said about the glacial motifs of the canonised glacier poetry gathered from the Norwegian literary histories, it appears that the glaciers themselves or their function as lyric motifs are rarely the reason for the literary histories' interest. Of the eighteen poems mentioned, 'Isberget' is the only one where the ice itself is discussed as a central motif and, still, most literary histories indicate that the iceberg is a symbol of something else. In spite of their size, powers and paradoxical existence, glaciers don't seem to be of significant interest to literary historians. Often, they are not even acknowledged. Another way to look at the lack of focus on lyric glaciers is that literary historians have not been aware of their interest in them. Glaciers can be central or rather peripheral in the poem, and still have an impact, directly as a visible iceberg, but also less noticeable, as the effects of glaciers include meltwater, general cooling of surroundings, albedo effect, erosion of mountains, filling of valleys, the sounds of cracking ice and unforeseen and sudden floods, glacial outbursts, calvings, the colours and light in the glacier as it is kindled by the sun and more. Perhaps literary history has captured some of these effects in glacier poems without having recognised that they originate from the lyrical glaciers.

Note

This historiographical presentation is part of a doctoral thesis. The exploration forms the basis for examining Norwegian glacial poetry and the functions of lyrical glaciers in poems from various times and from various poets. The review of Norwegian literary histories gives an incentive to explore a range of poems by several poets, where the glaciers are perceived, understood, and represented in different, often contrasting, ways.

Bibliography

Secondary sources

Andersen, P.T. 2012. *Norsk litteraturhistorie* (2nd ed.). Oslo: Universitetsforlaget.
Beyer, H. And E. Beyer. 1996. *Norsk litteraturhistorie*. Oslo: Tano Aschehoug.
Bing, J. 1904. *Norsk litteraturhistorie*. Kristiania: Det Nordiske Forlag.
Birkeland, T., G. Risa and K.B. Vold. 2018. *Norsk barnelitteraturhistorie*. Oslo: Samlaget.
Bull, F., F. Paasche, A.H. Winsnes and P. Houm. 1955–63. *Norsk litteraturhistorie*. Vols 1–7. Oslo: Aschehoug.
Dahl, W. 1981–1989. *Norges litteratur*. Vols 1–3. Oslo: Aschehoug.
Elster, K. 1934–35. *Illustrert norsk Litteraturhistorie*. Vols 1–2. Oslo: Gyldendal.
Engelstad, I., J. Hareide, I. Iversen, T. Steinfeldm and J. Øverland. 1988–90. *Norsk kvinnelitteraturhistorie*. Vols 1–3. Oslo: Pax.
Fidjestøl, B., P. Kirkegaard, S. Aarnes, A. Aarseth, L. Longum and I. Stegane. 1996. *Norsk litteratur i tusen år: teksthistoriske linjer*. Cappelen Damm Akademisk.
Gaski, H. 1998. *Skriftbilder: samisk litteraturhistorie*. Kárášjohka/Karasjok: Davvi Girji.
Hagen, E.B., J. Haaberg, J.M. Sejersted, T. Selboe and P. Aaslestad. 2007–09. *Den norske litterære kanon*. Vols 1–2). Oslo: Aschehoug.

Glacier Poetry in Norwegian Literary Historiography

Hambro, C.J. 1912. *Kortfattet norsk litteraturhistorie: med illustrasjoner.* Oslo: Gyldendal.

Havnevik, I. 2002. *Dikt i Norge: lyrikkhistorie 200–2000.* Oslo: Pax.

Jæger, H. 1896. *Illustreret norsk literaturhistorie.* Vols. 1–3. Kristiania: Hjalmar Biglers Forlag.

Liestøl, K. and E. Stang. 1938. *Norges litteraturhistorie.* Gyldendal.

Moretti, F. 2013. *Distant Reading.* Verso Books.

Nærup, C. 1905. *Illustreret norsk litteraturhistorie: sidste tidsrum 1890–1904.* Kristiania: Det Norske Aktieforlag.

Rottem, Ø. 1996. *Norges litteraturhistorie: etterkrigslitteraturen.* Vols 1–3. Oslo: Cappelen.

Sørbø, J.I. 2018. *Nynorsk litteraturhistorie.* Oslo: Samlaget.

Poetry

Bjørnson, B. 1858. *Arne.* Bergen: H.J. Geelmuydens Enkes Forlag.

Bjørnson, B. 1870. *Digte og sange.* København: Den Gyldendalske Boghandel.

Eggen, A. 1951. *Eld og is.* Oslo: Tiden Norsk Forlag.

Garborg, A. 1895. *Haugtussa.* Oslo: Aschehoug.

Hauge, O.H. 1951. *Under bergfallet.* Oslo: Noregs Boklag.

Hofmo, G. 1978. *Det er sent.* Oslo: Gyldendal.

Jacobsen, R. 1951. *Fjerntog.* Oslo: Gyldendal.

Lie, J. 1867. *Digte.* Christiania: Jac. Dybwads Forlag.

Munch, A. 1850. *Nye digte.* Christiania: Chr. Tønsberg.

Reiss-Andersen, G. 1946. *Dikt fra krigstiden.* Oslo: Gyldendal.

Sivertson, S. 1839. 'Norafjeld med Jøkel blaa'. In H. Wergeland (ed.) *Samling af Sange og Digte for den norske Sjømand.* Kristiania: H.T. Winters forlag. pp. 13–16.

Uppdal, K. 1920. *Altarelden.* Oslo: Gyldendal.

Welhaven, J.S. 1834. *Norges Dæmring: et polemisk digt.* Christiania: Johan Dahl.

Welhaven, J.S. 1848. *Halvhundrede Digte.* Kjøbenhavn: Universitetsforlager C.A. Reitzel.

Wergeland, H. 1833. *Digte af Henrik Wergeland: Anden Ring.* Christiania: F. Steen.

Wergeland, H. 1844. *Den engelske Lods.* Christiania: B.C. Fabritius.

Ørjasæter, T. 1953. *Ettersommar.* Oslo: Olaf Norlis Forlag.

Øverland, A. 1929. *Haustavler: digte.* Oslo: Fram Forlag.

The Author

Kristine Kleveland is a Ph.D. student in Nordic literature at the University of Oslo. Her research focuses on glaciers as motifs in Norwegian poetry. Kleveland is a former lecturer in Literature at the Western Norway University of Applied Sciences.

Chapter 13

THROUGH A MIRROR, DARKLY: BRINGING DEEP ENVIRONMENTAL HISTORY INTO THE MUSEUM

Felix Riede

Abstract

Around 13,000 years ago – at the tail end of a prolonged cold spell and towards the end of the Ice Age – the Laacher See volcano located in present-day western Germany erupted cataclysmically. Abrasive and poisonous ash from this eruption was transported in a vast swath across Europe with a primary fallout cloud stretching towards the north-east. Environments, climate, and contemporaneous human populations were affected in a variety of ways; the Nordic area specifically was affected – indirectly – by serving as a refugium for small bands of prehistoric disaster refugees. Despite the enormity of this event, and the threat that this merely dormant volcano still poses today, it remains poorly known to the public, not least in the Nordic region. I here explore how the narrative qualities of such an archaeological scenario can be transformed into a museum exhibition whose aim it is to (i) put societal vulnerability into a deep historical perspective and to (ii) highlight how contemporary societies, too, are vulnerable to processes of climatic change and extreme environmental events plentifully documented in the past.

Introduction: Learning from the Past – but How?

From David Hume to George Santayana, from Niccolò Machiavelli to Winston Churchill, philosophers, statesmen and historians have suggested that we can – that we should – learn from the past to inform our actions in the present. As humanity is being swept along in the Great Acceleration (cf. Steffen et al. 2015) of its own making towards the potentially 'ghastly future' (Bradshaw et al. 2021) of an Anthropocene – canonical geological epoch or not – in which ever-more planetary boundaries are irrevocably transgressed (Ripple et al. 2024), historians have become specifically and acutely concerned with what we may be able to learn from past interactions between climate, the environment and the fates of human societies (e.g. Chakrabarty 2009; Brooke 2014; Carey et al. 2014; Degroot et al. 2021).

In the tradition of counterfactualist David Staley (2002), historians are – so he argues at least – those best placed to make informed prognoses of future societal trajectories. Yet, the late David Lowenthal (1985) famously argued that the past is akin to a foreign country – one that becomes increasingly quixotic and difficult to understand the farther back in time one goes. This challenges the notion that history can be used to inform the future in well-founded and evidence-based ways: The longer a time interval is considered, the more different climate-environment-society interactions can be studied, yet the opaquer and less resolved these cases usually become. There is no doubt that past societies – like present ones – were to some degree dependent on climate (Degroot et al. 2022). It is unclear, however, how *precisely* the interactions between climate and, for instance, the fate of the Roman Empire, of the Maya cities or of even earlier polities and communities – all characterised by livelihoods, beliefs, kinship structures and values very different from contemporary ones – can inform concrete decision-making processes in the present, especially those affecting policy (see Lane 2015). While such possibilities are being explored by climate historians in ever more focused ways (Kaufman, Kelly and Vachula 2018; Riede and Sheets 2020; Allen et al. 2022; Izdebski, Haldon and Filipkowski 2022; Haldon et al. 2024), I here pursue a different pathway for how climate history may impact contemporary climate action. I do so specifically in the setting of the Nordic region where the large majority of the population lives in Western, Educated, Industrialised, Rich and Democratic (WEIRD) ways (Henrich, Heine and Norenzayan 2010). Consequently, lessons from historical ecology (Armstrong et al. 2017) or traditional agriculture (Guttmann-Bond 2010) are difficult to effectuate or remain largely disconnected from the overwhelmingly urban populations and their lifestyles of the present. Instead, I focus on the role of museums of culture history as fora for communication, debate, learning and action in relation to climate history and climate change – past and present.

Drawing on a long-term interdisciplinary research project investigating the societal impacts of the cataclysmic Laacher See volcanic eruption that occurred in the closing centuries of the last ice age, I here explore how the narrative qualities of such an archaeological scenario can be transformed into a museum exhibition whose aim it is to put societal vulnerability into a deep historical perspective and to highlight how contemporary societies, too, are vulnerable to processes of climatic change and the sorts of extreme environmental events plentifully documented in the past.

Felix Riede

Climate Change Archaeology Narratives and Where to Tell Them

Archaeology – the subject concerned with the material remains of human deep history – has a long tradition, not least in the Nordic region, of studying the environment alongside its focus on past societal factors (see Gron and Rowley-Conwy 2018). Variable preservation makes this view on the societies of the deep past often through a more or less darkened looking glass (cf. Rowley-Conwy 2000), but the evidence is nonetheless rich and multi-faceted. Often in collaboration with ecologists and geologists, archaeologists use a variety of proxies from their own excavations and from nearby 'off-site' contexts to better understand the particular environmental conditions under which people were living, and how these changed over time (Riede and Mannino 2024). Specific techniques such as dendrochronology, dendroclimatology and pollen analysis were pioneered in the Nordics and remain staples of many contemporary projects. Alongside these traditional approaches, recent work on ice cores as well as on a swath of biogeochemical proxies for past environments has practically revolutionised our understanding of past climates – not only for the Common Era (Ljungqvist, Seim and Huhtamaa 2021) but also much further back in time (cf. Burroughs 2005).

Archaeology is an epistemologically schizophrenic discipline, however. While most archaeologists engaged in 'climate change archaeology' (Van de Noort 2013) tend to operate more in the mindset of the Natural Sciences, just as many archaeologists draw on Social Theory, insights from anthropology or philosophy – and they are not commonly concerned with climate or the environment. Some have explored the narrative quality of archaeological interpretations, even suggesting that much of archaeological writing is closer to fiction than to science (Hodder 1993; Van Dyke and Bernbeck 2015; Mostafa 2019). Even among those concerned with contemporary climate change, this 'two-culture' divide (cf. Snow 1959) remains evident (Pétursdóttir and Sørensen 2023). While there are substantive theoretical and methodological differences between these alternative archaeologies, they all produce – implicitly or explicitly – *narratives* of past culture change under conditions of climate change. And narratives are among the most powerful tools of climate history (Cronon 1992).

The claim that narratives concerned with nature, the environment and climate can influence values and actions has been variously investigated by literary scholars (Schneider-Mayerson 2020; Schneider-Mayerson, Weik von Mossner and Małecki 2020; Schneider-Mayerson et al. 2023) and psychologists (Morris et al. 2019). Archaeological interpretations of past culture change during the many 'moments of crisis' (Tipping et al. 2012: 9) precipitated by past climate change share vital narrative qualities with climate fiction. Both are evidence-based writing about another time, and, in this way, archaeological narratives can be said to have

Through a Mirror, Darkly

the same affective qualities as climate fiction. Moreover, psychological research also indicates, albeit not conclusively (Ettinger et al. 2021), that narratives with negative valence are more likely to catalyse action on environmental matters (Morris et al. 2020). This, in turn, leads to a very specific genre of climate history – the apocalyptic. While there is little doubt that such narratives with acute tension are not only powerful ways of telling but also lucrative ways of selling a given history (cf. Pomeroy 2008), they do provide very particular analytical and rhetorical affordances. Volcanic eruptions, for instance, with their sudden onset and often widespread ash-fall make it possible – via stratigraphic analysis and other dating techniques – to clearly distinguish a 'before' and 'after'. In the original meaning of apocalypse as a revelation or turning point in a narrative, then, a focus on how such critical events unfolded and how past societies may have been affected by and responded to them allows us to readily transform mere chronicles into powerful stories.

Such stories can be told in many ways and in many places – in scientific books and articles, in popular magazines and newspapers and in the media. As the discipline concerned with deep history – including deep climate history (Caseldine and Turney 2010; Dukes 2013; Hussain and Riede 2020) – archaeology has an additional and very particular platform, and some unique exhibition objects: the culture history museum filled with the artefacts – the 'belongings' (Pitblado et al. 2025) – of past peoples. The museum sector at large has begun to engage strongly with issues of climate change and with what and how museums – as institutions and as loci of learning – can contribute in meeting the challenges associated with it (Cameron, Hodge and Salazar 2013; Cameron and Neilson 2015; Newell, Robin and Wehner 2016; Lyons and Bosworth 2019; Davis 2020). Museum scholars have argued that museums, as trusted spaces for dialogue and curiosity, may serve as democratic platforms for debating climate change (Cameron and Deslandes 2011) but also that they can catalyse concrete action (Rees 2017; Sutton and Robinson 2020).

This research has, however, almost exclusively focused on science centres and natural history museums. The degree to which museums of culture history have engaged with issues of anthropogenic climate change, societal responses to past climate change and the long arc of the Anthropocene that has its beginnings way back in earlier geological epochs (Foley et al. 2013; Svenning et al. 2024) has been limited, albeit with exceptions also in the Nordic region (see https://www.historiskmuseum.no/english/exhibitions/collapse/index.html, and Bjerregaard 2019a). Vitally, climate change and its articulation with societal change is not only a subject matter for the natural and technical sciences. And, notably, museums of culture history tend to attract a considerably larger number of visitors across all ages, thus providing a particularly powerful platform for dissemination and dialogue (Figure 1).

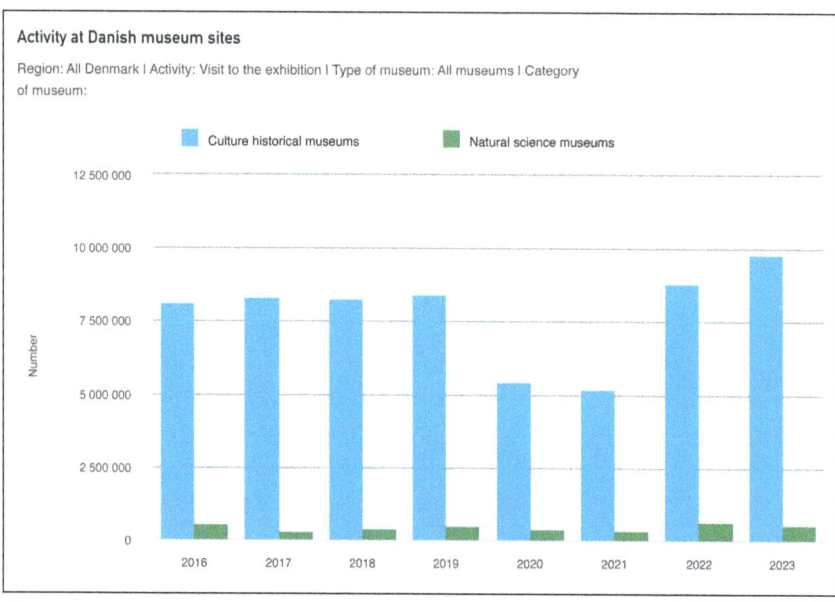

Figure 1. Visitor numbers for museums and cultural and natural history sites in Denmark for 2016 to 2023 (from Statistics Denmark, 'MUS3: Activity at Danish museum sites by region, category of museum, type of museum and activity', https://www.statbank.dk/MUS3*). The data for other Nordic countries look similar:* https://www.egmus.eu/nc/sl/statistics/complete_data/

Grounded in these observations, I designed a research project with a focus on the interaction between climate change, the eruption of the Laacher See volcano, and contemporaneous forager societies in which an exhibition at Moesgaard Museum outside of Aarhus in Denmark was an integral part. The exhibition was meant to represent a major deliverable showcasing the project's many results but also to serve as an experimental platform for how the complex research findings from environmental history may be transformed into an impactful exhibition, and for novel research (cf. Bjerregaard 2019b). The ambition was to directly mirror past and future climate, environments and societies – to let a particular and empirically rich case study inform how we may think of contemporary and future climate-society dependencies. The ambition was also to let the aesthetic and affective qualities of the apocalyptic narrative and its visual and auditory display qualities address both the hearts and minds of visitors. In the following, I summarise first the research findings of the project 'Apocalypse

Climate Change, the Laacher See Eruption and Societal Change at the End of the Last Ice Age

Then?' (see Riede 2017b) and then how these were transformed into exhibition components. In describing these, I focus on the intellectual co-creation of both the scientific results and the actual exhibition space and its content.

Following the so-called Last Glacial Maximum around 22,000 years ago, the geological epoch of the Pleistocene – the proverbial ice age – began to wane fitfully (Burroughs 2005). From around 15,000 years ago, temperatures rose so much that plants, animals and eventually humans began to settle in higher latitudes of Europe, including the Nordic area (Riede 2014). After some initial failures, forager societies had established at least a seasonal presence in what today is Denmark and Southern Sweden by around 14,000 years ago (Figure 2). These people likely had southern roots biologically (Posth et al. 2023) and culturally belonged to what archaeologists label the Federmessergruppen or Penknife Groups. This was a European-wide phenomenon whose characteristic material culture included flint tools such as arrow- and spearpoints, scrapers and burins for working various materials, fishhooks and -spears, art produced in amber and other rare raw materials, as well as domesticated dogs (Grimm et al. 2020) – although very few sites boast all these material characteristics.

The presence of these Federmessergruppen people coincides with stable warming, albeit punctuated by repeated pulses of cooling and storminess (Crombé et al. 2020). This period is identified in the Greenland ice-core record of northern hemisphere climate and labelled Greenland Interstadial 1c. Four centuries later, a cooling trend set in. Recognised from the Alps (Ammann et al. 2013) to Northern Europe (Fiłoc et al. 2024), this temperature depression is termed Greenland Interstadial 1b or the 'Gerzensee Oscillation'. Analyses relating temperature to hunter-gatherer demography and social organisation have found generalisable patterns where lower and more variable temperatures result in lower population densities (Tallavaara and Seppä 2012) and social networks that are more strained (Whallon 2006).

Just as temperatures were rising again at the end of the Gerzensee Oscillation – at 13,006±9 years before present (taken, by convention to mean 1950 CE; Reinig et al. 2021) – the Laacher See volcano located in present-day western Germany erupted. Today, the Laacher See is a placid lake set in a rich cultural and natural landscape that is part of the East Eifel Volcanic Zone (Figure 3). In the late spring or early summer 13,000 years ago, it was a dramatic ground zero when this cataclysmic eruption terminated a long phase of intermittent volcanism in the area (Schmincke 2004). Fuelled by a mantle plume, a station-

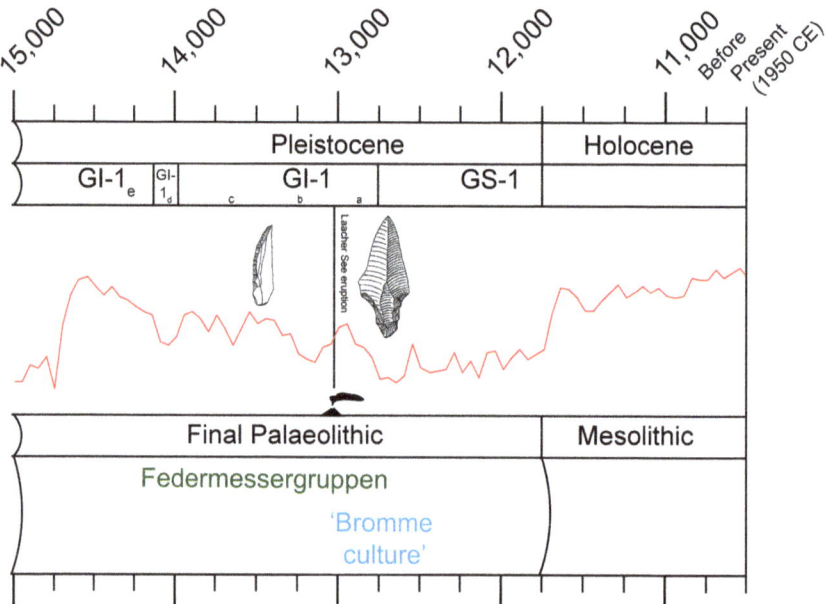

Figure 2. A climate historical timeline of the waning ice age. Oxygen isotope measurements in the Greenland ice cores provide a highly resolved temperature record for the Northern Hemisphere and offer a view on weather and climate change nested within broader geological epochs (i.e. the Pleistocene and Holocene). Human presence in higher latitudes during the Palaeolithic in general (Hublin and Roebroeks 2009) and the Final Palaeolithic in particular (Riede 2014) waxes and wanes with climate warming and stability. Source: Felix Riede.

ary pocket of magma at a relatively shallow depth (Zhu et al. 2012; Kreemer, Blewitt and Davis 2020), this eruption ejected more than six cubic kilometres of material from the Earth interior to the surface and covered an area upwards of 1,400 square kilometres with thick layers of fallout, ignimbrites and lahar deposits. These voluminous ejecta dammed the nearby River Rhine – Europe's third-largest river – leading to the rapid formation of a lake and widespread upstream flooding far up the Rhine and its tributaries (Park and Schmincke 1997, 2020). In turn, this dam broke during the waning stages of the eruption, sending flood waves laden with debris down the Lower Rhine reaches all the way to what today is the English Channel.

Figure 3. The Laacher See today. The crater remnant is heavily eroded and massively changed by historical land-use and water engineering projects. Source: Florian Sauer (author's project).

This eruption was devastating in its near-field but, in this devastation, contributed substantially to understanding the lives and livelihoods of contemporaneous forager populations. The ash (=tephra) and eruption debris covered, sealed and thereby preserved a late ice age landscape with the burnt remains of plants (Street 1986) and animal tracks (Baales and von Berg 1997) possibly including humans and their canine companions (Poschmann and von Berg 2020). Evidence of human presence in the region prior to the eruption is rich and includes both very small (von Berg 1994) and very large archaeological sites (e.g. Baales 2006). The frequency and size of these sites do indicate a reduction in settlement intensity during the cool Gerzensee Oscillation, followed by an abandonment of the region following the Laacher See eruption until some decades or even centuries later when people returned, albeit briefly, from a western direction (Baales and Jöris 2002).

The eruption impacted societies further afield too. As the massive ash column began to drift in the prevailing winds, it was transported initially towards the northeast and later towards the south as wind directions changed from spring to summer circulation (Niemeier, Riede and Timmreck 2021). Tephra-fall was uneven but still massive in many places. Our interdisciplinary research beginning back in 2007 (Riede 2007, 2008), has focused on using hypotheses derived from recent and historical eruptions and their impacts to evaluate whether and how societies living well away from the eruptive centre may still have been affected by it (Riede 2018). Specifically, we set out to better understand (i) if the emergence of a cultural phenomenon specific to southern Scandinavia, the so-called 'Bromme culture' can be causally linked to the Laacher See eruption; and (ii) if post-eruptive demographic fluctuations, abandonment or cultural responses can be detected in the regions affected by volcanic fall-out. With distance, tephra-fall thickness declines but the composition of the material – its grain size and surface-to-volume ratio – changes. The impact pathways and dynamics become more subtle but no less relevant as these intersect with changing ecological and demographic baselines that are conditioned in part by historical process as well as by latitudinal and ecological gradients. As part of this work, we explored the likely effects of grossly increased tooth wear among prey animals in humans as they ingested and chewed on ash-laden vegetation (Riede and Wheeler 2009); we also explored how dangerous – or at least how very irritating – the tiny ash particles would have been at ever-greater distance from the volcano (Riede and Bazely 2009); and we explored how the high doses of poisonous fluoride that were attached to the ash may have impacted animals and humans alike (Riede and Kierdorf 2020). We also conducted fieldwork designed specifically to test the hypothesis of a regional mid-field abandonment following the eruption (Riede 2016) using the time-marker of the ash as a critical analytical and narrative terminus (Riede 2012; Riede, Sauer and Hoggard 2018; Sauer, Stott and Riede 2019; Riede et al. 2024). Driven by fundamental culture- and climate-historical hypotheses, this research was interdisciplinary in multiple ways. The project's palette of approaches was co-designed with biologists, geographers, materials scientists, geologists and climate scientists, our results published chiefly in scientific journals and a final summary monograph (Riede 2017b).

From Science to Story, From Monograph to Museum

Emerging from this interdisciplinary research was a storyline in which the coupling of detrimental climate change – the Gerzensee Oscillation – and an unforeseen extreme event – the Laacher See eruption – impacted these ancient communities. Much of this work was conducted during a period when Icelandic volcanism was reawakening, likely due to rising global temperatures (Pagli and Sigmundsson 2008; Sigmundsson et al. 2010; Schmidt et al. 2013), reminding Europeans of the threat that even distant volcanoes pose to our present-day lifeways (e.g. Birtchnell and Büscher 2011; Lund and Benediktsson 2011) and infrastructure (Donovan and Oppenheimer 2018; Mani, Tzachor and Cole 2021). Despite clear evidence for the continued presence of magma underneath the Laacher See (e.g. Goepel et al. 2015), it is today chiefly seen as a touristic asset and not as a major threat (Erfurt-Cooper 2010).

Motivated by translating our insights from 'matters of fact' to 'matters of concern' (Stewart and Lewis 2017) but also mindful of the challenges of making such very remote periods of deep history relevant in the present (Riede 2017a), we set out to create a museum exhibition that directly articulates our empirical findings with contemporary quandaries around climate change and vulnerability.

Through a Mirror, Darkly

Creating a museum exhibition requires considerable resources and a co-creative process very different from scientific work. This exhibition with the title 'After the Apocalypse' was installed in a small (140 square metres) room dedicated to research-based thematic temporary exhibitions at Moesgaard Museum (https://www.moesgaardmuseum.dk) near Aarhus in Denmark (see Price 2015). At first, filling 140 square metres of empty exhibition space with the mostly minuscule and decidedly unspectacular artefacts of the Stone Age seemed terrifying – the curatorial equivalent of the writer's empty page syndrome. But working with the professional exhibition architects, curators and digital content creators – but also the stage painters, light and sound technicians and craftspeople of the museum's workshop (Figure 4) – led to the crystallisation of an overall vision: the exhibition was to be divided into two halves, where the deep past should mirror the near future. A central video installation inspired by graphic novels should serve as a dividing line; on the one side, this film would follow a Stone Age forager family experiencing the eruption and its effects from afar, on the other side, a near-future family would experience the same sequence but set in the context of a climate-changed contemporary. A heavy and uncanny rumbling derived from actual recordings of volcanic explosions provided the soundtrack to the experience. The overall ambience of the exhibition just as much as the object labels and panel texts were a co-created compromise between the ambition to include, on the one hand, as much research-based information as possible and to produce, on the other, an attractive and palatable exhibition that also would appeal to museumgoers.

Figure 4. 'After the Apocalypse' in the making. Source: Moesgaard Museum (author's project).

On opposing walls, massive (six by three metres) posters framed the overall narrative and the articulation of the eruption with the trajectories of climate and culture history. Here, using the powerful climate stripe visualisation originally developed by Ed Hawkins (https://showyourstripes.info/), we juxtaposed a timeline of warming and cooling during the closing millennia of the Pleistocene with the global warming that occurred since 1850 CE and into

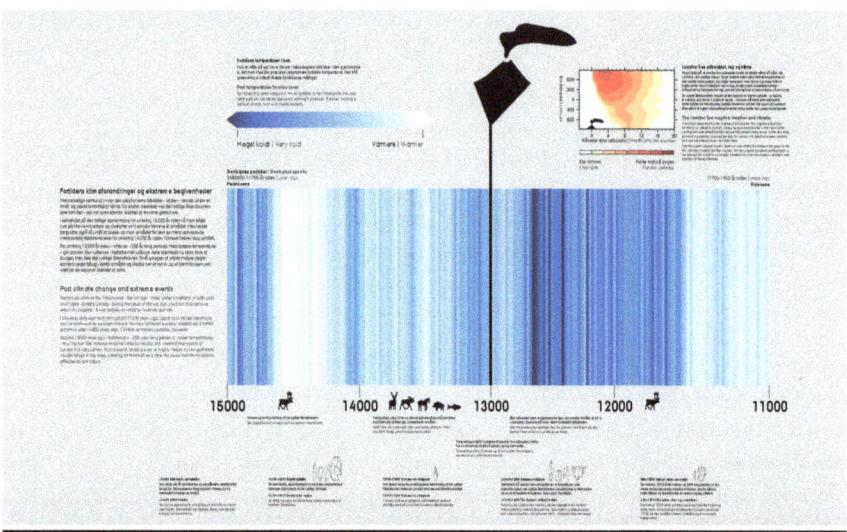

Figure 5. A. Poster content, as positioned in the exhibition: a climate historical timeline of the closing millennia of the Pleistocene and its transition into the Holocene, and major culture-historical events that occurred at this time in southern Scandinavia. The black vertical line marks the Laacher See eruption. Source: Moesgaard Museum (author's project).

Through a Mirror, Darkly

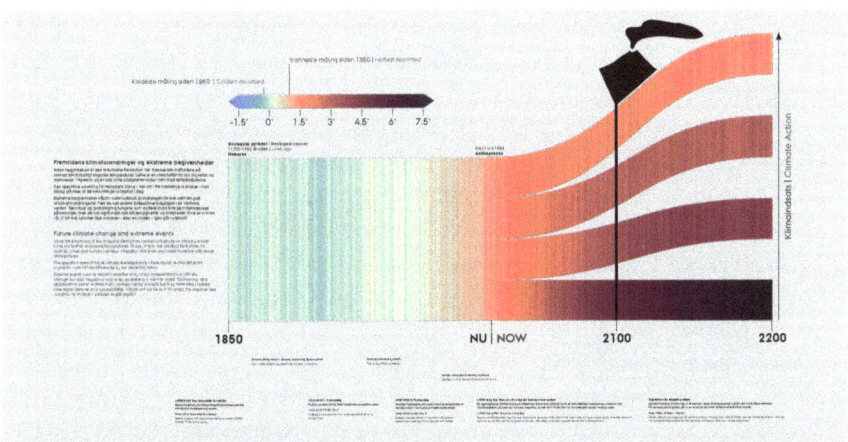

Figure 5. B. Poster content, as positioned in the exhibition: a climate historical timeline for the transition from the Holocene to the Anthropocene, including major climate-impacting technological innovations and near-future projections under variable scenarios of climate action, an idea inspired by Alexander Radtke (see https://grist.org/article/we-are-here-new-climate-design-shows-us-our-future-in-red-hot-stripes/). Source: Moesgaard Museum (author's project).

the Anthropocene – and how this is projected to develop towards 2100 CE under different scenarios of climate action. Volcanic markers place the actual past and speculative future eruptions of the Laacher See volcano at 11,000 BCE and 2200 CE respectively (Figure 5).

The choice of this particular visualisation format was grounded in the basic need for timelines in anchoring and structuring the overall narrative, the necessity to include visuals in communicating climate change as well as extreme events (Juneja and Schenk 2014; O'Neill and Smith 2014), and informed by analyses regarding their accessibility and immediacy (Windhager, Schreder and Mayr 2019). The background colour of the room was dark, and the floor painted in earth colours meant to represent ash-fall. Painted on the floor were animal tracks and human footprints – taken from the archaeological evidence uncovered near the Laacher See itself – meant to guide visitors from past to future in the otherwise quite open exhibition room. As one passed from the Pleistocene part of the exhibition space into its Anthropocene mirror image, these tracks took on modern shoe-sole formats and changed from those of wild animals to domestic ones. The four corners of the room housed interactive video screens where visitors could engage with experts on volcanology, archaeology, human-volcano interactions and contemporary risks. Finally, six cabinets – three dedicated to the past, three to the future – displayed the 'vibrant matter' (Bennett 2010) of past and present vulnerability and resilience, the artefacts that told salient thematic stories about how society in the deep past likely was affected by the compound pressures of climate change and a volcanic eruption, and how similar pressures may impact societies in the future (Figure 6).

All the displayed artefacts were selected to support particular key messages of the exhibition. On the archaeological side, the objects were intended to create an intimacy between visitors and the ancient peoples, to illustrate on what empirical basis and how we can make inferences about how climate change and an extreme event back then impacted lives and livelihoods. All these objects came from archaeological sites and themes that played a major role in the research project itself. On the future-oriented side of the exhibition space, the objects were selected to mirror the project's key themes – social network importance, health impacts, economic impacts and imaginations of apocalypse – and to do so with reference to our present. These objects – an ash-covered telephone, a face mask, bones deformed by ash-borne poison – alluded to negative impacts, while crafts and fiction relating to volcanic eruptions and similar calamities in the recent past also supported a story of the creative energy that catastrophes can generate. Many of these objects were either kindly supplied by colleagues near and far, sourced via the curator's network, or came directly from the project's collection and library. Notably, the exhibition ended with reference to iconic fiction author Ursula K. LeGuin – incidentally the daughter of an archaeolo-

Through a Mirror, Darkly

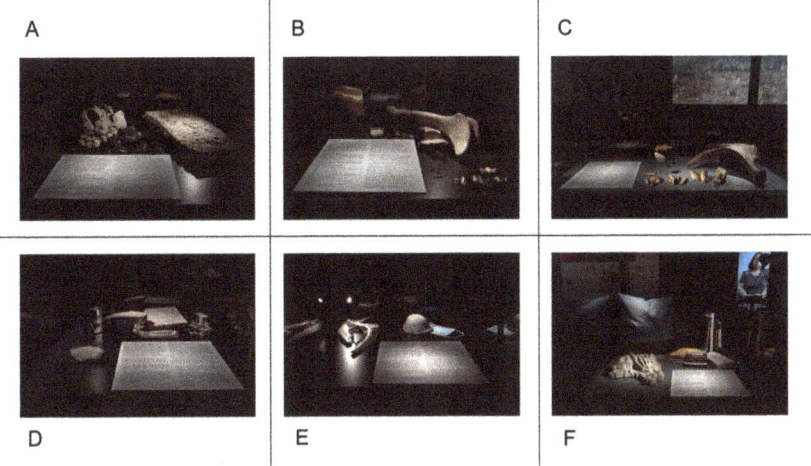

Figure 6. A collage of images from the cabinets in the 'After the Apocalypse' exhibition. (A): geological evidence from the near-vent area. (B): ecological impacts at distance. (C): far-field impacts on people and societal responses in southern Scandinavia. (D): the effects of volcanic ash-fall on contemporary infrastructure and economy as exemplified by an ash-covered telephone from Montserrat and bottled tephra from the 2010 Eyjafjallajökull eruption. (E): the effects of ash-fall on animals and health exemplified by the facemasks that were so ubiquitous at the time of the exhibition and animal teeth that had been strongly affected by industrial fluoride poisoning. (F): the impacts of volcanic activity on human imagination and sociality, both as catalyst for creativity and future imagination, as is found in myriad climate fiction narratives and in art and craft production inspired by volcanism. Photos: Moesgaard Museum (author's project).

gist – and her musings on how dystopian visions also contain kernels of utopia. The exhibition was quite deliberately a dark one but did close on a note of hope.

Conclusions: Climate History is Culture History, Also in the Museum

Museums offer a robust and important platform for communicating climate history to a diverse public – visitors of all ages, locals as well as foreigners. They certainly count among the impactful pathways for giving this research traction in the present – museums educate, inform and invite discussion and dialogue. While museums of natural history and science have become strongly engaged in recent debates on climate change, culture history museums have remained relatively silent. Yet, climate and environmental history *is* culture history and, thus, climate and the environment may claim their place in these galleries. Given the urgency of the contemporary climate crisis, such topics arguably should play a greater role in standard narratives of our past. As research within the psychology of climate action demonstrates that not factual but emotional engagement changes minds and catalyses action, the combination of scientific methods and patently relatable culture-historical human content makes for a particularly powerful blend.

The exhibition in focus here was only possible because of generous and consistent funding from the Independent Research Fund Denmark, and the opportunity of staging such a professionally produced, research-driven yet public-facing display at a museum with visitor numbers approaching half a million per year. Many intellectual and creative compromises are needed to turn research into an exhibition. This co-creative process alone, however, produces new knowledge. It simplifies but also clarifies and rarifies the narrative. Research findings informed the curatorial decisions, from the narrative structure and tone to the scientific and affective content, as well as the climate historical framework and its visualisation. The exhibition's implementation was further enriched by the design acumen of the many museum staff that became involved. While small, this exhibition was the result of co-creation and the coming together of human actors – the scientists, curators, craftspeople – and nonhuman actants – the machines, materials and artefacts – that together made it happen. This was not always an easy process, however. This exhibition was expensive. The burden of financing rested entirely on the research project, the museum partner provided the space and the exhibition expertise. This left the roles and responsibilities sometimes poorly defined, creating delays and undue difficulties. The curatorial vision for the exhibition was – in draft form at least – quite clear already early in the process. Achieving the final product took rather more negotiation in terms of scheduling, responsibilities and capacities: making museum exhibitions is a complex process, all the more so as our message had several facets: disasters are not natural in any meaningful sense but emerge in the interplay between climatic pressures and the vulnerabilities created by particular, historically constituted ways of life. We sought to highlight both the role of climate change – natural cooling at the end of the Pleistocene and anthropogenic warming that marks the beginning of the Anthropocene – and the existential risks associated with extreme, sudden and unexpected events (cf. Denkenberger and Blair 2018; Donovan and Oppenheimer 2018). A central concern was to use the past as an analytical mirror and the historical-archaeological perspective as a method to divert focus from dramatic societal impacts to the critical vulnerabilities that put the societies in focus – including our own – at risk in the first place.

Through a Mirror, Darkly

With regular evening lectures as well as advertisements on the backs of buses and in local magazines, the exhibition had a strong quotidian presence. Ultimately, however, the short six-month period during which it was in place – from September 2020 to May 2021 – cannot but have had only a limited impact on the total debate on climate, volcanism and societal vulnerability. Moreover, a Covid caveat came into play. The opening of the exhibition and its reach were impacted by the Covid pandemic in the untimeliest manner, the kind of *force majeure* to which museum venues are particularly vulnerable; there were no resources to convert the exhibition to an immersive online experience. While the exhibition could be set up, its opening ceremony was curtailed by numerous restrictions and the museum remained closed during much of the time during which the exhibition was in place. This unfortunate and itself catastrophic intervention also prevented real-time evaluations or visitor surveys such that the actual impact of this exhibition on the hearts and minds of visitors must remain speculative. History and prehistory have much to offer in terms of climate-historical narratives, apocalyptic or otherwise. The urgency of the climate crisis arguably puts into place a strong mandate for culture-historical museums to emphasise these aspects – a shift that would be especially salient if climate were to play a greater role in the permanent core exhibitions and across all time periods.

Bibliography

Allen, K. et al. 2022. 'Coupled insights from the palaeoenvironmental, historical and archaeological archives to support social-ecological resilience and the Sustainable Development Goals'. *Environmental Research Letters* **17** (5): 055011. https://doi.org/10.1088/1748-9326/ac6967.

Ammann, B. et al. 2013. 'Vegetation responses to rapid warming and to minor climatic fluctuations during the Late-Glacial Interstadial (GI-1) at Gerzensee (Switzerland)'. *Palaeogeography, Palaeoclimatology, Palaeoecology* 391: 40–59. https://doi.org/10.1016/j.palaeo.2012.07.010.

Armstrong, C.G. et al. 2017. 'Anthropological contributions to historical ecology: 50 questions, infinite prospects'. *PLoS ONE* **12** (2): e0171883. https://doi.org/10.1371/journal.pone.0171883.

Baales, M. 2006. 'Final Palaeolithic environment and archaeology in the Central Rhineland (Rhineland-Palatinate, western Germany): conclusions of the last 15 years of research'. *Anthropologie* **110** (3): 418–444.

Baales, M. and A. von Berg. 1997. 'Tierfährten in der allerödzeitlichen Vulkanasche des Laacher See-Vulkans bei Mertloch, Kreis Mayen-Koblenz'. *Archäologisches Korrespondenzblatt* 27: 1–12.

Baales, M. and O. Jöris. 2002. 'Between North and South – a site with backed points from the final Allerod: Bad Breisig, Kr. Ahrweiler (Central Rhineland, Germany)'. *L'Anthropologie* **106** (2): 249–267.

Bennett, J. 2010. *Vibrant Matter: A Political Ecology of Things*. Durham, NC: Duke University Press.

von Berg, A. 1994. 'Allerödzeitliche Feuerstellen unter dem Bims im Neuwieder Becken (Rheinland-Pfalz)'. *Archäologisches Korrespondenzblatt* 24: 355–365.

Birtchnell, T. and Büscher, M. 2011. 'Stranded: An eruption of disruption'. *Mobilities* **6** (1): 1–9.

Bjerregaard, P. 2019a. 'Exhibition-making as aesthetic inquiry'. In P. Bjerregaard (ed.) *Exhibitions as Research*. New York: Routledge. pp. 95–108.

Bjerregaard, P. (ed.) 2019b. *Exhibitions as Research: Experimental Methods in Museums*. London: Routledge. https://doi.org/10.4324/9781315627779.

Bradshaw, C.J.A. et al. 2021. 'Underestimating the challenges of avoiding a ghastly future'. *Frontiers in Conservation Science* **1**: 9. https://doi.org/10.3389/fcosc.2020.615419.

Brooke, J.L. 2014. *Climate Change and the Course of Global History: A Rough Journey*. Cambridge: Cambridge University Press. https://doi.org/10.1017/CBO9781139050814.

Burroughs, W.J. 2005. *Climate Change in Prehistory. The End of the Reign of Chaos*. Cambridge: Cambridge University Press.

Cameron, F., B. Hodge and J.F. Salazar. 2013. 'Representing climate change in museum space and places'. *WIREs Climate Change* **4** (1): 9–21. https://doi.org/10.1002/wcc.200.

Cameron, F.R. and A. Deslandes. 2011. 'Museums and science centres as sites for deliberative democracy on climate change'. *Museum and Society* **9** (2): 136–153.

Cameron, F.R. and B. Neilson (eds). 2015. *Climate Change and Museum Futures*. London: Routledge.

Carey, M. et al. 2014. 'Forum: Climate change and environmental history'. *Environmental History* **19** (2): 281–364. https://doi.org/10.1093/envhis/emu004.

Caseldine, C. and C.S.M. Turney. 2010. 'The bigger picture: towards integrating palaeoclimate and environmental data with a history of societal change'. *Journal of Quaternary Science* **25** (1): 88–93. https://doi.org/10.1002/jqs.1337.

Chakrabarty, D. 2009. 'The climate of history: Four theses'. *Critical Inquiry* **35** (2): 197–222. https://doi.org/10.1086/596640.

Crombé, P. et al. 2020. 'Repeated aeolian deflation during the Allerød/GI-1a-c in the coversand lowland of NW Belgium'. *CATENA* 188: 104453. https://doi.org/10.1016/j.catena.2020.104453.

Cronon, W. 1992. 'A place for stories: Nature, history, and narrative'. *Journal of American History* **78** (4): 1347–1376.

Davis, J. 2020. 'Museums and climate action: a special issue of Museum Management and Curatorship'. *Museum Management and Curatorship* **35** (6): 584–586. https://doi.org/10.1080/09647775.2020.1842535.

Degroot, D. et al. 2021. 'Towards a rigorous understanding of societal responses to climate change'. *Nature* 591 (7851): 539–550. https://doi.org/10.1038/s41586-021-03190-2.

Degroot, D. et al. 2022. 'The history of climate and society: a review of the influence of climate change on the human past'. *Environmental Research Letters* 17 (10): 103001. https://doi.org/10.1088/1748-9326/ac8faa.

Denkenberger, D.C. and R.W. Blair. 2018. 'Interventions that may prevent or mollify super-volcanic eruptions'. *Futures* 102: 51–62. https://doi.org/10.1016/j.futures.2018.01.002.

Donovan, A.R. and C. Oppenheimer. 2018. 'Imagining the unimaginable: Communicating extreme volcanic risk'. In C.J. Fearnley et al. (eds), *Observing the Volcano World: Volcano Crisis Communication*. Cham: Springer International Publishing. pp. 149–163.

Dukes, P. 2013. 'Big history, deep history and the Anthropocene'. *History Today* 63 (11): 4–5.

Erfurt-Cooper, P. 2010 'The Vulkaneifel in Germany. A destination for geotourism'. In P. Erfurt-Cooper and M. Cooper (eds), *Volcano and Geothermal Tourism: Sustainable Geo-resources for Leisure and Recreation*. London: Earthscan. pp. 281–285.

Ettinger, J. et al. 2021. 'Climate of hope or doom and gloom? Testing the climate change hope vs. fear communications debate through online videos'. *Climatic Change* 164 (1): 19. https://doi.org/10.1007/s10584-021-02975-8.

Fiłoc, M. et al. 2024. 'Late-Weichselian (Vistulian) environmental changes in NE Poland – Evidence from Lake Suchar Wielki'. *CATENA* 234: 107546. https://doi.org/10.1016/j.catena.2023.107546.

Foley, S.F. et al. 2013. 'The Palaeoanthropocene – the beginnings of anthropogenic environmental change'. *Anthropocene* 3: 83–88. http://dx.doi.org/10.1016/j.ancene.2013.11.002.

Goepel, A. et al. 2015. 'Volcano-tectonic structures and CO_2-degassing patterns in the Laacher See basin, Germany'. *International Journal of Earth Sciences* 104 (5): 1483–1495. https://doi.org/10.1007/s00531-014-1133-3.

Grimm, S.B. et al. (eds). 2020. *From the Atlantic to Beyond the Bug River: Finding and Defining the Federmesser-Gruppen/Azilian*. Heidelberg: Propylaeum. https://books.ub.uni-heidelberg.de/propylaeum/catalog/book/575.

Gron, K.J. and P. Rowley-Conwy. 2018. 'Environmental archaeology in southern Scandinavia In E. Pişkin, A. Marciniak and M. Bartkowiak (eds), *Environmental Archaeology: Current Theoretical and Methodological Approaches*. Cham: Springer International Publishing, pp. 35–74.

Guttmann-Bond, E.B. 2010. 'Sustainability out of the past: how archaeology can save the planet'. *World Archaeology* 42 (3): 355–366. https://doi.org/10.1080/00438243.2010.497377.

Haldon, J. et al. 2024. 'Past answers to present concerns. The relevance of the premodern past for 21st century policy planners: Comments on the state of the field'. *WIREs Climate Change* e923. https://doi.org/10.1002/wcc.923.

Henrich, J., S.J. Heine and A. Norenzayan. 2010. 'The weirdest people in the world?' *Behavioral and Brain Sciences* **33** (2–3): 61–83. https://doi.org/10.1017/S0140525X0999152X.

Hodder, I. 1993. 'The narrative and rhetoric of material culture sequences'. *World Archaeology* **25** (2): 268–282. https://doi.org/10.1080/00438243.1993.9980243.

Hublin, J.-J. and W. Roebroeks. 2009. 'Ebb and flow or regional extinctions? On the character of Neandertal occupation of northern environments'. *Comptes Rendus Palevol* **8** (5): 503–509.

Hussain, S.T. and F. Riede. 2020. 'Paleoenvironmental humanities: Challenges and prospects of writing deep environmental histories'. *WIREs Climate Change* **11** (5): e667. https://doi.org/10.1002/wcc.667.

Izdebski, A., J. Haldon and P. Filipkowski (eds). 2022. *Perspectives on Public Policy in Societal-Environmental Crises: What the Future Needs from History*. Cham: Springer. https://link.springer.com/book/10.1007/978-3-030-94137-6.

Juneja, M. and G.J. Schenk. 2014. *Disaster as Image. Iconographies and Media Strategies across Europe and Asia*. Regensburg: Schnell & Steiner.

Kaufman, B., C.S. Kelly and R.S. Vachula. 2018. 'Paleoenvironment and archaeology provide cautionary tales for climate policymakers'. *The Geographical Bulletin* **59** (1): 5–24.

Kreemer, C., G. Blewitt and P.M. Davis. 2020. 'Geodetic evidence for a buoyant mantle plume beneath the Eifel volcanic area, NW Europe'. *Geophysical Journal International* **222** (2): 1316–1332. https://doi.org/10.1093/gji/ggaa227.

Lane, P.J. 2015. 'Archaeology in the age of the Anthropocene: A critical assessment of its scope and societal contributions'. *Journal of Field Archaeology* **40** (5): 485–498. https://doi.org/10.1179/2042458215Y.0000000022.

Ljungqvist, F.C., A. Seim and H. Huhtamaa. 2021. 'Climate and society in European history'. *WIREs Climate Change* **12** (2): e691. https://doi.org/10.1002/wcc.691.

Lowenthal, D. 1985. *The Past is a Foreign Country*. Cambridge: Cambridge University Press.

Lund, K.A. and K. Benediktsson. 2011. 'Inhabiting a risky earth: The Eyjafjallajökull eruption in 2010 and its impacts'. *Anthropology Today* **27** (1): 6–9. https://doi.org/10.1111/j.1467-8322.2011.00781.x.

Lyons, S. and K. Bosworth. 2019. 'Museums in the climate emergency'. In R.R. Janes and R. Sandell (eds), *Museum Activism*. London: Routledge. pp. 174–185.

Mani, L., A. Tzachor and P. Cole. 2021. 'Global catastrophic risk from lower magnitude volcanic eruptions'. *Nature Communications* **12** (1): 4756. https://doi.org/10.1038/s41467-021-25021-8.

Morris, B.S. et al. 2019. 'Stories vs. facts: triggering emotion and action-taking on climate change'. *Climatic Change* **154** (1): 19–36. https://doi.org/10.1007/s10584-019-02425-6.

Morris, B.S. et al. 2020. 'Optimistic vs. pessimistic endings in climate change appeals'. *Humanities and Social Sciences Communications* **7** (1): 82. https://doi.org/10.1057/s41599-020-00574-z.

Through a Mirror, Darkly

Mostafa, J. 2019. 'In search of lost time: Fiction, archaeology, and the elusive subject of prehistory'. *Australian Humanities Review* 65: 3.

Newell, J., L. Robin and K. Wehner. 2016. *Curating the Future: Museums, Communities and Climate Change*. London: Routledge.

Niemeier, U., F. Riede and C. Timmreck. 2021. 'Simulation of ash clouds after a Laacher See-type eruption'. *Climate of the Past* **17** (2): 633–652. https://doi.org/10.5194/cp-17-633-2021.

O'Neill, S.J. and N. Smith. 2014. 'Climate change and visual imagery'. *Wiley Interdisciplinary Reviews: Climate Change* **5** (1): 73–87. https://doi.org/10.1002/wcc.249.

Pagli, C. and F. Sigmundsson. 2008. 'Will present day glacier retreat increase volcanic activity? Stress induced by recent glacier retreat and its effect on magmatism at the Vatnajökull ice cap, Iceland'. *Geophysical Research Letters* **35** (9): L09304. https://doi.org/10.1029/2008gl033510.

Park, C. and H.-U. Schmincke. 1997. 'Lake formation and catastrophic dam burst during the late Pleistocene Laacher See eruption (Germany)'. *Naturwissenschaften* 84: 521–525.

Park, C. and H.-U. Schmincke. 2020. 'Multistage damming of the Rhine River by tephra fallout during the 12,900 BP Plinian Laacher See Eruption (Germany). Syn-eruptive Rhine damming I'. *Journal of Volcanology and Geothermal Research* 389: 106688. https://doi.org/10.1016/j.jvolgeores.2019.106688.

Pétursdóttir, Þ. and T.F. Sørensen. 2023. 'Archaeological encounters: Ethics and aesthetics under the mark of the Anthropocene'. *Archaeological Dialogues* 30 (1): 50–67. https://doi.org/10.1017/S1380203823000028.

Pitblado, B.L., S. Thomas, A. Wessman and S. Woodward. 2025. 'Retheorizing archaeological "artefacts" as "belongings"'. *Archaeologies*. https://doi.org/10.1007/s11759-025-09523-1.

Pomeroy, A.J. 2008. *Then it was Destroyed by the Volcano. The Ancient World in Film and on Television*. London: Duckworth.

Poschmann, M. and A. von Berg. 2020. 'Spuren in der Asche'. *Archäologie in Deutschland* 2020 (1): 60–61.

Posth, C. et al. 2023. 'Palaeogenomics of Upper Palaeolithic to Neolithic European hunter-gatherers'. *Nature* **615** (7950): 117–126. https://doi.org/10.1038/s41586-023-05726-0.

Price, N. 2015. 'The new MOMU: Meeting the family at Denmark's flagship Museum of Prehistory and Ethnography'. *Antiquity* **89** (344): 478–484. https://doi.org/10.15184/aqy.2015.16.

Rees, M. 2017. 'Museums as catalysts for change'. *Nature Climate Change* 7: 166. https://doi.org/10.1038/nclimate3237.

Reinig, F. et al. 2021. 'Precise date for the Laacher See eruption synchronizes the Younger Dryas'. *Nature* **595** (7865): 66–69. https://doi.org/10.1038/s41586-021-03608-x.

Riede, F. 2007. 'Der Ausbruch des Laacher See-Vulkans vor 12.920 Jahren und urgeschichtlicher Kulturwandel am Ende des Alleröd. Eine neue Hypothese zum Ursprung der Bromme Kultur und des Perstunien'. *Mitteilungen der Gesellschaft für Urgeschichte* 16: 25–54.

Riede, F. 2008. 'The Laacher See-eruption (12,920 BP) and material culture change at the end of the Allerød in Northern Europe'. *Journal of Archaeological Science* 35 (3): 591–599. https://doi.org/10.1016/j.jas.2007.05.007.

Riede, F. 2012. 'Tephrochronologische Nachuntersuchungen am endpaläolithischen Fundplatz Rothenkirchen, Kreis Fulda. Führte der Ausbruch des Laacher See-Vulkans (10966 v. Chr.) zu einer anhaltenden Siedlungslücke in Hessen?'. *Jahrbuch des nassauischen Vereins für Naturkunde* 133: 47–68.

Riede, F. 2014. 'The resettlement of northern Europe'. In V. Cummings, P. Jordan and M. Zvelebil (eds), *Oxford Handbook of the Archaeology and Anthropology of Hunter-Gatherers*. Oxford: Oxford University Press, pp. 556–581.

Riede, F. 2016. 'Changes in mid- and far-field human landscape use following the Laacher See eruption (c. 13,000 BP)'. *Quaternary International* 394: 37–50. http://dx.doi.org/10.1016/j.quaint.2014.07.008.

Riede, F. 2017a. 'Past-forwarding ancient calamities. Pathways for making archaeology relevant in disaster risk reduction research'. *Humanities* 6 (4): 79. https://doi.org/10.3390/h6040079.

Riede, F. 2017b. *Splendid Isolation. The Eruption of the Laacher See Volcano and Southern Scandinavian Late Glacial Hunter-Gatherers*. Aarhus: Aarhus University Press.

Riede, F. 2019. 'Doing palaeo-social volcanology: Developing a framework for systematically investigating the impacts of past volcanic eruptions on human societies using archaeological datasets'. *Quaternary International* 499: 266–277. https://doi.org/10.1016/j.quaint.2018.01.027.

Riede, F. et al. 2024. 'Geoarchäologisch motivierte Grabung am spätpaläolithischen Fundplatz Mühlheim-Dietesheim'. *hessenArchäologie* 2023: 65–69.

Riede, F. and Bazely, O. 2009. 'Testing the "Laacher See hypothesis": a health hazard perspective'. *Journal of Archaeological Science* 36 (3): 675–683. https://doi.org/10.1016/j.jas.2008.10.013.

Riede, F. and U. Kierdorf. 2020. 'The eruption of the Laacher See volcano (~13,000 years BP) and possible fluoride poisoning amongst contemporaneous wildlife and human foragers – Outline of a hypothesis and the way to test it'. *International Journal of Osteoarchaeology* 30 (6): 855–871. https://doi.org/10.1002/oa.2916.

Riede, F. and M. Mannino. 2024. 'Environmental archaeology'. In T. Rehren and E. Nikita (eds), *Encyclopedia of Archaeology*. 2nd ed. Oxford: Academic Press. pp. 260–266. https://doi.org/10.1016/B978-0-323-90799-6.00219-6.

Riede, F., F. Sauer and C. Hoggard. 2018. 'Rockshelters and the impact of the Laacher See eruption on Late Pleistocene foragers'. *Antiquity* 92 (365): e2. https://doi.org/10.15184/aqy.2018.217.

Riede, F. and P. Sheets (eds). 2020. *Going Forward by Looking Back: Archaeological Perspectives on Socio-Ecological Crisis, Response, and Collapse*. New York: Berghahn Books.

Riede, F. and J.M. Wheeler. 2009. 'Testing the "Laacher See hypothesis": tephra as dental abrasive'. *Journal of Archaeological Science* 36 (10): 2384–2391. https://doi.org/10.1016/j.jas.2009.06.020.

Ripple, W.J. et al. 2024. 'The 2024 state of the climate report: Perilous times on planet Earth'. *BioScience* 74 (12): 812–824. https://doi.org/10.1093/biosci/biae087.

Rowley-Conwy, P. 2000. 'Through a taphonomic glass, darkly: the importance of cereal production in prehistoric Britain'. In J.P. Huntley and S. Stallibrass (eds), *Taphonomy and Interpretation*. Oxford: Oxbow. pp. 43–53.

Sauer, F., D. Stott and F. Riede. 2018. 'Search for new final Palaeolithic rock shelter sites in the Federal State of Hesse'. *Journal of Archaeological Science: Reports* 22: 168–178. https://doi.org/10.1016/j.jasrep.2018.09.021.

Schmidt, P. et al. 2013. 'Effects of present-day deglaciation in Iceland on mantle melt production rates'. *Journal of Geophysical Research: Solid Earth* 118 (7): 3366–3379. https://doi.org/10.1002/jgrb.50273.

Schmincke, H.-U. 2004. *Volcanism*. Berlin: Springer.

Schneider-Mayerson, M. 2020. '"Just as in the book"? The influence of literature on readers' awareness of climate injustice and perception of climate migrants'. *ISLE: Interdisciplinary Studies in Literature and Environment* 27 (2): 337–364. https://doi.org/10.1093/isle/isaa020.

Schneider-Mayerson, M. et al. 2023. 'Environmental literature as persuasion: An experimental test of the effects of reading climate fiction'. *Environmental Communication* 17 (1): 35–50. https://doi.org/10.1080/17524032.2020.1814777.

Schneider-Mayerson, M., A. Weik von Mossner and W.P. Małecki. 2020. 'Empirical ecocriticism: Environmental texts and empirical methods'. *ISLE: Interdisciplinary Studies in Literature and Environment* 27 (2): 327–336. https://doi.org/10.1093/isle/isaa022.

Sigmundsson, F. et al. 2010. 'Climate effects on volcanism: influence on magmatic systems of loading and unloading from ice mass variations, with examples from Iceland'. *Philosophical Transactions of the Royal Society A: Mathematical, Physical and Engineering Sciences* 368 (1919): 2519–2534. https://doi.org/10.1098/rsta.2010.0042.

Snow, C.P. 1959. *The Two Cultures and the Scientific Revolution*. New York: Cambridge University Press.

Staley, D.J. 2002. 'A history of the future'. *History and Theory* 41 (4): 72–89. https://doi.org/10.2307/3590669.

Steffen, W. et al. 2015. 'The trajectory of the Anthropocene: The Great Acceleration'. *The Anthropocene Review* 2 (1): 81–98. https://doi.org/10.1177/2053019614564785.

Stewart, I.S. and D. Lewis. 2017. 'Communicating contested geoscience to the public: Moving from "matters of fact" to "matters of concern"'. *Earth-Science Reviews* 174: 122–133. https://doi.org/10.1016/j.earscirev.2017.09.003.

Street, M. 1986. 'Ein Wald der Allerödzeit bei Miesenheim, Stadt Andernach (Neuwieder Becken)'. *Archäologisches Korrespondenzblatt* **16** (1): 13–22.

Sutton, S. and C. Robinson. 2020. 'Museums and public climate action'. *Journal of Museum Education* **45** (1): 1–4. https://doi.org/10.1080/10598650.2020.1722513.

Svenning, J.-C. et al. 2024. 'Defining the Anthropocene as a geological epoch captures human impacts' triphasic nature to empower science and action'. *One Earth* **7** (10): 1678–1681. https://doi.org/10.1016/j.oneear.2024.08.004.

Tallavaara, M. and H. Seppä. 2012. 'Did the mid-Holocene environmental changes cause the boom and bust of hunter-gatherer population size in eastern Fennoscandia?'. *The Holocene* **22** (2): 215–225. https://doi.org/10.1177/0959683611414937.

Tipping, R. et al. 2012. 'Moments of crisis: climate change in Scottish prehistory'. *Proceedings of the Society of Antiquaries of Scotland* 142: 9–25.

Van de Noort, R. 2013. *Climate Change Archaeology: Building Resilience from Research in the World's Coastal Wetlands*. Oxford: Oxford University Press.

Van Dyke, R.M. and R. Bernbeck. 2015. *Subjects and Narratives in Archaeology*. Boulder, CO: University Press of Colorado. https://www.jstor.org/stable/j.ctt14bthnr.

Whallon, R. 2006. 'Social networks and information: Non-"utilitarian" mobility among hunter-gatherers'. *Journal of Anthropological Archaeology* **25** (2): 259–270.

Windhager, F., G. Schreder and E. Mayr. 2019. 'On inconvenient images: Exploring the design space of engaging climate change visualizations for public audiences'. Paper presented at the Workshop on Visualization in Environmental Sciences (EnvirVis). Porto: Eurographics. https://doi.org/10.2312/envirvis.20191098.

Zhu, H. et al. 2012. 'Structure of the European upper mantle revealed by adjoint tomography'. *Nature Geoscience* **5** (7): 493–498. http://www.nature.com/ngeo/journal/v5/n7/abs/ngeo1501.html#supplementary-information.

The Author

Felix Riede is German-born and British educated with a Ph.D. in archaeology from Cambridge University. Inspired by evolutionary and ecological theory and methods, he seeks to understand human-environment relations past, present and future. His work focuses on major tipping point episodes such as the end of the Pleistocene, extreme environmental events such as volcanic eruptions, earthquakes and tsunamis, novel ecosystems, and on the archaeology of the Anthropocene. After leaving Cambridge for UCL and then Aarhus University, Felix is now Professor, affiliated both with the Departments of Archaeology and Heritage Studies, and of Biology. At Aarhus, he founded the Centre for Environmental Humanities; he was also Visiting Professor at the Oslo Centre for Environmental Humanities and Visiting Scholar at the Oeschger Centre for Climate Change Research. Felix brings a distinct perspective on deep time and material relations to environmental history.

Chapter 14

BACK TO THE FUTURE: WEAVING CLIMATE HISTORY INTO NORDIC NATIONAL MUSEUM NARRATIVES

*Natália Nascimento e Melo, Bergsveinn Þórsson,
Felix Riede and Stefan Norrgård*

Abstract

Museums, perceived as trusted institutions, have significant potential for fostering public understanding of climate change. This study examines the integration of climate narratives in the permanent exhibitions of five Nordic national museums. The analyses focused on human-climate relations and the museums' role as societal reflection and change agents. Despite the growing academic emphasis on the importance of climate narratives in museum exhibitions, the analysis reveals that such narratives were scarce in the studied museums. When present, the narratives were fragmented, isolated themes rather than cohesive and integrated elements of historical storytelling. Additionally, this study explores barriers that prevent the integration of climate narratives in permanent exhibitions and proposes practical curatorial strategies for reframing existing historical narratives. The strategies aim to inspire collective action and critical engagement that position national museums as dynamic platforms for addressing climate challenges. Finally, the article highlights the necessity for curatorial practices to evolve and integrate inclusive and forward-looking narratives that empower audiences to confront the climate crisis.

Introduction

Adapting to human-induced climate change is a clear and vital societal challenge. It demands an urgent, multi-scalar response that simultaneously heeds processes that unfold at global, national and local levels. For people to feel motivated to engage and act, they need to relate to the topic personally and with affection, but also intellectually and based on knowledge. This could be achieved by linking

Natália Nascimento e Melo, Bergsveinn Þórsson, Felix Riede and Stefan Norrgård

the climatic past and present with the future, and by highlighting how personal and community experiences relate to global outcomes. Rees and Filho (2018) argued for a transformation in the way climate challenges are communicated, noting that, just as scientific evidence of climate change has been a key focus for policymakers, there is a need for public engagement. This requires rethinking communication channels and models that stress the complexity, extent and urgency of climate action. It requires that perspectives on the climate crisis and climate actions are made accessible and relatable to a diversity of stakeholders. While communicating the science is important, Allen and Crowley (2017) have argued that climate education needs to establish frames of relevance with the public through participation and engagement in their own communities. This requires that community-based knowledge and scientific knowledge are on equal footing (Honwad, Coppens, DeFrancis, Stafne and Bhattarai 2020). Innovative and creative, as opposed to purely intellectual and rational, approaches have regularly been contended as more effective ways to tap into the affective and personal domain of learning (Newell, Robin and Wehner 2017; McGhie, Mander and Underhill 2018).

The academic discourse increasingly emphasises the role of museums in addressing the climate crisis (Davis 2020; Þórsson 2020; Harrison and Sterling 2021). Subsequently, natural history museums have started to rethink their role in communicating climatic information; art museums have started providing spaces for creative and imaginative climate narratives; while museums devoted to climate and climate change are founded around the world (Oliveira et al. 2020; Newell 2020). Museums hold great potential to become important actors in raising awareness and encouraging climate change action but, as Robert Janes (2020) argues, they have not done enough compared to the gravity of the situation. Examples of dissatisfaction with museums' commitment to the climate crisis include criticism of oil sponsorship and, in recent years, the increasing number of climate activists' interventions inside museums (Demos 2023; Sharp 2022).

This paper addresses perceived gaps between the ideals rapidly emerging in contemporary museological literature and the presence and valence of climate change implications in permanent exhibitions of cultural history museums in the Nordics. It examines the presence of climate narratives in permanent exhibitions of Nordic national museums, i.e., the Historisk Museum in Oslo, the National Museum of Iceland, the National Museum of Finland[1], the Swedish History Museum and the National Museum of Denmark. While many specialised

---

1. In 2024, the National Museum of Finland was undergoing a renovation and expansion project, expected to end in 2027. The new museum will provide additional facilities for temporary exhibitions, events and other public activities (see www.kansallismuseo.fi/en/exhibitions/tulevaisuuden-kansallismuseo).

museums, such as the Klimahuset, the Finnish Museum of Natural History and the Nordiska Museet, address climate issues, national museums play a distinct role in shaping collective identity and historical narratives. By exploring how climate narratives are presented in current exhibitions, this study aims to stimulate reflections on the necessary efforts and possibilities to bring climatic narratives into focus within the permanent exhibitions of national museums.

National Museums' Narratives

According to Simon Knell (2011), national museums serve as cultural pillars for a shared sense of identity. National museums are places where representations of a shared history foster a collective consciousness. Based on the idea of shared knowledge, rituals and historical symbols, national museums create a sense of unity between individuals who have never met (Knell 2011; Anderson 2006 [1983]; Macdonald 2003). Thus, national museums display the political, economic, cultural and social history of a nation, but why are they not displaying how climate has impacted societal challenges and historical development?

Often originating in the eighteenth or nineteenth centuries, when nationalism swept through Europe and long before climate warming became topical, the task of national museums was to construct and support national narratives. The purpose was to highlight the relationship between the nation as an ethnic unity and its territory, and to underline the temporal depth of these relations. National museums were often charged with managing the past by caring for the tangible and intangible heritage of their nations. While history plays an important role in constructing national narratives, Simon Knell (2011: 9) argued that national museums are not strictly bound to the 'requirements of rigorous historiography' but navigate somewhere between the public's understanding of nationhood and the work of experts. However, historians tend to be resistant to embracing the full participation of other stakeholders as a legitimate alternative to history (Knell 2011).

Storytelling is quite firmly anchored in the permanent exhibitions of national museums and the narrative provides a sense of permanence, manifested through the materiality of collections accumulated over a long time. Although permanent exhibitions of national museums are not completely uncontested (Kjartansdóttir and Schram 2008), they remain impervious to change. This is partially because of the established narratives in the permanent exhibitions, but also because of the heavy financial, professional and political investments put into these nation-building projects.

The position of national museums makes them ideal places to raise awareness and initiate discussions on climate action. By linking to national or regional climate narratives, as opposed to global ones, the museums have the possibility to make otherwise abstract climatic issues relatable. However, if exhibitions continue to focus on traditional historical themes that do not expose visitors to climate narratives, they miss salient opportunities to engage with contemporary environmental realities and challenges.

Current Representation of Climate Narratives in Nordic National Museums

An analysis of the permanent exhibitions in the five studied museums reveals a noticeable absence of the physical climate and historical climate-society interactions. Only the Historical Museum of Oslo includes significant references to the physical climate and human-climate relations in both its permanent and temporary exhibitions. In the other museums, the focus is predominantly on human relationships framed by social, political, economic and religious perspectives. The importance of the past is a defining element in the studied museums and historical perspectives often take precedence over contemporary concerns in the permanent exhibitions. The museums emphasise chronological narratives and traditional themes that guide visitors through the historical development of the nation.

The almost twenty-year-old exhibition *Making of a Nation* at the National Museum of Iceland, presents its historical collection with the aim of answering the question: what makes a nation? The exhibition divides the history of Iceland into distinct periods in a chronological order from settlement during the Viking Age, to Christianity, being under Norwegian and Danish rule, to independence. Starting from the ninth and tenth centuries, the presence of environmental factors is noticeable. Pointing to archaeological findings on vegetation, the exhibition attempts to paint a picture of the landscape when the first settlers arrived. While the environment is present, the focus is mostly on nature as a resource and the bountiful landscape as an attraction for human settlement. Little is stated about climate conditions or the impact that human settlements had on the environment. As the story progresses, more emphasis is put on societal and governance structures. During the transition from paganism to Christianity, the exhibition focuses on power structures, while giving a glimpse into the peasant society and means of food production and export. From the eleventh to the sixteenth centuries, the focus is on changes in governance, and the nineteenth century is presented as the age of struggle for independence. When entering the twentieth century, technology, industrial advancements and the rise of consumer culture take central stage.

Some of the museums stimulate comparisons between the past and the present, inviting visitors to reflect on their own lives compared to the lives of

Back to the Future

people in the past. The descriptions of prehistory exhibitions at the National Museum of Finland and the Swedish Museum of History are examples of this approach. The Swedish History Museum's exhibition *Prehistories* poses a series of questions around the daily life and social customs of prehistoric communities. Questions such as 'What were relationships like between men and women?' and 'How did Roman objects end up in Swedish graves?'[2] invite the visitors to imagine the lives of early humans through the lens of social and cultural understanding. While the questions do not address human-environment interactions or the specific nexus of human-climate relationships, they could easily do so and embrace a broader context of environmental and ecological interrelations.

In addition, the analysis revealed a (subtle) engagement with forward-looking considerations and concerns about the future. These narratives, although limited, offer glimpses of a future-oriented perspective and make connections to contemporary issues through the lens of cultural heritage, sometimes inviting reflection on future challenges and values.

The temporary exhibition *Future Fragments* at the National Museum of Iceland stands out as a case where art, historical artefacts from the museum's collection and environmental themes come together to explore the human-environment relations. Featuring artworks by Þorgerður Ólafsdóttir, the exhibition results from her research on conceptions of cultural and natural heritage, using art to explore what the museum describes as 'planting the seeds of a future collection'. By displaying artefacts and ecofacts together, the exhibition aims to challenge conventional boundaries between culture and nature.[3] This approach encourages reflection on how heritage might be perceived in the future, emphasising an integrated view of natural and cultural elements that could foster bridges between past heritage and future ecological concerns.

Similarly, at the end of its permanent exhibition *History of Sweden*, the Swedish Museum of History incorporates elements that gesture towards the future implications of change by introducing a cabinet of curiosities from the 1740s. The text introducing this part of the exhibition on the museum's website is entitled 'A new interest in science'. It refers to the rise of industrialism and subsequent social changes, including socio-economic gaps and early feminist movements. These references may indicate an awareness of social change from historical antecedents and suggest visitor's reflections on the future by examining past changes. The description refers to 'reflections on the contemporary scene', connecting the past and present. However, despite the title, the text does not deal explicitly with science or nature-related topics, as might be expected in

---

2. https://historiska.se/utstallningar/prehistories/
3. https://www.thjodminjasafn.is/en/syningar/brot-ur-framtid

a cabinet of curiosities, nor does it refer to the contemporary environmental context or the climate crisis.[4]

The National Museum of Finland's current permanent exhibition highlights the political, economic and social history of Finland and the birth of the welfare state. While the museum is closed for renovation until 2027, its vision is to integrate future-oriented reflections more directly into its exhibitions. It seeks 'to strengthen cultural sustainability, to respond to the ever-growing demand for culture and to make the common cultural heritage even more accessible and usable'.[5] The museum also documents contemporary phenomena and issues relevant to environmental and societal shifts, such as 'environmental impact of plastic, civic activism related to climate change, the costume culture of Romani in Finland, and the impact of the COVID-19 pandemic on society'.[6] While these efforts recognise that cultural institutions must capture the evolving social and environmental realities, references to ecological and future-oriented issues, such as the impact of climate warming, are missing from the descriptions of the museum's permanent exhibitions. Likewise, it is unclear whether and to what extent the results of this documentation have been, or will be, integrated into the museum's permanent exhibitions.

The Historical Museum in Oslo provides some of the most explicit examples of permanent exhibitions that link the past with the present and incorporate future-oriented narratives. The exhibition *Heritage – Our Place in History* examines human relationships with heritage and displays historical and contemporary heirlooms to highlight the commonalities in human relationships with the past. The narrative encourages reflections on how choosing to elevate or forget heritage affects contemporary communities and the ways in which this determines how society addresses the future.[7]

The past is also linked to the future in the exhibitions *Control – Attempting to Tame the World* and *Collapse – Human Beings in an Unpredictable World*. These exhibitions explore the relationship between humans and the environment. They highlight questions that reflect upon the consequences of humanity's quest to control nature and provide examples of human adaptability in the face of environmental challenges. By focusing on human-animal relationships, *Control* explores the environmental and social implications of the attempt to control nature. Using the relationship between humans and cows as an example, the exhibition explores domestication, forms of interaction, religious beliefs, power

---

4. https://historiska.se/utstallningar/history-of-sweden/
5. https://www.uusikansallinen.fi/
6. https://www.kansallismuseo.fi/en/collections/kansallismuseo-dokumentoi-nykypaei-vaeae
7. https://www.historiskmuseum.no/english/exhibitions/heritage/index.html

Back to the Future

and futures. The themes invite visitors to ask, 'What are the consequences of seeing the world as under human control?' and whether this perspective has contributed to contemporary environmental crises such as deforestation and endangered species, but also to what extent industrial agriculture has impacted the climate crisis.[8]

Collapse juxtaposes historical and contemporary examples of human responses to environmental crises. It invites visitors to 'meet people who, in various ways and at different times, face forces of nature that threaten to collapse their world.' With stories ranging from the Stone Age pioneers of the Oslo Fjord to historical and contemporary human-environment relationships in Polynesia, the exhibition offers a broader historical perspective on human adaptability and vulnerability in the face of natural forces.[9] Using Polynesia as an example, the last part of the exhibition is meant to provide an alternative to cartesian dualisms between humans and nature, and highlight adaptation and integration both in the past and present. However, using an example from a distant region with an unfamiliar climate can create distance between visitors and the narrative.

Of the analysed exhibitions, *Collapse* and *Control* provided the most explicit references to climate change and human-climate relations. However, these references are situated within a broader narrative of human-environment interactions that often perpetuate a sense of antagonism, instead of connectedness, between humans and nature. While some museums express an intention to establish connections between past, present and future, this connection is limited and the focus of permanent exhibitions remains on the past. The limited engagement with climate narratives and sustainability highlights the untapped potential of museums as platforms for addressing pressing issues such as the climate crisis.

Barriers to Integrating Climate Narratives in Permanent Exhibitions

It is challenging to integrate climate histories and human-environment relations into permanent exhibitions. Limited resources (human and monetary), spatial constraints and a lack of specialisation make it difficult to implement substantial changes to longstanding displays. In contrast, temporary exhibitions often make such engagements readily feasible due to their lower costs and structural complexities. By their nature, however, temporary exhibitions are smaller and short-lived, meaning they have a limited outreach, especially when considering shared frames of reference across multiple visitor generations. Temporary exhibitions provide opportunities for innovative approaches and experimental

---

8. https://www.historiskmuseum.no/english/exhibitions/control/index.html
9. https://www.historiskmuseum.no/english/exhibitions/collapse/index.html

storytelling, but their transient nature prevents them from leaving a lasting imprint on the narrative framework of museums. As a result, temporary exhibitions fail to contribute to the long-term preservation of narratives and debates of societal concerns. This ephemerality jars with the urgency of the climate crisis. The long-term changes and consequences of climate warming require sustained and inter-generational efforts to raise awareness and inspire action over time.

It is worth considering whether the lack of climatic perspectives stems from the perception that climatic themes fall outside historical museums' field of expertise. Relegating the topic to natural history or science museums leads to the absence of climatic narratives and human-environment interactions in the museums' permanent displays. Overcoming this perception requires a shift in how historical museums perceive their role in fostering societal engagement with global challenges. It demands that museums recognise climate change as deeply intertwined with the history and future of human societies. Over the last decades, the rise and rapid growth of fields such as climate history, environmental history and environmental archaeology (Olsen, Farstadvoll and Godin 2024; White, Pfister and Mauelshagen 2018) highlight the relevance of climate in history and prehistory, which provides the means to bridge entrenched disciplinary boundaries. Growing awareness of the complex interplay between human societies, climate and the environment adds pressure on cultural history museums to integrate this knowledge into their constructed narratives.

National museums arguably carry the historical burden of the strict conceptual and institutional division between *nature* and *culture* that permeates Western academic thought and organisations. Even when permanent exhibitions explicitly address human-climate relations, the narrative tends to frame human-environment interactions in terms of the dichotomy between control and resistance. It presents nature as either an adversary to be dominated or an uncontrollable force that threatens the human ways of life. The presence of human-environment relations is more apparent in early historical periods, reflecting similar biases in school curricula (Riede 2022). This framing perpetuates an antagonistic Cartesian view that overlooks both the evident structuring role that climate plays in unfolding human social history as well as more nuanced and cooperative past human-environment relationships (Hussain 2024; Hussain and Baumann 2024; Hussain and Riede 2020). Given the significant challenges – both practical and conceptual – of transforming permanent exhibitions, it is necessary to consider whether historical museums can explore alternative narratives.

Cultural-historical museums often emphasise the preservation and interpretation of material culture and heritage as a means of representing human experiences through the preservation of the tangible traces of the past. Simon

Back to the Future

Knell (2011) stated that museums compress time and often present human subjects indirectly through the artefacts with which they have interacted. Knell used the open camera lens as a metaphor to exemplify how museums view the world. While the lens remains open, the contours of human presence become blurry and fade out leaving behind the material vestiges that remain the focus of museological representation. Focusing on the preservation of material culture can lead to a reluctance to address living, evolving issues such as climate change, which require a dynamic forward-looking and human-centred approach (Knell 2011).

Knell (2011) also noted that history, or more broadly the engagement with the past, remains central to the construction of national identity in national museums. However, the traditional and established idea that museums are custodians of the past limits their ability to engage with contemporary and dynamic subjects. As an urgent and unfolding issue, the climate crisis challenges the conventional museological emphasis on timeless preservation and demands a reimagining of historical narratives to include connections to the present and future.

Historical museums are well-positioned to present complex, multi-dimensional narratives. By integrating multiple perspectives and promoting critical reflection, based on past human-climate relations, historical museums could present a broader understanding of the complex impacts of contemporary climate change and inspire complementary approaches to the climate crisis. To achieve this, it is crucial to interrogate how perceptions of the role of museums as guardians of the past, as well as the antagonistic framing of culture and nature, shape decisions about what stories are told. Understanding these dynamics helps to identify opportunities to integrate climate narratives in ways that resonate with the missions of national historical museums. Such narratives could highlight moments of cooperation, conflict and decision-making in response to climate anomalies and extreme weather events, providing historical context to inform current and future responses. By taking a more integrated approach to human-environment relations, these museums could contribute to meaningful discussions about climate change, inspiring collective action and reflection grounded in historical perspectives.

Integrating Climate History: Six Curatorial Strategies

Incorporating climate themes does not require a complete transformation of permanent exhibitions. Instead, cultural history museums can adopt innovative strategies to revise historical narratives and embed climatic perspectives into their storytelling, while still preserving their core mission. Nonetheless, this process requires a conceptual shift in how exhibitions are framed and how climate narratives are integrated into historical and cultural contexts. We propose six climatic strategies as approaches to climate change to help museums in their attempts to incorporate climatic narratives into their permanent exhibitions. The strategies provide a conceptual and practical framework that aims to incorporate climatic perspectives into the museums' storytelling practices.

1. Mindful of Space and Time

National museums have the prerogative of covering the history of a specific geographical region over a long period of time. In showcasing the impact of climate warming and the role of historical climate variability and extreme weather, the examples should be from their respective countries. A Finnish museum should use examples from Finland, as this is the climate type that Finnish visitors are familiar with, can relate to and can understand. It might be tempting to use more dramatic examples from other parts of the world, but little is gained if the examples derive from places, cultures and climates of which the visitors have little prior knowledge. Thus, Nordic museums should use national, regional or local narratives. Prehistoric examples can be used, but examples from more recent history are recommended as it may be easier for visitors to identify with or understand changes in recent history than those from the distant past. It is important and advisable to create a climatic bond between then and now. This can only be done if the museum visitors can identify or at least recognise the examples in both space and time.

2. Highlighting Climate

Climate is often considered highly abstract and therefore visitors often connect more easily with the tangible and immediate concept of weather and its impact. As humans also have become more urban and alienated themselves from climate, a historical perspective should illuminate societal change and climatic impact over time. The Nordic countries, albeit not all, are unique from a European climatic perspective in that they all have a shared history of being impacted by ice and snow and they have four clearly distinguishable seasons. A climatic perspective could highlight seasonal change and how the short growing season during summer has affected agriculture or how snow and ice during winters, or naturally occurring spring floods, have impacted societies economically, culturally and socially, but also how societies slowly adapted to these limiting factors or tried to mitigate their impact. Comparative perspectives, then and now, especially if considering national museums, would let visitors learn how climate impact has affected nation-building parallel to technological and societal development. For

Back to the Future

example, icebreakers provided the means for Finland to engage in trading all year round, but this process of adapting to the ice waters of the Baltic Sea took over 100 years. Perspectives like these might demand a new type of research, or revisiting old research with a fresh perspective, and joint collaboration between the humanities and natural sciences.

3. Highlighting Weather

Extreme weather events could be effective tools for illustrating the impacts of climate warming (Cameron, Hodge and Salazar 2014), especially if seen from a long-term perspective and by including changes in frequency and intensity. By emphasising historical human responses to extreme weather events, museums can ground their exhibitions in relatable, real-world contexts. By focusing on past societies and adaptation strategies, mitigation attempts, resilience and policymaking, museums can highlight how societies in the past have been impacted by and dealt with extreme events and how future climate warming will change societies' vulnerability. In this volume, see for example Norrgård's article on ice jam events in Turku 1739–2024. An exhibition could discuss the occurrence of ice jams, depict them as extreme events and discuss their economic impact. Moreover, there is a cultural aspect as ice breakups and ice jams have been deeply rooted in past societies' climatic memory. The sounds, the excitement, the poems on ice breakups are why the citizens gathered in the thousands waiting to be entertained and experience the true force of nature. By linking historical responses to contemporary climate challenges or weather extremes, museums can inspire visitors to perceive climate issues as immediate and relatable, fostering a shared sense of responsibility and action towards climate justice. Each country, however, should try to focus on climatic variables or extreme events that are representable for that country, as the challenges vary from country to country.

4. Promoting Conversation, Not (Just) Information

Rather than focusing solely on providing information about climate change, museums should prioritise fostering a culture of communication by positioning themselves as spaces for community storytelling and active listening (Rees and Leal Filho 2018). By highlighting shared narratives, from stories of overcoming contemporary climate challenges to those of past cooperation in the face of extreme weather, museums can promote collective understanding among audiences. This approach reframes climate narratives as opportunities to engage communities in discussions about citizenship and collective responsibility. By emphasising conversation, museums can inspire action rooted in shared values and preferences.

Natália Nascimento e Melo, Bergsveinn Þórsson, Felix Riede and Stefan Norrgård

5. Integrating Local and Personal Stories

Climate warming is a societal challenge and research fields within the humanities need to engage more rigorously in crafting stories that foster engagement and understanding. By incorporating localised narratives – stories rooted in specific places and personal experiences – the humanities would help audiences to perceive climate challenges as immediate and familiar concerns rooted in everyday human experience instead of making them appear as distant abstractions (Daniels and Endfield 2009).

Memories and experiences of past extreme weather events and their impacts have shaped how communities understand and respond to climate uncertainty. This has resulted in diverse coping strategies that foster both conflict and cooperation in adapting to environmental challenges (Daniels and Endfield 2009). Including memories and experiences also exemplifies how extreme weather events may provoke long-lasting cultural trauma (Scarlett, Rothenberg, Riede and Holmberg 2023; Sztompka 2000). By linking historical experiences to contemporary climate issues, museums can offer valuable insights into human-environment interactions. Exhibitions that emphasise historical adaptability, resilience and vulnerability enable audiences to reflect on lessons from the past, and provide a foundation for reflections on future challenges and strategies.

Transitioning from rigid, professionalised historical narratives to more personal and communal storytelling fosters perceptions of relevance, emotional connection and deeper engagement with climate issues. Museums can incorporate these localised narratives and personal anecdotes, whether drawn from the past or the present, into their exhibitions. Such relatable, human-scale stories can resonate more deeply with audiences than broad crisis narratives, fostering connections between climate change and their own lives and local contexts.

6. Incorporating Future-Oriented Perspectives

Although seemingly paradoxical, the field of heritage studies – under which museum studies fall – is future-oriented (Holtorf and Högberg 2020). To promote climate resilience, museums can integrate narratives that connect history and experiences of the past with forward-looking perspectives. Daniels and Endfield (2009) argue for climate change narratives framed in historical-geographical contexts, using imaginative scenarios based on past stories and future possibilities. Museums can reframe permanent exhibitions by creating imaginative displays connecting history, memory and future-oriented narratives related to climate resilience. This invites visitors to reflect on the applicability of knowing about the past in imagining the future and taking collective action. The 'Great Acceleration' towards the Anthropocene can be tracked by the dramatic

Back to the Future

increase of certain materials and material flows in much the same manner as the previous transitions from stone to bronze and to iron are used to mark the onset of new ages. Extrapolations of these material flows and their climatic correlations make it clear that the world will look different in the decades to come. The materiality of these changes makes them patently exhibitable and the provocative addition of galleries dedicated to the Plastic Age also allows national museums to proactively include dimensions otherwise covered by the emerging climate museums (Newell 2020).

While transforming entire permanent exhibitions is challenging, small but impactful changes and innovative approaches can be implemented to intro-

Proposals for changes
How to include climate narratives without remodelling permanent exhibitions

REDESIGNING LABELS AND VOCABULARY
Adjusting exhibition texts to include terms and themes related to climate, contextualizing the environmental impacts and relations of historical objects.

EXPERIMENTING WITH INTERPRETIVE NARRATIVES
Introducing climate themes through interpretive labels or micro-stories embedded within existing exhibits, explaining the environmental context of objects.

ADDING LOCAL CONTEXTS
Incorporating regional or community-based climate stories alongside exhibits to build relatability. These additions can ground abstract climate issues in tangible, relatable experiences.

USING DIGITAL MEDIA
Integrating climate narratives into permanent exhibitions through digital supplements like videos, interactive maps, or augmented reality can enhance storytelling and provide greater context.

LINKING HISTORICAL THEMES TO THE PRESENT
Connecting past acts of resilience or crises with present-day climate challenges. This approach highlights the continuity between historical and present-day responses to environmental crises.

BUILDING ON EXISTING THEMES AND FRAMEWORKS
Introducing climate change as a topic related to other phenomena already discussed in exhibitions, such as war or migration, drawing parallels between historical events and contemporary challenges.

Figure 1. National museums can take a practical and incremental approach to integrating climate narratives into permanent exhibitions, making small changes without the need for major renovations.

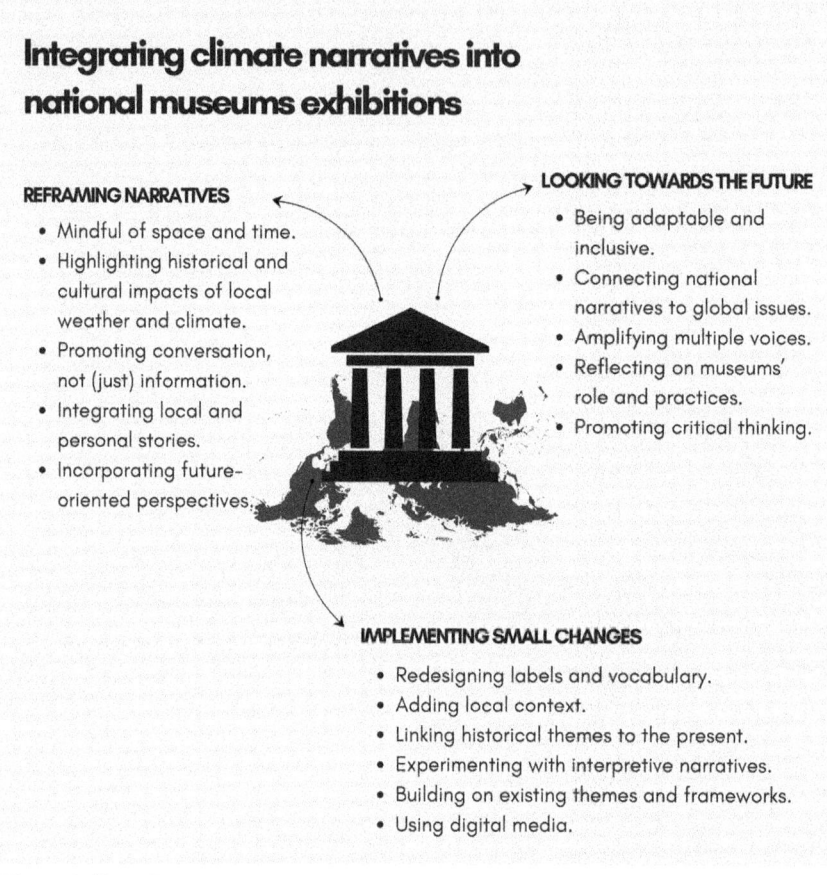

Figure 2. By reframing narratives, implementing small changes and linking history to a look towards the future, national museums can foster critical reflection, community engagement and informed action for climate justice.

duce climate narratives. Museums can focus on practical, low-cost, incremental interventions that integrate climatic narratives into the existing themes and exhibitions without requiring large-scale remodelling. It can be as simple as redesigning labels and refreshing the vocabulary to adding local context and linking historical themes to the present, or using digital media and experimenting with narratives and micro-stories.

Museums can combine small-scale approaches with a broader sense of reframing the narratives, using multiple strategies at once. By reframing how objects and stories are presented, national museums can bridge historical human-environment interactions with contemporary climate challenges, foster-

Back to the Future

ing a connection between past, present and future. By embracing narratives that inspire resilience, adaptability and climate awareness, they can act as catalysts for critical thinking and action. These incremental adjustments pave the way for national museums to serve as dynamic spaces for critical reflection and community engagement, highlighting the urgency and relevance of climate discourse.

Looking towards the future, national museums should be adaptable and inclusive, connecting local and national narratives to global warming while amplifying diverse voices. Museums should challenge conventional practices, provoke reflection on their own role and encourage audiences to engage with contemporary complexities through a lens informed by history and oriented towards the future.

Final Considerations

National museums have an important position as custodians of cultural heritage and agents of societal transformation. However, an analysis of Nordic cultural history and national museums reveals a significant gap between the narratives addressed in their permanent exhibitions and contemporary climatic, environmental and societal challenges. Human-climate references are largely absent and, even when present, the climatic narratives are fragmented, isolated stories or pieces of information rather than comprehensive themes that weave through the exhibitions. The current content of permanent exhibitions in the studied museums lags behind both academic discussions and societal needs. As institutions, museums need to evolve and take a more active role in addressing climate change. The museums need to rethink the narratives of their permanent exhibitions. They need state-of-the-art research and attempts to create dialogues that engage audiences. However, changing entire permanent exhibitions requires time and resources that museums often lack. With their limited capacities, should museums focus on changing permanent exhibitions to include climate narratives, or would alternative approaches (such as educational programmes, public events and temporary exhibitions) be more effective when engaging with complex and pressing societal issues? While practical constraints may limit the complete transformation of permanent exhibitions, the potential for national museums to serve as dynamic spaces for promoting climate awareness and action remains vast. Different approaches must be combined, and museums can take a positive direction in incorporating climate stories into permanent exhibitions by reframing already existing narratives, implementing small changes and looking towards the future.

Reimagining the role of national museums in the action for climate justice requires a shift from a singular, celebratory narrative of national history to one that embraces interdisciplinary collaborations, a tapestry of collective memory, inclusive storytelling and participatory dialogue. By combining the preservation of cultural heritage with contemporary narratives, museums are better equipped to promote understanding of pressing societal challenges. Including societal challenges transforms museums into dynamic spaces for public reflection and action. By weaving climate narratives into their core exhibitions, national museums can move beyond preservation and become active platforms for exploring how the past informs future action. This approach can position museums as critical spaces for reflecting on how historical and cultural legacies can inspire action and shape a more sustainable and resilient future.

Further research into curatorial practices and the inclusion of forward-looking narratives is crucial to support this transformation. While history is at the heart of national museums, their approach to history needs to evolve to meet current and future challenges. By embracing inclusive and forward-looking narratives, national museums can fulfil their potential as agents of societal change, empowering communities to confront climate issues with a sense of shared purpose and adaptability.

Bibliography

Anderson, B. 2006 [1983]. *Imagined Communities: Reflections on the Origins and Spread of Nationalism*. London: Verso.

Allen, L.B. and K. Crowley. 2017. 'Moving beyond scientific knowledge: Leveraging participation, relevance, and interconnectedness for climate education'. *International Journal of Global Warming* 12 (3–4), 299–312. https://doi.org/10.1504/IJGW.2017.084781

Aronsson, P. and G. Elgenius (eds). 2015. *National Museums and Nation-Building in Europe 1750–2010: Mobilization and Legitimacy, Continuity and Change*. London: Routledge.

Cameron, H., B. Hodge and J.F. Salazar. 2014. 'Representing climate change in museum space and places'. *Wiley Interdisciplinary Reviews: Climate Change* 4 (1): 9–21. https://doi.org/10.1002/wcc.200.

Daniels, S. and G.H. Enfield. 2009. 'Narratives of climate change: Introduction'. *Journal of Historical Geography* 35 (2): 215–222. https://www.sciencedirect.com/science/article/pii/S0305748808001102.

Davis, J. 2020. 'Museums and climate action: a special issue of museum management and curatorship'. *Museum Management and Curatorship* 35 (6): 584–586. https://doi.org/10.1080/09647775.2020.1842235.

Demos, T.J. 2023. 'Is there a future for just stop oil's art protests?', artreview.com, 11 January. Accessed 3 June 2024. https://artreview.com/is-there-a-future-for-just-stop-oil-art-protests/.

Back to the Future

Díaz-Andreu, M. 2007. *A World History of Nineteenth-Century Archaeology. Nationalism, Colonialism and the Past.* Oxford: Oxford University Press.

Díaz-Andreu, M. and T. Championn (eds). 1996. *Nationalism and Archaeology in Europe.* Oxon: Routledge.

Harrison, R. and C. Sterling. 2021. *Rethinking Museums for the Climate Emergency.* London: UCL press.

Holtorf, C. and A. Högberg (eds). 2020. *Cultural Heritage and the Future.* London: Routledge. https://doi.org/10.4324/9781315644615.

Honwad, S., A.D. Coppens, G. DeFrancis, M. Stafne and S. Bhattarai. 2020. 'Weaving strands of knowledge: Learning about environmental change in the Bhutan Himalayas'. *Nordic Museology* 30 (3): 62–73.

Hulme, M. 2008. 'The conquering of climate: Discourses of fear and their dissolution'. *Geographical Journal* 174 (1): 5–16. https://doi.org/10.1111/j.1475-4959.2008.00266.x.

Hussain, S.T. 2024. 'Feral ecologies of the human deep past: Multispecies archaeology and palaeo-synanthropy'. *Journal of the Royal Anthropological Institute* 0: 1–23. https://doi.org/10.1111/1467-9655.14152.

Hussain, S.T. and C. Baumann. 2024. 'The human side of biodiversity: Coevolution of the human niche, palaeo-synanthropy and ecosystem complexity in the deep human past'. *Philosophical Transactions of the Royal Society B: Biological Sciences* 379 (1902): v20230021. https://doi.org/10.1111/j.1475-4959.2008.00266.x.

Hussain, S.T. and F. Riede. 2020. 'Paleoenvironmental humanities: Challenges and prospects of writing deep environmental histories'. *Wiley Interdisciplinary Reviews: Climate Change* 11 (5): 1–18. https://doi.org/10.1002/wcc.667.

Knell, S. 2011. 'National Museums and the National Imagination'. In S. Knell, P. Aronsson and A. Amundsen (eds), *National Museums.* London: Routledge, pp. 3–28

Janes, R.R. 2020. 'Museums in perilous times'. *Museum Management and Curatorship* 35 (6): 587–598. https://doi.org/10.1080/09647775.2020.1836998.

Kjartansdóttir, K. and K. Schram. 2008. 'Re-negotiating identity in the National Museum of Iceland'. In K. Goodnow and H. Akman (eds), *Scandinavian Museums and Cultural Diversity.* New York: Berghahn Books. pp. 221–229. https://doi.org/10.1515/9781789204049-025.

Kohl, P.L. and C. Fawcett (eds). 1995. *Nationalism, Politics and the Practice of Archaeology.* Cambridge: Cambridge University Press.

Macdonald, S. 2003. 'Museums, national, postnational and transcultural identities'. *Museum and Society* 1: 1–16.

Madsen, P.K. and L. Jørgensen (eds). 2010. *The National Museums in a Globalised World: A Conference on the Bicentenary of The National Museum of Denmark, Copenhagen May 21–22 2007.* Copenhagen: National Museum of Denmark.

Natália Nascimento e Melo, Bergsveinn Þórsson, Felix Riede and Stefan Norrgård

McGhie, H., S. Mander and R. Underhill. 2018. 'Engaging people with climate change through museums'. In W. Leal Filho, E. Manolas, A. Azul, U. Azeiteiro and H. McGhie. (eds), *Handbook of Climate Change Communication: Vol. 3. Climate Change Management*. Springer, Cham. https://doi.org/10.1007/978-3-319-70479-1_21.

Newell, J. 2020. 'Climate museums: powering action'. *Museum Management and Curatorship* **35** (6): 599–617. https://doi.org/10.1080/09647775.2020.1842236.

Newell, J., L. Robin and K. Wehner. 2017. *Curating the Future: Museums, Communities and Climate Change*. New York: Routledge.

Oliveira, G., E. Dorfman, N. Kramar, C.D. Mendenhall and N.E. Heller. 2020. 'The Anthropocene in natural history museums: A productive lens of engagement'. *Curator: The Museum Journal* **63** (3): 333–351.

Olsen, B.J., S. Farstadvoll and G. Godin (eds). 2024. *Unruly Heritage: Archaeologies of the Anthropocene*. New York, NY: Bloomsbury Academic.

Parkins, H. 1997. 'Archaeology and nationalism: Excavating the foundations of identity'. *Nations and Nationalism* **3**: 451–458. https://doi.org/10.1111/j.1354-5078.1997.00451.x.

Riede, F. 2022. 'Deep history curricula under the mandate of the Anthropocene. Insights from interdisciplinary shadow places'. *FECUN* **1**: 172–185. https://doi.org/10.7146/fecun.v1i.130246.

Rees, M. and W. Leal Filho. 2018. 'Disseminating climate change: The role of museums in activating the global public'. In W. Leal Filho, E. Manolas, A. Azul, U. Azeiteiro and H. McGhie (eds), *Handbook of Climate Change Communication: vol. 3. Climate Change Management*. Springer, Cham. pp. 319–328. https://doi.org/10.1007/978-3-319-70479-1_20.

Scarlett, J.P., M.A.W. Rothenberg, F. Riede and K. Holmberg. 2023. '"Dark heritage" – landscape, hazard, and heritage'. In R. Jigyasy and K. Chmutina (eds), *Routledge Handbook on Cultural Heritage and Disaster Risk Management*. London: Routledge. pp.225–243. https://www.routledge.com/Routledge-Handbook-on-Cultural-Heritage-and-Disaster-Risk-Management/Jigyasu-Chmutina/p/book/9781032274805#.

Sharp, C-M. 2022. '"Shell is proud to present… *The Spirit Sings*": Museum sponsorship and public relations in oil country'. *Museum & Society* **20** (2): 172–189. https://doi.org/10.29311/mas.v20i2.3780.

Sztompka, P. 2000. 'Cultural trauma: The other face of social change'. *European Journal of Social Theory* **3** (4): 449–466. https://doi.org/10.1177/136843100003004004.

Þórsson, B. 2020. 'Curating climate – museums as contact zones of climate research, education and activism'. *Nordic Museology* **20** (3): 2–13. https://doi.org/10.5617/nm.8625.

White, S., C. Pfister and F. Mauelshagen (eds). 2018. *The Palgrave Handbook of Climate History*. Palgrave Macmillan, London.

Back to the Future

The Authors

Natália Nascimento e Melo is a researcher at the University of Évora and a collaborator at the Institute of Contemporary History (IHC|IN2PAST). She holds a Ph.D. in History and Philosophy of Science with a specialisation in Museology. Her research explores the intersections of climate, museum narratives and public engagement, with a focus on how the Anthropocene, climate change, and human-climate relations are represented in museums. She is interested in the role of arts in fostering dialogues about science and societal issues, and how material culture shapes public perceptions of environmental change and human-environment relations. She also works on projects related to public history, placemaking and transdisciplinary approaches to citizen participation.

Bergsveinn Þórsson holds a Ph.D. in Museology and is an Associate Professor at Bifröst University in Iceland and Programme Manager for Public Administration. He teaches courses on cultural management, sustainability and the sociology of climate change. His research focuses on the Anthropocene in museums, climate and sustainability implementation in cultural organisations, and speculative future thinking. With a strong interdisciplinary approach, he explores how cultural institutions navigate uncertainty and address global challenges. He is affiliated with the CoFutures Research Group and is currently a Fellow on the project *The Nordic Little Ice Age (1300–1900) Lessons from Past Climate Change (NORLIA)* at the Centre for Advanced Study at the Norwegian Academy of Science and Letters. In addition, he serves as Editor-in-Chief of Nordic Museology and co-curated the Beyond Barcode exhibition at the Interkultural Museum in Oslo, which explored locally generated future scenarios for the city.

Felix Riede is German-born and British educated with a Ph.D. in archaeology from Cambridge University. Inspired by evolutionary and ecological theory and methods, he seeks to understand human-environment relations past, present and future. His work focuses on major tipping point episodes such as the end of the Pleistocene, extreme environmental events such as volcanic eruptions, earthquakes and tsunamis, novel ecosystems, and on the archaeology of the Anthropocene. After leaving Cambridge for UCL and then Aarhus University, Felix is now Professor, affiliated both with the Departments of Archaeology and Heritage Studies, and of Biology. At Aarhus, he founded the Centre for Environmental Humanities; he was also Visiting Professor at the Oslo Centre for Environmental Humanities and Visiting Scholar at the Oeschger Centre for Climate Change Research. Felix brings a distinct perspective on deep time and material relations to environmental history.

Stefan Norrgård is a senior researcher and climate historian at the Department of History at Åbo Akademi University in Turku, Finland. Subsequent to reconstructing climate in West Africa during the 1700s, his research interests have centred on riverine ice breakups in Finland. He has reconstructed spring ice breakups for

Natália Nascimento e Melo, Bergsveinn Þórsson, Felix Riede and Stefan Norrgård

both the Aura River (Turku) and the Kokemäki River (Pori) between the 1700s and 2000s. He has several publications on ice breakups but his research field also covers historical climate adaptation processes and meteorological observations in Finland and Sweden in the 1700s. His ongoing research project, founded by the Kone Foundation, investigates climate, culture and society in Finland in the 1700s.

INDEX

A

abandonment of farms or settlements 9, 11, 12, 16, 77, 85, 87, 89–91, 93–97, 107, 116, 136–37, 141, 301
ACCORD project 193, 195, 202
actor-network theory 149
adaptation; adaptability 2–3, 10, 13–15, 20, 47–48, 59, 68–70, 77–78, 80, 94–97, 103, 104, 113–15, 118, 123–146, 147–164, 170, 181, 216, 218, 228, 317, 322–23, 326–28, 331
agriculture *see also* farm 4, 9–11, 15, 16, 18, 63, 83, 85, 86, 88, 92–95, 108, 130, 139, 169, 176, 181, 190, 209–36, 266–70, 29, 323, 326
Akershus, Norway 151, 158, 224–27, 229
Andersen, P.T. 278, 290
annals 103, 109, 110, 117–18, 189–90
Anthropocene 294, 297, 305–06, 308, 328
anthropogenic influence 1, 15, 16, 19, 47, 49, 61, 67, 124, 125, 127, 170, 176, 237, 239, 297, 308
archaeology 3, 10, 60–62, 67, 80, 116, 125–27, 150, 231, 263, 266, 296, 306, 324
archives 2, 17
architecture 16, 59, 70, 95, 96, 123, 126, 137, 139, 141, 147–64
Arctic 8, 30, 43, 79, 189, 192–93, 203
  Circle 29, 30
artefact 68, 69, 85, 92, 116, 297, 303, 306, 308, 321, 325
ash *see also* tephra 104, 105, 110, 114–17, 249–50, 294, 298, 301–02, 306–07
Aura river, Finland 15, 237–58
Aurasilta bridge, Turku, Finland 237, 238, 240, 245–51, 253
awareness 16, 105, 109, 114, 115, 127, 213, 238, 261, 265, 271, 318, 319, 321, 324, 331

B

Baltic Sea 40, 176, 239, 327
barley 85, 214, 216, 218, 219, 228
barometer 31, 36
Bárðarbunga eruption, Iceland 104, 115
baseline, climatic / ecological 1, 17, 302
Beda Venerabilis 112, 113
Bergen 40, 68, 151, 161
  Museum 266, 267
Bergþórsson, Páll 192–93
Berlin, Johan Daniel 33, 34, 39
Bjerknes, Vilhelm 266
Bjørkvik, Halvar 268
Black Death 8, 12, 92
Borgarfjörður, Iceland 110
Brahe, Tycho 132
brick 130, 133, 134, 136, 147, 150, 152, 154, 156–58, 162
bridge 135, 237, 239, 240, 243, 248–50, 253–54
Bronze Age 60, 64, 66, 70, 81, 266, 270
Bugge, Alexander 265, 266
Bugge, Niels 129
Bugge, Thomas 34, 39
built environment 15, 16, 96, 132, 147–50
Bull, Edvard 7, 266, 267, 268

C

Carrard, Philippe 264, 270
carrying capacity 61, 65
castle 134, 148, 150–53, 158
catastrophe *see also* disaster 11, 118, 306
causation; causality 7, 12, 13, 20, 61, 62, 67, 181, 209, 230, 262–64, 266, 301
cellar 15, 147, 148, 150, 152–54, 156–62
Celsius, Anders 6, 31, 33, 39
Celsius thermometer 29, 32, 33
cemetery 79, 81, 83, 85, 87, 92, 96, 97
Chambers, Robert 103, 104, 118
Childe, V.G. 61
Christian IV, King of Denmark 132, 158
Christiania (Oslo) 32, 37,
Christianity 92, 103, 106–08, 113, 117, 118, 320

Index

church 83, 103, 113–16, 118, 152, 155, 160, 161, 282
climate change *see also* global warming 1, 2, 4, 6, 7, 10–12, 14, 16–18, 30, 43, 46–49, 59, 60, 79, 124–27, 147–52, 158, 162, 189, 193, 213, 218, 230, 261–71, 276, 291, 296–300, 302–03, 306–08, 317–18, 322–29, 331
climate crisis 11, 12, 16, 124, 125, 217, 308, 309, 317, 318, 322–25, 328
Climatic Research Unit, University of East Anglia 8, 193
climatology 1, 3, 4, 6, 7, 8, 10, 39, 104
historical 1, 7, 8, 10
coast 3, 4, 14, 15, 30, 40, 46, 65, 68, 69, 77, 78, 81, 83, 85, 87, 114, 128, 132, 141, 186–87, 189, 192, 194, 196, 202, 218, 227
colonialism 1
comfort 147, 148, 151, 152, 156, 158, 162
commons 169
communication 2, 3, 20, 83, 85, 87, 88, 93, 94, 96, 124, 224, 295, 306, 307, 318, 327
compensation 137, 169, 179
computer modelling 1
cooling 4, 9, 12, 41–43, 47, 60, 66–67, 77–79, 93–94, 174, 201, 202, 209, 210, 217, 218, 221, 292, 299, 303, 308
Copenhagen 29, 34, 37, 39, 44, 46–48, 137, 210
court records 167–69, 171, 174–81
Covid-19 309, 322
crisis *see also* climate crisis 8, 11, 12, 15, 16, 20, 78, 80, 95, 114, 132, 213, 217, 223–31, 296, 328
culture 2, 3, 11, 12–14, 18, 19, 20, 61, 62, 67, 68, 77, 80, 103, 118–19, 124, 126, 127, 134, 136–37, 139–41, 149, 266, 275, 291, 295–99, 301–03, 307–09, 319, 321–22, 324–28, 331–32
curator; curation 303, 306, 308, 317, 325, 332

D

Danish Meteorological Institute (DMI) 8, 37
Danish Scientific Society 33–36
dataset 1, 43, 44, 47, 60–65, 193, 213–15, 217, 220
declensionism 20
demography; demographics *see also* population 7, 9–12, 15, 59–62, 65–69, 71, 93, 210, 224–25, 275, 299, 301–02
Denmark 3–4, 9, 12, 13, 15, 30, 34–37, 39, 46, 48, 71, 118, 123–46, 190, 225, 298, 299, 303, 308, 318
determinism 18, 263, 265, 271
development, economic / societal 11, 12, 17, 18, 92, 93, 147, 150, 162, 319, 326
disaster *see also* catastrophe 13–16, 20, 67, 80, 103, 104, 114, 116, 167–71, 178–82, 230, 294, 308
disease 7, 9, 12, 79, 108, 202, 210, 213, 216, 227, 228, 229
Domkirkeodden, Norway 152–53, 155, 159–61
drain 147–56, 162
drainage 15, 84, 86, 88, 95, 128, 130, 239
drought 4, 10, 14, 93, 174, 176, 181, 227, 264
Dybdahl, Audun 9
dynamite 250–51, 255

E

East Greenland Current 6, 46, 187, 192–93
ecology 16, 80, 126–27, 295
economy 7, 10, 12, 14, 18, 29, 30, 59, 60, 68, 71, 80, 85, 92–96, 116, 136, 167, 169–70, 176, 179, 215, 221, 227, 229, 241, 306, 307, 319, 320, 322, 326, 327
Eddic poetry 79, 107
Ehrenheim, F.V. von 7
Eidsberg, Norway 34, 35, 212–14, 225–26, 228
Einarsson, Oddur 6
El Niño 132

Index

Eldgjá eruption, Iceland 103, 106–08, 118, 119
Elster, K. 277, 283
Enlightenment 7, 31, 33, 210, 212
epidemic *see* disease
Eriksdotter, Gunhild 148
Eriksson, Magnus 178
Ersland, Geir Atle 269
Esmark, Jens 40, 265
Eura, Finland 175
excavation 64, 81, 83–88, 90, 94, 115, 116, 123, 130–37, 139, 141, 151–56, 296
exchange *see also* trade 81, 85, 96, 16, 181
exhibition 16, 17, 20, 294–95, 297–99, 302–09, 317–32
export 168, 224, 227, 320
extreme weather 14, 30, 108, 132, 174, 200, 227, 229, 325–28

F

Fahrenheit, Daniel Gabriel 32
Fahrenheit temperature scale; thermometer 29, 33–34
Falk, Oren 108
famine 9, 12, 60, 61, 79, 108, 132, 199, 200, 202, 210, 225, 227
farm; farming *see also* agriculture 9, 12, 15, 16, 59–76, 85–86, 92, 95, 107, 110, 115, 116, 128, 139, 169, 180, 181, 209–36
Faroe Islands 37
Federmessergruppen people 299
feedback 14, 127, 170
Fimbulwinter 79
Finland 3, 4, 9, 10, 12, 15, 30, 32, 33, 35, 38–40, 46, 167–85, 237–57
 National Museum of 319, 321–22, 326, 327
Finnish Meteorological Institute 5
Finnish Society of Sciences and Letters 38, 40
Finnmark, Norway 40
Finnsson, Hannes, bishop 6
fire 14, 15, 39, 40, 104, 105, 107–19, 152, 167–85, 199, 240, 288
fireplace 158

fishing; fisher 68–69, 81, 83, 85, 88, 92, 95, 189, 190, 196, 198, 228, 229, 240
FitzRoy, Robert, 36
Flateyjarbók 111
Fleischer, Baltzar Sechmann 228
flood 13, 14, 84, 108, 113, 114, 115, 123, 137, 140, 170, 201, 211, 225, 237–41, 244–48, 250, 254, 264, 292, 300, 326
fluctuation
 climate 4, 41, 42–46, 67, 174, 187, 230
 demographic 59–63, 301
foraging; forager 59–76, 81, 83, 85, 298, 299, 301, 303
forecast 1, 17, 36–38
forest 86, 116, 167–85, 284
forest fire ordinance 177
forestry 215, 227
Fram Strait 193
Franklin, Benjamin 266
frost 4, 13, 14, 48, 150, 158–59, 161, 197, 212, 219, 221, 225, 285
Frydendahl, Knud 8
future 1, 2, 14, 16, 17, 19–21, 34, 49, 187, 194, 212, 238, 254, 262, 294–95, 298, 303–07, 318, 321–28, 330–32

G

Gammelå, Denmark 129–30, 134–37, 140–41
Gerzensee Oscillation 299, 301, 302
Ginges, Andreas 35
Gísladóttir, Guðrún Alda 116
glacier 2, 16, 38, 43, 106, 187, 192, 200, 264, 267, 275–293
global warming *see also* anthropogenic influence; climate change 1, 17, 19, 43, 46, 47, 49, 67, 303, 324, 331
governance 3, 320
government 37, 38, 126, 168, 169, 177, 190, 224, 228, 231, 262
Grágás 103–04, 113–15, 118
grain 10, 116, 169, 210, 214–19, 221, 224, 227–29, 284
Gran, Norway 155, 156
Grannavollen, Norway 152, 160

Index

grass; grassland 86, 216
grazing 85, 88, 90, 95, 96, 229
Great Acceleration 294, 328
Great Famine 12, 132
Greenland 30, 35–37, 40, 43, 105, 106, 186, 187, 196, 266, 267
  ice-cores 299–300
Grindavík, Iceland 196
Growing Degree Days (GDD) 16, 209–10, 217–26, 229, 230
growing season 4, 61, 93, 209, 214, 217–19, 221, 326

H

Hagen, Anders 267
Halinen Dike, Aura River, Finland 240, 247, 248, 253
Hallgrímsson, Jónas 36
Hallsson, Gizurr 117
Hällström, G.G. 39
Hamar, Norway 152–53, 155–61
Hammerfest, Norway 40
Hanko, Finland 40
Hansteen, Christoffer 40
Haparanda, Sweden 38
harbour 77, 84, 88, 94, 95, 196, 240–42, 246, 253
Hardangervidda plateau, Norway 66
harvest 4, 9–10, 211, 213–17, 219, 224, 226, 228, 285
  failure 9–10, 12, 16, 108, 209, 210, 212, 217, 221, 224, 225–28, 230–31
Hasund, Sigvald 8, 9, 266, 267, 270
Hauge, Olav H. 278, 280, 288
Hauksbee thermometer 29, 32
Havnevik, I. 278, 285, 289
hazard *see also* risk 10, 13–15, 20, 107, 113–15, 119, 127, 169, 170
Heitbréf Eyfirðinga 114
Hekla eruption, Iceland 103, 104, 109–10, 115–18
Helsinki 38, 39, 40, 44–48
Herstad, John 227, 229
Hinsch, Erik, 68–69
historical climatology *see* climatology

historiography 6, 261, 262, 265, 270, 271, 275–93, 319
history, public 16, 261–74
Hoff, Norway 160
Hofmo, Gunvor 275–78, 280–81, 287–88, 291
Holocene 3, 15, 59–76, 92, 209, 304–05
Holocene Thermal Maximum 265–67
Horrebow, Peder 34
Huhtamaa, Heli 10, 170, 181
Huittinen, Finland 176
Hungrvaka 118
hunting 66, 81, 107, 229
Huntington, Ellsworth 265–66

I

ice 14, 43, 67, 81, 106, 108, 276, 278, 286, 291, 326
  breakup 237–57, 327
  core 2, 79, 104, 105, 106, 296, 299–300
  jam 15, 237–57, 327
  lake 2
  sea 6–8, 14, 46, 60, 79, 186–208, 266
  river 237–57
iceberg 279–80, 286–92
icebreaker 250, 254, 327
ICEHIST project 193–94
Iceland 6, 7, 8, 10, 12, 14, 15, 29, 30, 33–37, 40, 46, 103–122, 186–208, 266, 267, 302
  National Museum of 318, 320, 321
Icelandic Meteorological Office (IMO) 38, 39
Icelandic Society of Letters 33
identity 138, 139, 170, 261, 275, 276, 319, 325
Ilulissat, Greenland 31, 36, 37, 42, 43
imaginary; imagination 20, 306, 307, 318, 328
import 136, 224, 227
Indigenous knowledge 10
Indre Østfold, Norway 211, 212, 218, 219, 221, 222–24, 229–30
Industrial Revolution 61, 167,
industrialisation 241, 321

Index

instrument, meteorological 6, 8, 29–55, 187, 191, 195, 214
interdisciplinarity 2, 3, 9, 10, 17, 18, 20, 103, 295, 301, 302, 332
Intergovernmental Panel on Climate Change (IPCC) 48, 238
Irminger Current 46, 48, 187
Iron Age 15, 66, 70, 83–85, 88, 94, 217, 265–66
Isidore of Seville 111

J

Jochumsson, Mattías 189, 202
 Hafísinn 202
Jónsson, Arngrímur 6, 196
Jónsson, Björn 196
Jónsson, Jón, the elder 199
Jónsson, Páll, Bishop 117
Jónsson, Trausti 8, 192
journal, weather 30, 32–34, 49, 237, 241
Joys, Charles 267
Julin, Johan 33
Jurin, James 32, 33, 49
Jutikkala, Eino 9
Jutland, Denmark 123–146, 210

K

Kajaana, Finland 40
Kalajoki, Finland 174, 176, 180
Kalm, Pehr 34, 266
Karelia 30
Karl XI, King of Sweden 179
Karlberg, Jakob 33
Katla eruption, Iceland 104, 106
Kelly, R.L. 61
Ketilsson, Magnús 35
King's Mirror 103, 111–14, 119
Kington, J.A. 31
Kink, Pehr 32
Knell, Simon 319, 325
Koch, Lauge 8, 192–94, 201
Kokemäki, Finland 55, 175
Kristiansund, Norway 4, 87
Kuopio, Finland 40

L

Laacher See volcano 294–95, 298–99, 301–06
Laki fissure, Iceland 105, 200, 219
LALIA (Late Antique Little Ice Age) 4, 11, 261, 264, 265, 270
Lamb, Hubert 8, 268
Landnám Tephra Layer, Iceland 105–06
Landnámabók 105, 107, 108, 116
Late Medieval Agrarian Crisis 8, 9, 12, 132, 267, 269, 270
latitude 2, 3, 30, 46, 48, 77, 218, 219, 299, 300
Latour, Bruno 149
law; legal *see also* legislation 20, 177–79, 181
 Icelandic 103, 104, 109, 111, 113–15, 118
lava 104–08
Lebesby, Norway 40
Leche, Johan 33, 49, 241, 244, 247
legislation 167–70, 177–79, 181
Lemvig, Jutland, Denmark 210
Levetzau, Albert Philip 229
LIA-like 60, 62, 64–67, 71
Lievog, Rasmus 35
Lilja, Sven 9
Lillehammer, Arnvid 268
literature 3, 16, 33, 212, 262, 263, 265, 275–93
Little Ice Age (LIA) 3–5, 8–11, 14–15, 19, 59–60, 67, 132, 147–64, 168, 186, 202, 203, 209, 261, 264, 265, 267–71
livestock 81, 110, 115, 132, 214–16, 228, 229
Lofthus rebellion 232
Longyearbyen, Svalbard 31, 32, 43
Løvlien, Astrid 227
Lowenthal, David 295
Lower Satakunta, Finland 168–69, 171, 175, 176, 179, 180, 181
Lucretius 112–13
Lund, Sweden 32–33
Lunden, Kåre

Index

M

Magnús lagabætir, King of Norway 111, 113
Magnus, Olaus 1, 6
Magnússon, Gísli 197
Malthus, Thomasx 61
Mann-Kendall test 29, 43, 47
Mannheim, Germany 34, 212, 213
Mariehamn, Åland islands 40
marginality 2, 10, 17, 79, 220, 267, 270
maritime climate 30, 40, 46, 48
Marmøy, Reidar 267
Martinsilta Bridge, Turku 240, 246
material culture 2, 3, 126, 134, 137, 139, 140, 149, 296, 299, 324–25
MCSPD (Monte Carlo summed probability density method) 62–64
medicine 7, 210
Medieval Climate Anomaly 3, 4
Medieval Warm Period 4, 8, 150, 261, 264–66, 268–70
Mediterranean 11
Mesolithic 59, 60, 63, 65, 66, 68, 70
meteorology 6, 7, 8, 14, 16, 29–55, 189, 209–13, 217, 219, 229, 230, 241
migration 11, 12, 13, 14, 71, 93, 105, 107, 229, 265
Migration Period 81, 85, 87–88, 93, 96, 97
military 11, 13, 37, 92, 93, 111, 227
mitigation 13, 20, 95, 103, 110, 115, 124–26, 136, 139, 237, 238, 241, 250, 254, 326, 327
Mjøsa, Lake, Norway 152, 155
mobility 59, 62, 71
Moesgaard Museum, Aarhus 298, 303, 305, 307
Mohn, Henrik 37, 40
Munch, Andreas 275–83
museums 3, 16, 294–36
Musschenbroek, Pieter van 211
mythology 79, 107, 118, 119, 285

N

Nansen, Fridtjof 8
Napoleonic wars 231
narrative 11, 16, 17, 20, 103, 108–09, 116, 118, 124, 231, 261–74, 275–76, 294–98, 303, 306–09, 317–36
nationalism 1, 271
Neolithic 59, 60, 63, 65–71, 266, 270
Nissum Fjord, Denmark 128, 130, 133–35, 140–41
Nordli, Øyvind 8
Norlind, Arnold 7
Nørre Vosborg, Denmark 123–45
North Atlantic 6, 7, 8, 48, 104, 193, 268
Current 4, 30, 187–88
North Ostrobothnia, Finland 168–71, 174–77, 179–81
North Sea 8, 13, 128, 140, 141
Norway 3, 4, 7, 8, 9, 15, 16, 30, 32–35, 37, 40, 46, 49, 59–76, 77–102, 109, 147–64, 209–36, 261–74, 275–93
Norwegian Meteorological Institute (MET Norway) 37, 39
Nuuk, Greenland 35, 37

O

oats 85, 214–21, 228–30
Oddaverjaannáll 118
Ólafsdóttir, Þorgerður 321
Öræfajökull eruption, Iceland 104, 115
Oripää, Finland 239
Ørland, Norway 77, 78, 83–85, 88, 95
Orning, Hans Jacob 263, 269
Oslo 32, 37, 40, 44, 46, 47, 48, 50, 151, 155, 156, 158, 160, 161
Fjord 59, 65, 66–70, 323
Historical Museum 318, 320, 322
Observatory 40
University 3, 265, 267, 269
Østfold, Norway *see also* Indre Østfold 211, 216, 224, 229
Oulu, Finland 33, 35, 40
Øverland, Arnulf 275–76, 278–79, 281, 284
Øvre Borgesyssel, Norway 225, 227

P

palaeoclimatology 9, 264, 270
palaeodemography 59, 60, 68

Index

Páls saga byskups 117
Pálsson, Gísli 108
Pálsson, Sveinn 35
pathway 3, 14, 15, 17–20, 179, 295, 302, 307
patriotism 275, 279–81, 285
Paus, Johan Ludvig Christian Ludvigsen 228
peasant 15, 139, 167–85, 320
permafrost 43
Pettersson, Otto 7
plate, tectonic 104
Pleistocene 81, 299, 300, 303–08
Pliny the Elder 112
plurality 2, 15
poetry 2, 3, 14, 16, 79, 107, 189, 196, 202, 275–93, 327
politics; political 11–14, 18, 19, 29, 30, 80, 92–96, 105, 156, 168, 230, 231, 319, 320, 322
population *see also* demography 2, 10, 14, 15, 16, 59–63, 79, 85, 92, 93, 95, 97, 108, 132, 168, 169, 176–79, 181, 192, 200, 210, 213, 228, 231, 239, 291, 294, 295, 299, 301
   growth 61–63, 65–70, 167, 225, 227, 229
precipitation 4, 30, 37, 77, 132, 197, 214, 217–21, 228, 239, 246, 251–54
   gauge 36
prehistory 3, 60–61, 77, 81, 217, 227, 261, 264–66, 268–70, 294, 309, 321, 324, 326
pre-industrial society 4, 14, 17, 218, 229, 230
proxy 2, 59–67, 69, 71, 77, 105, 106, 209, 296

Q

Qaqortoq, Greenland 37
Quensel, Conrad 32

R

racism 1
radiocarbon dating 62, 63, 68, 81, 85, 86, 88–90, 94, 105, 106, 116, 156
Reaumur thermometer 33, 35
reductionism 18, 20
relief *see also* compensation 14, 178, 182
religion *see also* Christianity 11, 15, 103, 107, 110, 111, 114, 119, 213, 262, 320, 322
resilience 10, 13–15, 19, 61, 77, 78, 80, 92, 94, 96, 97, 103, 116, 119, 123–24, 126–27, 129, 134–4, 167, 168, 170, 177, 181, 182, 228, 267, 270, 280, 306, 327, 328, 331, 332
Reykjavik 29, 31, 35, 36, 40, 44, 45–48, 190, 198
Rhine, River 85, 92, 300
Ringsaker, Norway 160
risk *see also* hazard 1, 13, 15, 17, 19, 20, 103, 104, 115, 118, 127, 132, 169, 170, 176, 221, 225, 227, 229–30, 250, 306, 308
Roman 84, 263, 266, 31, 322
   Climatic Optimum 92–93
   Empire 82, 92–93, 295
   period 85, 92–93, 265
   Warm Period 264, 265, 270
romanticism 1, 249, 275, 279, 280–83
Rømer, Ole 31–32
Rottem, Ø 279–80, 288
Royal Danish Academy of Sciences (Danish Scientific Society) 33–34
Royal Netherlands Meteorological Institute 36
Royal Norwegian Society of Science and Letters 33, 50
Royal Society, London 32, 49
Royal Swedish Society of Sciences 33, 34, 50
Russia 38, 168
Russian Revolution 38
Ryan, Siobhan Moira 212
rye 214, 218, 219

S

Sagas (Icelandic) 103, 108–10, 118
Saloinen, Finland 177
Sandvik, Hilde 269
Scheel, Hans Jacob von 191
Schenmark, Nils 34

Index

Schøning, Gerhard 227–28
season; seasonality *see also* growing season
    5, 12, 19, 30, 48, 187, 194, 196, 198,
    201, 211, 216, 220, 225, 249, 299,
    326
Second World War *see* World War Two
seismic activity 38, 111, 113
settlement 6, 11, 14, 15, 16, 43, 65, 68–70,
    77–102, 103, 106, 107, 108, 115,
    116, 118, 119, 141, 167, 168, 181,
    192, 201, 265, 266, 268, 301, 320
Shetelig, Haakon 266, 270
shock, climatic or environmental 2, 10, 11,
    15, 16, 103, 104, 118, 230, 231
Sicily 11
Sigurðsson, Jón Viðar 268
Skálholt diocese, Iceland 110, 116, 117
Skaftá river, Iceland 105, 199
Skaftártunga, Iceland 107, 108
skating 237, 240, 24–50, 253
Skoglund, M. 10
slash-and-burn 169, 176, 181
snow 1, 30, 113, 114, 132, 147, 148, 150,
    162, 198–200, 225, 251, 282–86,
    291, 326
    melt 148, 211, 239, 247
socio-economic 10, 14, 19, 167, 168, 178,
    181, 190, 202, 231, 321
Sogn, Norway 147
Sogner, S. 227
soil 77 88, 94, 105, 116, 152, 160, 216–17,
    219–20, 228, 230, 249, 250
SPD (summed probability density) 62–64,
    86, 90
Speerschneider, C.I.H. 7
Spöring, Hermann 32
Spydeberg, Norway 34, 35, 209–36
Staley, David 295
Steenberg, Axel 9
Steinvikholm, Trøndelag, Norway 151–52,
    158
Stockholm 4, 34, 35, 38–40, 44–48
Storå, river, Denmark 128–29, 135
Storm surge 15, 84, 123–24, 130–41
stratigraphy 105, 106, 297
stress, climatic 2, 15, 16, 20, 77, 209, 231
Sturluson, Snorri 109

Stykkishólmur, Iceland 31, 36, 4, 42, 191,
    192
stone 15, 130, 133–36, 139–40, 147–62,
    237, 239, 240, 243, 245–47, 329
Stone Age 70, 303, 323
Storøya, Tyrifjorden, Norway 160
stove 15, 134, 137, 147, 150, 158, 162
summer 2, 4, 5, 6, 10, 12, 30, 116, 132, 168,
    172–76, 181, 187, 194, 196–98,
    200, 210, 212, 218, 221, 224, 226,
    228, 229, 246, 249, 267, 281, 284,
    299, 301
surface pressure 214
Svalbard 30, 31, 43
Sveinsson, Brynjólfur, Bishop 110
Sweden 3, 4, 6, 7, 9, 10, 12, 30, 32, 35,
    38–40, 46, 71, 148, 177, 225, 230,
    267, 299
Swedish History Museum 321
Swedish Meteorological and Hydrological
    Institute (SMHI) 37, 39

T

tar 168, 169, 174, 176, 178, 181
tax 12, 92, 167–69, 175, 180–82
techno-fix 20
technology 20, 59, 147–51, 158, 160, 162,
    305, 320, 326
telegraph system 37, 40
temperature *see also* variability, climatic 2,
    4–6, 10, 15, 30, 32, 35, 37, 40–50,
    59–76, 79, 132, 147, 148, 158, 159,
    172–73, 187–208, 215, 218–21,
    225, 230–31, 239, 241, 245, 246,
    251–54, 264, 285, 299, 300, 302
    air 186, 187, 193, 202, 214
    reconstruction 60, 67, 174
    sea 187, 188, 193
    series 29, 34, 38, 39, 41, 43, 191
tephra *see also* ash 103–08, 115, 19, 301,
    307
thermometer 31–36, 39
Thorlacius, Árni 36
Thoroddsen, Thorvaldur 7, 192, 193, 266,
    268
Thorsteinsson, Jón 36

Index

Tornio, Finland 33, 35, 238
Torricelli, Evangelista 31
trade *see also* exchange 62, 111, 190, 210, 224, 227
Trandilsson, Gaukr 115
tree-ring 2, 4, 5, 12, 19, 79
trend, climatic 2, 4, 6, 19, 29, 41–49, 60, 63, 67, 93, 174, 201, 239, 299
Trøndelag 151–52, 225, 228
Trondheim, Norway 33–35, 39–42, 87, 158 Fjord 83
Tuomiokirkkosilta bridge, Turku, Finland 240, 242–43, 245, 248, 250, 251
Turku, Finland 15, 32–35, 39–42, 239–57, 327

U

Upernavik, Greenland 37
Uppdal, Kristofer 275–81, 289–91
Uppsala, Sweden 6, 31–33, 35, 39, 41–42, 49
urbanisation 132, 168, 291
Utterström, Gustaf 9, 12

V

variability, climatic 3, 5, 6, 8–12, 14, 19, 41, 46–48, 79, 168, 170–71, 174, 176, 178, 181, 186–187, 193, 201, 202, 209, 210, 217, 220–22, 227, 230, 326
Vessas, Tarjei 290
Veyne, Paul 264, 270
Vilhjálmsson, Vilhjálmur Örn 116
Vik, Norway 77–102
Viking 6, 104, 115, 116, 265, 270, 320
Vinjefjord, Norway 77, 78, 87–90
Vinjeøra, Norway 77–102

volcano *see also* individual volcano names 38, 103–22, 294, 298, 299, 302, 306
vulnerability 12–17, 77–80, 92, 94, 97, 103–22, 124, 127, 136–40, 162, 167–85, 187, 209–36, 267, 270, 280, 294, 295, 302, 306, 308, 309, 323, 327, 328

W

Wargentin, Pehr 34
Wergeland, Henrik 277, 279, 280, 282
Westman Islands, Iceland 199
wheat 214, 216, 218, 219
wildfire *see* fire
Wilse, Jacob Nicolai 34, 209–36
winter 3, 4, 5, 6, 13, 30, 32, 38, 48, 79, 108, 109, 110, 117, 118, 132, 148, 158, 160–62, 174, 187, 194, 196–201, 219, 224, 228, 229, 238, 239, 242–43, 245, 249–51, 253, 254, 284, 285, 326
Winter, Johan 241
Wishman, Erik 8, 224
witchcraft 11, 12, 108
wood 106, 111, 116, 151–56, 159, 177, 189, 227, 239, 247–47, 250, 282
World War Two 242, 265
Worm, Ole 196, 197

Þ

Þórhallsson, Magnús 110
Þjórsárdalur valley, Iceland 104, 115, 116, 118
Þorsteinsson, Klængur, Bishop 117
Þorðarson, Sturla 109
Þórðarson, Jón 110

www.ingramcontent.com/pod-product-compliance
Lightning Source LLC
Chambersburg PA
CBHW040745020526
44114CB00049B/2929